Monetizing Entertainment

Monetizing Entertainment: An Insider's Handbook for Careers in the Entertainment & Music Industry offers a thorough, guided exploration of the current state of the industry, with an emphasis on trends in copyright, digital streaming, and practical advice for developing a career as an artist, technician, or industry executive.

This book investigates a variety of topics within the entertainment and music industry, ranging from traditional and emerging business models to intellectual property rights to the creative destruction happening currently. The book strategically outlines the existing gaps that make being successful as an artist a dynamic interaction between creativity and business.

This book includes the following:

- An overview of the creative destruction process that has destroyed some of the old business models and created a number of career options.
- A look at innovative, entrepreneurial career options.
- A step-by-step examination for both creative and business professionals of the administrative and financial structures of the industry.
- Detailed analysis of trends and topics shaping the current entertainment and music industry drawn from insiders' perspectives and other contemporary resources.

An accompanying website (www.routledge.com/cw/wacholtz), hosting case studies, videos, data, infographics, and blog posts on business models, is the perfect companion to this authoritative resource.

Dr. Larry Wacholtz is a professor of entertainment and music business at Belmont University in Nashville. A consultant to industry professionals, he is a member of NARAS, USASBE, MEIEA, IFBPA, and the Academy of Entrepreneurship. He is the author of eight other bestselling books on the entertainment and music business, including *Off the Record: Everything You REALLY Need to Know About the Music Business*, *Star Tracks*, *Principles for Success in the Music and Entertainment Business*, *How the Music Business Works*, and *Inside Country Music* (a Billboard book).

Dr. Beverly Schneller is an associate provost for academic affairs at Belmont University in Nashville. She is the author of *Anna Parnell's Political Journalism: Context and Texts*, *Writing About Business and Industry*, and *Writing About Science* (with Elizabeth Bowen) as well as numerous articles on literature and higher education. She serves as a Teagle Scholar through the Wabash College Center of Inquiry.

MONETIZING ENTERTAINMENT

An Insider's Handbook for Careers in the Entertainment & Music Industry

Larry Wacholtz

Edited by
Beverly Schneller

Routledge
Taylor & Francis Group

NEW YORK AND LONDON

First published 2017
by Routledge
711 Third Avenue, New York, NY 10017

and by Routledge
2 Park Square, Milton Park, Abingdon, Oxon OX14 4RN

Routledge is an imprint of the Taylor & Francis Group, an informa business

Library of Congress Cataloguing-in-Publication Data
Names: Wacholtz, Larry E., author.
Title: Monetizing Entertainment: an insider's handbook for careers in the
  entertainment and music industry / Larry Wacholtz.
Description: New York; London: Routledge, 2016. | "2016
Identifiers: LCCN 2016010404 | ISBN 9781138886049 (hardback) |
  ISBN 9781138886018 (paperback) | ISBN 9781315713922 (e-book)
Subjects: LCSH: Music trade—Vocational guidance. | Music entrepreneurship. |
  Cultural industries—Vocational guidance.
Classification: LCC ML3795. W23 2016 | DDC 780.23—dc23
LC record available at http://lccn.loc.gov/2016010404

ISBN: 978-1-138-88604-9 (hbk)
ISBN: 978-1-138-88601-8 (pbk)
ISBN: 978-1-315-71392-2 (ebk)

Typeset in Palatino
by Apex CoVantage, LLC

Visit the companion website: www.routledge.com/cw/wacholtz

Printed and bound in the United States of America by
Edwards Brothers Malloy on sustainably sourced paper

CONTENTS

AUTHOR'S ACKNOWLEDGMENTS

Without the support of Belmont University, I would have been unable to write this book. Special thanks to President Robert C. Fisher, Provost Thomas D. Burns, Associate Provost for Academic Affairs Beverly Schneller, my colleagues in the Mike Curb College of Entertainment and Music Business, and the university's Tenure, Leave, Promotion Committee for the sabbatical award.

Over the years, I have had the pleasure and privilege of working with many industry insiders, friends, and associates, who have graciously shared their collective wisdom and insights about the business of entertainment. In particular, the legendary Tony Conway, who generously read the manuscript and provided invaluable suggestions for its betterment. They have sustained my passion for the industry and richly informed my teaching and research.

This book is really for my students, past and present. They are the future of the industry. I have learned more from the questions they asked than anything I taught them.

Thanks to you all!

Blessings, L.

INTRODUCTION: TAKE A LOOK AT THIS!

The letters T, B, and M make me smile. They represent the *trillions, billions, and hundreds of millions* of dollars the global entertainment and media industries generate annually. Can I have just 1/10 of 1%? I'd never work again; or at least not at some of the crazy jobs I've had to survive! So what are these global entertainment and media industries?

Various research companies define the *global entertainment and media industries* differently. As an example, Pricewaterhouse-Coopers (PwC) describes the industries as 13 segments: *book*

Figure I.1 The many segments of the entertainment industry.

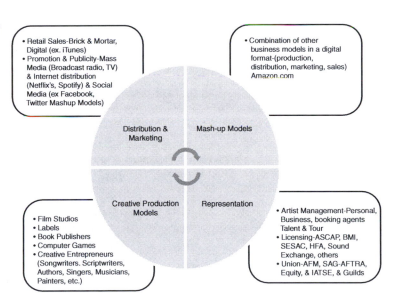

- Retail Sales-Brick & Mortar, Digital (ex. iTunes)
- Promotion & Publicity-Mass Media (Broadcast radio, TV) & Internet distribution (Netflix's, Spotify) & Social Media (ex Facebook, Twitter Mashup Models)

- Combination of other business models in a digital format-(production, distribution, marketing, sales) Amazon.com

Distribution & Marketing

Mash-up Models

Creative Production Models

Representation

- Film Studios
- Labels
- Book Publishers
- Computer Games
- Creative Entrepreneurs (Songwriters. Scriptwriters, Authors, Singers, Musicians, Painters, etc.)

- Artist Management-Personal, Business, booking agents Talent & Tour
- Licensing-ASCAP, BMI, SESAC, HFA, Sound Exchange, others
- Union-AFM, SAG-AFTRA, Equity, & IATSE, & Guilds

publishing, business to business, cinema, Internet access, Internet advertising, magazine publishing, music, newspaper publishing, out-of-home advertising, radio, TV advertising, TV and video, and *video games.* Other research companies combine different types of industry-related businesses, making the final numbers hard to crunch. Thus, there are many ways to figure the estimated "value" of the entertainment industry. Yet, using many sources, we can still find variations in what to count as part of these industries and some rather conservative estimates of the value of different sectors. No matter what is on "the list," it appears to represent revenue of about $1.8 trillion in 2016 with an expected increase to about $2.1 trillion by 2020.

To help us see what this looks like, I built Table I.1 from multiple sources, including PricewaterhouseCoopers (PWC), the Recording Industry Association of America (RIAA), the International Federation of Phonographic Industry (IFPI), Music Business Worldwide (MBI), Nielsen, Apple, TechCrunch, Statistica, *Forbes, The Financial Times*, and individual industry-related companies, such as Live Nation, Universal, Warner Bros. Entertainment & Music Groups, Sony Entertainment, and many others. Since these sources define each segment differently, more than one is needed to even get close to an accurate estimate of the relative value of each segment separately and combined. Research firms make their money by selling information to companies that want to market products that will be successful in the related entertainment and media industries to their customer base. So, they routinely release various parts of the "picture" and it's up to someone (such as me) to attempt to take the given data from a variety of sources to construct a reasonable "financial snapshot" of its value. Government sources from the U.S. Census and Economic Index also provide some validity to published resources. However we paint it, the entertainment and media (both traditional and social) generate about $1.8 trillion annually. I personally like the way the "t" rolls off the tongue. All the segments on a global scale generate revenues in the billions of dollars, and

Table I.1 The Ts, Bs, Ms. The table is modelled on accounting reports with the first category representing the total gross income from the global entertainment and media markets. This is currently $1.8 trillion and projected to be $2.14 trillion in 2020. The concern within the entertainment industry is that the business is dead. By showing the projected increase in gross revenues, we can see the industry is growing significantly based on new business models. The first subsection of the total, the second category, includes advertising, film, music industry, mass media, Internet access, digital games, sports, and publishing. The purpose of illustrating the subsections is to help you understand the entertainment industry is huge, and it is an aggregate of different types of businesses- all of which are still working together. Notice that it takes almost all the industries collaborating in various ways before any can be successful. As an example, film companies, labels, and publishing companies need to use advertisers for promotion and publicity. Another example is record labels who provide free master recordings to broadcast radio (media) for promotional purposes so that consumers can discover there is a new recording or a new act on the market. One last example, computer games that are becoming very popular through interactive Internet access. Without the media, the Internet, advertisers, and the rest of the industry the entertainment production companies' businesses would fail as they have no means to deliver, market, distribute through retail outlets, and promote their products to consumers.

Category	1st Subsection	2nd Subsection	3rd Subsection	Total Global Market 2016 Gross $	First Subsection 2016 Gross $	Second Subsection 2016 Gross $	USA Market Section 2016 $	Projected Gross 2020 $
Global entertainment & media market				$1.8 trillion				$2.14 trillion
	Advertising (in all traditional and digital media)				600 billion			717
		Entertainment advertising				272.8		350.98
			TV advertising				27.5	29
			Radio advertising				16	17.28
			Internet & other digital advertising				188	260.4
			Newspaper advertising				14.2	15.3
			Magazine advertising				15.1	16.6
			Outdoor advertising/ billboards				7.6	7.9

(Continued)

Table I.1 Continued

Category	1st Subsection	2nd Subsection	3rd Subsection	Total Global Market 2016 Gross $	First Subsection 2016 Gross $	Second Subsection 2016 Gross $	USA Market Section 2016 $	Projected Gross 2020 $
			All others, including cinema & directories				4.4	4.5
		Gaming advertising				136.5		167
			Video game revenues				70	85
			Traditional gaming revenue				50	60
			Social/casual revenue				16.5	22
		Other types of advertising (financial, healthcare, etc.)				190.7		195
	Film (in all production, uses, & licenses)				564 billion			679
		TV & subscriptions (digital & most countries for broadcasts)				221.9		259

Table I.1 Continued

	TV & video revenue	TV USA market		48	134.7	51.3
	Global box office revenues			26		15
		U.S. box office revenues			10	14
	Digital video (on-demand)			9.8	9.8	17
		TVOD (one time rental—e.g., Amazon.com)			2	2.3
		SVOD (monthly subscription—e.g., Amazon.com & Netflix)			5.3	7.5
		EST (cloud access to purchases—e.g., iTunes, Google, & Amazon.com)			2.3	3.9
		Other			0	3.3
	Other			259		336

(Continued)

Table I.1 Continued

Category	1st Subsection	2nd Subsection	3rd Subsection	Total Global Market 2016 Gross $	First Subsection 2016 Gross $	Second Subsection 2016 Gross $	USA Market Section 2016 $	Projected Gross 2020 $
	Music industry (production, recordings, uses, licenses)				54.4			59.1
		Digital music streaming (e.g., Spotify, Apple Music)				4.5		5.3
		Live music ticket sales (e.g., concerts, festivals)				25		28.5
		Live music sponsorship				2		
		Physical recorded music (CDs, etc.)				8.4		5.2
		Digital music downloading (iTunes, Amazon, Google)				2.4		8.6
		SoundExchange (use of recordings in digital media streaming)				0.75		

Table I.1 Continued

Category	Subcategory				
Music publishing	Performance rights			11.4	7.55
	Mechanical rights				1.32
	Private transactions mechanicals				0.77
	Grand & sync rights				1.7
	Other			5	
Mass media			63.1		
	Radio (minus online, all broadcasts radio)			44	72
					50
	Radio USA market				18.1
	Online radio (all online transmissions)			0.8	
	Digital video/music			9.781	13.698
	Other			14	

(Continued)

Table I.1 Continued

Category	1st Subsection	2nd Subsection	3rd Subsection	Total Global Market 2016 Gross $	First Subsection 2016 Gross $	Second Subsection 2016 Gross $	USA Market Section 2016 $	Projected Gross 2020 $
	Internet access						510	634.8
		High-speed (30+Mbps)				136.7		344
		Medium (between 4 and 30 Mbps)				279		265.3
		Low-speed (4 and below Mbps)				95		24.7
	Digital games (all production, sales, licenses)				52.4			66.8
		Mobile games				24.293		31.302
		Download (full version)				11.192		14.763
		Other				16.6		19.2
	Sports (all professional sports—all revenue streams)				145.3			150
		North American	Ticket revenues			67	18.6	

Table I.1 Continued

Category	Item			
	Media rights	18.4		
	Sponsorships	16.4		
	Merchandise	13.8	78.3	
Rest of the world				150
Publishing (books, magazines, e-books, newspapers, etc.)		147		
Global publishing (print & electronic)			130	92.1
	Books USA market			27
	Newspapers USA market			13
	Magazines USA market			8
E-publishing			16.3	28.9
	E-books	11		20
	E-magazines	1		2.9
	E-papers	3		6
Other			0.07	

Note: With the exception of the first row that captures the totals in trillions, the remainder is in billions of dollars.

when we break it down into additional subsections we'll often find the numbers in the hundreds of millions of dollars. Want to be part of this show-me-the-money world?

PricewaterhouseCoopers (PwC) (2016) claims *publishing* revenues (defined as all types of consumer and educational printed books and e-books) total about $147 billion annually. It also estimates the publishing segment will slow for printed books, but because of the 11.7% increase expected in the revenues of e-books, the entire publishing industry will increase about 1.7% until 2020. The popular media has been saying for years books are dying—that seems not to be the case. What is important is that the delivery method for reading is changing and e-books, considered "new" a few years ago, have stabilized as an entertainment media revenue source in more ways than you probably knew.

Cinema is defined somewhat narrowly by PwC (2016) as box office and advertising gross revenues. Cinema revenue has a global impact of about $26 billion. And if we add in subscription TV (in many countries of the world TV broadcast is licensed), plus production, video-on-demand, and many other film-based or related sources for entertainment, the value of the cinema business jumps to about $564 billion.

The *music industry* is the first of the entertainment segments to suffer the financial slide of the creative destruction process caused by the advances in technology, negative consumer behaviors, and the lack of enforcement of the copyright laws. These three factors have damaged the overall value and potential value of the music industry as the sale of recorded products (CDs, etc.) has lost about 70% of its revenues over the last few years. Published sources claim the revenues have been dropping steadily from about $16 billion to less than $5 billion. Safe harbors and the unwillingness of the court system to actually support copyright holders as the owners of entertainment products have placed the "value" of the industry into question and forced labels more into marketing, branding, and promotion business models instead of the recording companies trying to sell CDs. Still, as consumers shift away from stealing music online to streaming it, the industry has been able to generate hundreds of millions of dollars from the streaming companies. These alternative revenue sources are just starting to cover the losses from album sales and other dormant revenue streams. When we combine digital streaming with live music ticket sales, sponsorships, CD, digital downloads, and music publishing royalties the industry generates about $54 billion, with a predicted increase of 2.1% to about $59 billion in 2020.

Advertising is the big enchilada, with revenue in the neighborhood of $600 billion in 2016. Entertainment advertising on TV, radio, the Internet, print, and other types of social and traditional media currently grosses about $272 billion. The figure is estimated to rise significantly to over $350 billion by 2020.

Computer and digital games account for about $52.4 billion now but revenues are expected to rise to over $66 billion by 2020. The global *Internet access* market rocks at about $510 billion being generated now and around $634 billion in 2020. The growth in revenue appears to correlate with the improvement in the availability of faster downloading speeds, which will also help the streaming of movies, music, and games. The North American (U.S. and Canada) professional sports market totals about $67 billion currently from ticket sales, media rights licensing, sponsorship, and merchandise receipts. The world market tops out at about $145 billion with the addition of football (soccer), the Olympics, and other types of professional sports. Table I.1 provides some insight into the size of the global entertainment and media markets.

Technological Revolutions

The surprise to many consumers, students, and even career professionals (in the industry) is the degree to which different segments of the industry must work together if any are to be financially successful. Without the great songwriters, the musicians, artists, labels, radio, and other forms of media, entertainment ventures are dead in the water. On the other hand, without initial and long-term financial support and industry connections, songwriters and their music publishers usually fail to sell or license any of their material. Thus, if you really want a great gig in this business, you need to start by understanding the big picture of how it all works together. On top of that, we are living in a time when the status quo is being disrupted on a daily basis. According to Bothun and Sviokia in their article "You're a Media Company. Now What?" (2016),

> *Nowhere are these porous and evolving borders more evident than in the entertainment and media (E&M) industry. The past 20 years have brought a wave of disruptions to distribution, formats, technologies, and consumption patterns . . . There has always been an intimate and complex relationship among consumer and industrial companies . . . The 1950s-era daytime serials were known as "soap operas" because they were sponsored by the companies that made soap. In 1940, at the dawn of the radio era, listeners tuned in to the Texaco Metropolitan Opera broadcasts. The Wonderful World of Disney, the television show that debuted in 1969, integrated media, experience, entertainment, and merchandising. Hello Kitty was born in Japan in 1975 as a way to cute-ify merchandise, and then developed into television series, comics, and video games. Still, through the 20th century, most brands relied on the creativity and expertise of the media, advertising, and entertainment companies to create content and deliver audiences.*[1]

But now consumers using digital devices can avoid buying entertainment products in various platforms at retail stores (e.g., DVDs and CDs) and just stream free or steal it outright. Say good-bye to retail brick-and-mortar stores. Entertainment has always been an "experience" we'd have to acquire by buying a copy, such as piano rolls for the player piano (look it up), vinyl records, cassettes, CDs, or tickets to the local theater to watch the latest Hollywood movie. TV changed some of that by allowing us to see visual programs in our homes, but we still have to tolerate those darn boring commercials. Radio did the same thing for music in our cars. Therefore, Netflix, Spotify, e-books, and all the other Internet-based programing are just the next step in the technical digital delivery revolution for acquiring information and entertainment products that started well over 600 years ago. Welcome to the Age of Enlightenment and the Gutenberg printing press with movable type. Let's pause and consider the impact the work of one inventor had in revolutionizing access to both educational and entertainment products with moveable type:

> *He who first shortened the labor of copyists by device of movable types was disbanding hired armies, and cashiering most kings and senates, and creating a whole new democratic world: he had invented the art of printing. (Thomas Carlyle, Sartor Resartus, 1833) . . . The first man to demonstrate the practicability of movable type was Johannes Gutenberg (1398–1468), the son of a noble family of Mainz, Germany. A former stonecutter and goldsmith, Gutenberg devised an alloy of lead, tin and antimony that would melt at low temperature, cast well in the die, and be durable in the press. It was then possible to use and reuse the separate pieces of type, as long as the metal in which they were cast did not wear down, simply by arranging them in the desired order. The mirror image of each letter (rather than entire words or phrases), was carved in relief on a small block. Individual letters, easily movable, were put together to form words; words separated by blank spaces formed lines of type; and lines of type were brought together to make up a page. Since letters could be arranged into any format, an infinite variety of texts could be printed by reusing and resetting the type.*[2]

So, that is how books and newspapers eventually made it into every home. Now, let's look at another common source of entertainment, and one we think of as free, in our homes and in cars. The word "radio" can be either the device we use to capture and hear a radio signal or the content of the signal being broadcast. The invention of radio is closely related to the invention of the telephone by Alexander Graham Bell (1847–1922), and the telegraph by Samuel Morse (1791–1872).

> *During the 1860s, Scottish physicist, James Clerk Maxwell (1831–1889), predicted the existence of radio waves; and in 1886, German physicist Heinrich Rudolph Hertz (1851–1894)*

demonstrated that rapid variations of electric current could be projected into space in the form of radio waves similar to those of light and heat . . . In 1866, Mahlon Loomis, (1826–1886), an American dentist, successfully demonstrated "wireless telegraphy." Loomis was able to make a meter connected to one kite cause another one to move, marking the first known instance of wireless aerial communication . . . Guglielmo Marconi (1874–1937), an Italian inventor, . . . sent and received his first radio signal in Italy in 1895 . . . By 1899 he flashed the first wireless signal across the English Channel . . . Nikola Tesla (1856–1943) is now credited with being the first person to patent radio technology; the Supreme Court overturned Marconi's patent in 1943 in favor of Tesla.[3]

World War II brought us the atom bomb, but it also created the first serious opportunity for electronic computers.

The earliest electronic computers were not "personal" in any way: They were enormous and hugely expensive, and they required a team of engineers and other specialists to keep them running. One of the first and most famous of these, the Electronic Numerical Integrator Analyzer and Computer (ENIAC), was built at the University of Pennsylvania to do ballistics calculations for the U.S. military during World War II. ENIAC cost $500,000, weighed 30 tons and took up nearly 2,000 square feet of floor space. On the outside, ENIAC was covered in a tangle of cables, hundreds of blinking lights and nearly 6,000 mechanical switches that its operators used to tell it what to do. On the inside, almost 18,000 vacuum tubes carried electrical signals from one part of the machine to another.[4]

The personal computer was developed with the inventions of transistors, integrated circuits, and microprocessors.

In 1948, Bell Labs introduced the transistor, an electronic device that carried and amplified electrical current but was much smaller than the cumbersome vacuum tube. Ten years later, scientists at Texas Instruments and Fairchild Semiconductor came up with the integrated circuit, an invention that incorporated all of the computer's electrical parts—transistors, capacitors, resistors and diodes—into a single silicon chip . . . But one of the most significant of the inventions that paved the way for the PC revolution was the microprocessor . . . They could run the computer's programs, remember information and manage data all by themselves.[5]

Former vice president Al Gore did not invent the Internet. This is how it really happened:

The first workable prototype of the Internet came in the late 1960s with the creation of ARPANET, or the Advanced Research Projects Agency Network. Originally funded by the U.S. Department of Defense, ARPANET used packet switching to allow multiple computers to communicate on a single network. The technology continued to grow in the 1970s after scientists Robert Kahn and Vinton Cerf developed Transmission Control Protocol and Internet Protocol, or TCP/IP, a communications model that set standards for how data could be transmitted between multiple networks. ARPANET adopted TCP/IP on January 1, 1983, and from there researchers began to assemble the "network of networks" that became the modern Internet. The online world then took on a more recognizable form in 1990, when computer scientist Tim Berners-Lee invented the World Wide Web. While it's often confused with the Internet itself, the web is actually just the most common means of accessing data online in the form of websites and hyperlinks.[6]

The bottom line is that now many nonentertainment product (shoes to the kitchen sink) companies have expanded into entertainment; the older traditional industry–related companies and media have had to reinvent their business models. Again, according to Bothun and Sviokia (2016),

The evidence is ubiquitous. Nike has become a major presence in social media, digital video, mobile apps, and e-commerce—witness the company's recently launched YouTube miniseries focusing on a fitness bet between two sisters. ANZ Bank, one of Australia's largest financial institutions, has built a finance news portal, Blue-Notes, which is staffed by well-known business journalists. Marriott has created a content studio, supported by Hollywood talent, to develop videos for distribution in social media and elsewhere, all with the business objective of increasing the hotel brand's appeal to millennials. FairPrice, a Singapore-based supermarket, maintains the highly popular food content platform Food for Life, which hosts 2,000 video assets in a range of languages. And the list goes on.[7]

This trillion-dollar industry has changed. Do you see how the parts both naturally and intentionally were made to work together? Adapting Bothun and Sviokia's (2016) unique concept for analyzing business models, I've revised it to fit the entertainment and media industry. The following combines the separate industries into four types of business models based on how they contribute to the system and the processes of creating, licensing, promoting, marketing, distribution, and sales of various entertainment products acquired through retail or streaming, and the theft associated with illegal downloads (which take away from the entire industry's bottom line as well as yours as a creative).

Entertainment production businesses create entertainment products based on intellectual properties they buy or license from the original owners. As an example, songs come from songwriters and music publishers license the rights to make a recorded copy of the song. Film companies purchase or lease the book rights from major authors and publishing companies to turn a popular book, and it does not have to be bestseller, into a movie. In the case of the book into film, the industry executives recognize how the storyline will connect emotionally with the movie goers they want to reach. The movie companies then hire scriptwriters to remake the book into a movie script that directors can follow to bring the book to life as a film with actors. See how this is all connected? Do you want a career in one of these parts of the entertainment development and delivery process? Part of this section of businesses, beyond the major labels and studios, are the entrepreneurial creative writers, authors, songwriters, painters, and others, who own (copyright) their own original creative works once they are placed in a tangible form (known as real property). Either they or an entertainment-based representation company can sell or license the creative works to the major production companies, which moves the creative work from a pleasing idea to a monetizable consumer product.

Distribution and marketing businesses are the companies that often distribute the creative products to consumers through purchase, subscription, licensing, or streaming. The distribution is either part of a production company, with a separate purpose (business model), or a totally separate company that releases entertainment products and provides promotion and publicity to consumers. The distributors usually have to obtain products by buying them wholesale or through licenses, for which they pay a percentage of the profits to the label or the studios, as examples. Examples include retail chain store outlets, everything from Target to Kroger. The disadvantage of retail brick-and-mortar stores is that they have to get the customers to come to the stores. That costs money in building, heating, lighting, employees, advertising, and so forth. Find many record stores around anymore? Not when consumers can acquire their music much cheaper online or by streaming.

The marketing of entertainment products includes the promotion and publicity that is used in both traditional media and social media. Advertisements for movie trailers, free clips on iTunes, and stories based on acts' images are used to alert potential consumers about new entertainment product releases. Promotion gives consumers a chance to "discover" the entertainment product while publicity provides the backstory of the act tied to their image. Both are very important as consumer acceptance of new entertainment products leads to the production companies' use of branding to increase profits. You will learn more about this in the chapters on record labels.

Representation is the term used to describe the companies that offer special technical types of products that require very little direct connection to consumers. Their clients are the businesses who need audio installations, video production studios, websites, and many different types of products that will help other businesses (business to business) improve their operations and increase revenues. These are talent agencies, personal management companies, and licensing companies, such as ASCAP, BMI, SESAC, Harry Fox, SoundExchange, and others discussed later in the book in the chapters on artist management and touring. You might find yourself working for one of these important entertainment industry businesses, giving career advice to artists, finding jobs for film, stage, and tours, or analyzing the finances, paying all the bills and taxes, and contributing to the success of both your company and the industry overall.

When I was in high school a long time ago, we had a basketball coach who would start to pull up his socks when things started to get weird. Then he'd jump up, prance around, and start yelling at just about everyone. By the way, he would later become a college coach and had Magic Johnson on his team when they won the Final Four!

Well, start to pull up your socks. *Mash-ups* are the new industry-related businesses that are driving the old established industry crazy. Many of them live in safe harbors that protect them and allow consumers to violate the copyright law all the time. You'll read about that stuff later in the book. However, they are allowed to exist so that technology may make advances to improve the quality of human life. The hard part is twofold: copyright owners are not often paid for their creative works and investments, as sometimes the safe harbor companies tend to get away with ignoring the law, and they may also push the privacy boundaries of individual citizens past established ethical and legal limits. The upside is that now we have YouTube, computer software, and many other forms of communication; these have forced the media and entertainment companies to reinvent their business models. Examples include Amazon.com, Spotify, and many other types of entertainment-based companies that merge the functions of the *creative producer models, distribution and marketing models, representation models,* and *mash-ups,* which all provide new opportunities for individual creative artists. The great products usually come from the professional workers and artists, and the major labels, publishers, and movie studios. What comes from the efforts of these entertainment industry novices (called wannabes by insiders) are uploaded products that are usually really poor to okay, and they have the accumulated effect of simply jamming up the Internet and making it harder for truly talented or promising artists to be heard or seen. Think about what you can do in the industry and where you might fit best based on your passion and creativity. Then think about what you need to know to be successful as you read this book and how you might apply it in ways that will contribute to the entertainment industry's growth and sustainability.

The world is quickly changing and you will hear lots of frustration from the old industry suits and creatives who are struggling to keep up with the industry to which they have given their lives. As you read through the chapters of this book, you'll also feel my frustration at the situation. However, embrace it! It is simply a waste of time to think things are going to return to the way they were before the Internet, technology, and social media. So let's look at the bigger picture. Honestly, the entire industry is based on the protection of creative works by copyright laws. And that is where the sock pulling, yelling, and frayed nerves are breaking out in all parts of the entertainment and media industries. What a time to have a perfect storm in the middle of a technological revolution. It's enough to threaten an entire industry—or, at the very least, to make anyone who wants to be a part of it realize how much the changes are creating new opportunities and new models for entertainment creation, marketing, sales, and distribution by fostering new and varied revenue streams.

Notes

1. Bothun, Deborah, and John Sviokia. "You're a Media Company. Now What?" *Strategy Business*. 2016. Accessed June 25, 2016. http://www.strategy-business.com/article/Youre-a-Media-Company-Now-What?gko=ac350.

2. Kreis, John S. "The Printing Press." *The History Guide, Lectures on Modern European Intellectual History.* September 11, 2014. Accessed June 25, 2016. http://www.historyguide.org/intellect/press.html.

3. Bellis, Mary. "The Invention of Radio—Wireless Telegraphy." About.com Inventors. October 21, 2015. Accessed June 25, 2016. http://inventors.about.com/od/rstartinventions/a/radio.htm.

4. History.com Staff. "Invention of the PC." History.com. 2011. Accessed June 25, 2016. http://www.history.com/topics/inventions/invention-of-the-pc.

5. History.com Staff. "Invention of the PC." History.com. 2011. Accessed June 25, 2016. http://www.history.com/topics/inventions/invention-of-the-pc.

6. Andrews, Evan. "Who Invented the Internet?" History.com. 2013. Accessed June 25, 2016. http://www.history.com/news/ask-history/who-invented-the-internet.

7. Bothun, Deborah, and John Sviokia. "You're a Media Company. Now What?" Strategy Business. 2016. Accessed June 25, 2016. http://www.strategy-business.com/article/Youre-a-Media-Company-Now-What?gko=ac350.

THE PERFECT STORM   1

I am a member of the pirate generation. When I arrived at college in 1997, I had never heard of an MP3. By the end of my first term I had filled my 2-gigabyte hard drive with hundreds of bootlegged songs. By graduation, I had six 20 gigabyte drives of music, nearly 15,000 albums' worth . . . I pirated on an industrial scale, but told no one . . . The files were procured in chat channels, and through Napster and BitTorrent; I haven't purchased an album with my own money since the turn of the millennium.[1]

—Stephen Witt

In the United States only, which is about one-third of the world's entertainment product market, record companies lose almost $3 billion per year from illegal downloads. The film industry loses almost $6 billion every year, which is twice as bad for investors and production companies (Figure 1.1).

The old music industry is dead. We're standing in the ruins of a business built on private jets, Cristal, $18 CDs and million-dollar

Figure 1.1 The annual loss to the music industry is about $2.7 billion in earnings (RIAA 2015), and about $6.1 billion in gross income to the film and movie industry (MPAA 2015), as stated in the article "Film Piracy, A Threat to the Entire Music Industry."

**U.S. Entertainment Industry Market Loss
Annually to Internet Downlods**

recording budgets. We're in the midst of the greatest music industry disruption of the past 100 years. A fundamental shift has occurred—a shift that Millennials are driving. For the first time, record sales aren't enough to make an artist's career, and they certainly aren't enough to ensure success. The old music industry clung desperately to sales to survive, but that model is long gone.[2]

—Honeyman

The sale of recordings is only a small part of the music industry; however, when we are talking billions, it still means real money (see Figure 1.2). The music industry has lost almost two-thirds of its markets to illegal downloads. Once again, the investors who put millions into the marketing and the recordings have lost the incentive to invest in prerecorded

Figure 1.2 Recorded music sales have dropped from $14.6 billion in the United States in 1999 to only $6.85 billion in 2014.
Source: RIAA (2015), IFPI (2015).

Gross U.S. Recorded Market

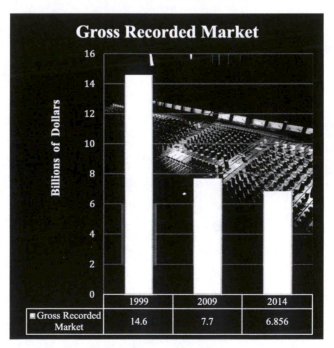

music. Now, the industry has forced the labels to refocus on branding and touring to recover their investments.

I Disagree

The music industry has hit the skids, but by changing its business model, it is starting to make money again. Because people still want to listen to great music, the industry will find a way to profit from its investments as long as laws can be enforced. However, it won't be easy! Selling albums at retail outlets is largely a thing of the past; digital downloads of albums or singles on Internet provider sites, such as iTunes or Amazon.com, reign now, but who can say what the next way we will consume and enjoy music will be. What might strike you as the uncertainty facing us in the entertainment industry is a very natural process of change that in the end will be creative, innovative, and profitable. Or at least, I hope it will be and you should, too!

As you read this book, you'll start to understand the process of creative destruction and the rebirth of the new music industry and you'll be able to see how these changes are also trends in the entire entertainment industry. We have to hope the U.S. Congress and others who control and regulate the entertainment industry will continue to re-evaluate the balance between free markets, the innovations technology brings to the creation and delivery of entertainment products, and the need to protect and insure the honest financial rewards creative individuals so richly deserve. Things are changing at all levels of the industry, from the types of recording artists labels sign to the production methods used to create entertainment products and the distribution of movies. The old industry isn't dead; it's just morphing into an opportunity to build the *new* industry based on the best of the past and the opportunities we'll discover in the future. However, there are some serious lessons about the uses of new technologies, software, entrepreneurship, ethics, creativity, and laws that we may need to understand if the new industry—whatever that will be—is to survive.

301 Special Report: The Office of the President of the United States

This chapter started with four versions of where the entertainment industry appears to be in the 21st century: illegal Internet downloads; significant lost revenue for traditional entertainment industry products and their creators; almost predictable drops in record sales as personal devices replace radios and record players in homes; and an awareness among artists that there has to be change in order for the entertainment industry to survive. Think about the implications of all these things coming together at one time in one industry. Are Witt, who seems at least to be an honest crook, and people like him, who have downloaded entertainment products without paying for them because they were "free," acting like thieves, pirates, or Robin Hood, stealing from the rich to support their passion to be entertained? Do you think they even consider the impact of their actions on the artists whose works they are so eager to have? And even if they do, who cares about the long-term impact of what they have taken? Did anybody really get hurt when Witt downloaded music in his home to listen to for his own enjoyment? And, if we were to ask him or anyone who has done the same thing, we might hear a response along the lines of "Well, it's out there, isn't it, and no one says I can't." Right! Why should anyone care if we shoot a video, mix in a major hit, and upload it to YouTube.com or any of the other thousands of worldwide sites? Besides, it was just for us anyway and we didn't sell the video with the original recording, so we didn't make any money.

If I had a buck for every time I've heard those types of statements, I'd own a sailboat in the Caribbean and could probably buy an island next to billionaire Sir Richard Branson. The fact is that when Witt illegally downloaded a song, the songwriter, music publisher, label, and recording artist each lost a few pennies. So you do the math—when a million or so people illegally download that same recording the creative careers of the individuals who wrote and sang the tracks and the company who invested its money into it are destroyed. Sadly, this is a world problem that must be addressed, with a growing significance to the extent that the president of the United States commissioned a specific report on the consequences of patent and trademark infringement as well as copyright infringement just to try to understand it better. *The Special Report From the Office of the President of the United States* (2015) states,

> *The problems of trademark counterfeiting and copyright piracy continue on a global scale and involve the mass production and sale of a vast array of fake goods and a range of copyright-protected content pirated in various forms. Counterfeited goods include semiconductors and other electronics, chemicals, automotive and aircraft parts, medicines, food and beverages, household consumer products, personal care products, apparel and footwear, toys, and sporting goods.*
>
> *. . . consumers, legitimate producers, and governments are harmed by trademark counterfeiting and copyright piracy . . . Producers and their employees face diminished revenue and investment incentives, an adverse employment impact, and loss of reputation.*[3]

Let's translate what the report says into the economic impact on jobs, profits, and lost taxes (Figure 1.3).

Figure 1.3 The cost of Internet illegal downloads is serious for several reasons, including the loss of over 71,000 jobs, $2.7 billion in earnings, and $422 million in taxes.
Source: RIAA (2015).

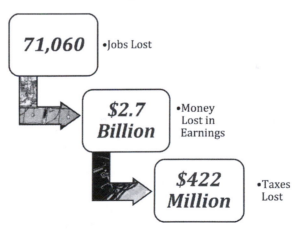

Cost of Illegal Downloads

71,060 •Jobs Lost

$2.7 Billion •Money Lost in Earnings

$422 Million •Taxes Lost

The Insider's Perspective

Honestly, many industry insiders are frustrated by the public's lack of understanding about how the entertainment and music industry works. Such is a gap I am hoping to fill with this book. In these pages, I am inviting you to think deeply and critically about the current and future states of the entertainment and music industry. I'm assuming you are reading this book because you, as I did, want to be part of the dynamic world of creating and valuing entertainment products that, put simply, make our lives a whole lot better. There are not many things we can say with certainty, but one thing is for sure—we love our movies, books, TV shows, and, above all, our music. So, one of the biggest challenges facing the entertainment and music industry today is how to keep making money, how to keep the business alive, in the face of the popularity of streaming, and other new forms of entertainment delivery. The culture of the entertainment industry is changing, and you need to know the processes, laws, and practices that have shaped it, where things stand now, and then use what you have learned to make the industries what you want them to be! Fasten your seatbelts and let's get moving. But first, here's a little about me.

I've spent most of my adult life meeting personalities and industry insiders, working as a bad producer and a passible audio engineer. I've gotten to know people all across the entertainment industry; some are friends and others are my colleagues and may one day become yours. All of us are especially interested in understanding and sustaining the entertainment industry. I'm always the student and the artists and entertainment leaders are always the professors. I've never asked for an autograph or picture. I'm simply not interested in the glitz and glamor. I'm a voting member of the Grammy Awards, and I have found most of the people in this industry to be bright, talented, often humble, passionate people. Sure, there are a few jerks, but they are usually not among superstars or top executives. In addition, most artists are also business people. Some great, some naïve, and some awful until they learn from the school of hard knocks that they better get it together if they want to get paid anything for their art. Because consumers can opt to "own" a CD or a DVD through downloading, artists have to think about how they can prevent or at least affect any amateur's (called wannabes in the industry) use of digital entertainment products that the artists have worked hard to create. Wannabes make it tougher for new and established artists to maintain economic success because the amateur's product, usually made with the same or similar recording tools, is rarely at the same level of quality as the professionals' works. Wannabes upload what they have created on the same platforms the professionals use, but they give it away to get noticed, creating a larger swamp of thick oak moss that makes it harder to find the quality products. The consumer may or may not have the experience or the interest to distinguish between the work of a professional and that of a wannabe, because the consumer wants the pleasure, the experience, of the creative work.

"Taking Care of Business"

As part of my professional training and experience in the music industry, I developed a psychographic research method that allowed me to provide image-consulting research for MCA and Word Records' recording artists. My interest is in understanding not what people buy or are attracted to as consumers but why they buy it or are drawn to it, by analyzing where people live, how much education they have, and how what other kinds of things they buy might influence why they buy a music or entertainment product. For instance, a person who buys tickets to an NFL season might also own a Volvo, or a Prius, or a Harley motorcycle. What do these things have in common? Such are the variables that make psychographic research challenging, as there are many things shaping why someone would buy something that do and don't compare with their other buying habits. In other words, you don't have to own cats or like the opera to admire the work of British composer Sir

Andrew Lloyd Webber. However, at some level the common consumer preferences come together and we discover that the guy who rode the Harley to the football game likes a certain type of beer and the rugged, outdoor, rebellious-attitude lifestyle "hero" or "villain" found in the images of heavy metal rock (when young and older 55+, as an example), and country music acts with a rebellious image when they are middle-aged. And there you have it, the connection between teenage males (heavy metal fans) to middle-aged male fans who relate to country music. We can also find out where they live by using computer-marketing software, down to their zip plus 4. There are lots of other methods and more detailed forms of market and lifestyle research available, but consumer research has value to businesses only when consumers buy or legally use entertainment products through subscription providers. It's useless information if people are just going to rip it off the Internet for free.

I am one of those people God missed when He was handing out the talent to sing or be a musician, but it does not mean I lost interest in the entertainment industry. In fact, my passion has been to try to figure out how entertainment products communicate with consumers and how the industry monetizes them (assesses their commercial value). I've lived my life with one foot in the industry and one in academia, working with a few other talented faculty to help many talented young people become world-famous acts and industry insiders in publishing, songwriting, film, and production, and successful business executives. I've been very blessed. However, here I am toward the late stages of my career realizing that many people think the entertainment industry is suffering and the music industry is dead. I disagree. But how did this misperception of an early "death" happen?

The Process of Change

Researchers discover natural phenomena and inventors develop devices we use based on their "discoveries." Energetic entrepreneurs and businessmen and women invest in the new "discoveries and devices" to create entertainment products, and the corresponding movie stars, recording artists, poets, authors, and gamers realize some sort of profit. Researchers often want to answer a problem bugging them; inventors often want to develop new devices that may help improve the quality of life on the planet; and investors usually want to use the devices to create an industry for profit. Business is business. Discover a combination of natural or synthetic elements that cure an illness and the world will throw money at you. Think about what Steve Jobs did at Apple. He created devices consumers want, generating billions in profits. Where would we be without our iPod, iPad, iPhone, and personal computer? We need to remember that Edison invented the phonograph in 1870, though I would bet some of you have never used, or maybe never even seen, a record player except in an antique shop or in a picture. And, hey, there is nothing wrong with that! The fact that things are changing does not mean anything is "dying" and innovation is to blame. In fact, we thrive on and promote innovations in transportation modes, waste management collection methods, housing, education, and almost everything else in our lives. So, what is there to worry about if the entertainment industry also undergoes a transformation? Let's take a look at the size of the entertainment industry as projected by growth in revenues by 2019.

Projected Entertainment Industry Growth

In Figure 1.4 and Table 1.1, we see the projected sales revenues across the businesses that encompass the entertainment industries. You will want to pay particular attention to how these industries are grouped as this is how data is collected for analysis within public and private sector sources.

The Reality of Innovation

Many events have occurred in human history that have provided a stimulant for changing the ways we perceive reality. As an example, blaming the millennial generation for

Figure 1.4 Pricewaterhouse-Coopers (2015) projects growth in various segments of the entertainment industry, including global music, and touring will rise to roughly $29 billion by 2019. *Source*: Global entertainment and media outlook (2015–2019), PWC, Ovum.

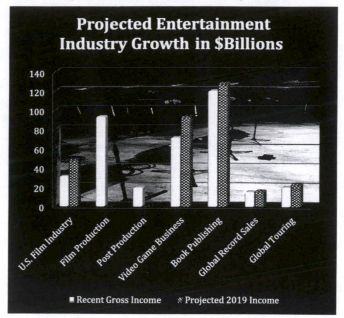

Table 1.1 Current and Projected Incomes from the Major Entertainment Segments.

Entertainment and Media Markets	Recent Gross Income (in $ billions)	2019 Projected Revenues (in $ billions)
U.S. Film Industry (major and independent studio distribution)	31	48.4
Film Production	93	N/A
Postproduction	19.5	N/A
Global Music Industry (including all recorded music and live touring)	26	29
Video Games (traditional, social, and video game advertising)	71	93.18
Book Publishing	120.13	128.34
Business-to-Business	193.9	234.4

the music and entertainment industry's decline is about as crazy as blaming Columbus and his crew of crooks, robbers, cutthroats, and heroes for proving the world is physically round. By that way of thinking it was his fault that everyone had to change the ways they "survive" and "make sense" of this "brave" new world. One day we build our life around "what we are sure we know" (called a frame of reference) and the next day we wake up and the world is a round ball, not flat and limited. That's the way life is—progressive and constantly moving forward through a window of time. Every time there's a technological improvement in devices, we'll usually buy the new device. When there are new acts, movies, or computer games to get excited about, we want them also. But how in the world did acquiring music free ever happen? And what will that mean to the future of the music industry and to entertainment as a whole?

The Digital "Round" World

Change is a bitch! But it is here and we've got to embrace it. It doesn't really matter if we lived at the time of Columbus or right now because change caused by events is a constant. It often alters our lives in ways we never imagined. Then, just when you get comfortable and think you know it all, the next thing you know, you get bit by something that flips your world again. Powerful thing, this process of evolution! There's one additional element of change that's really cool. When change happens, especially in industry, there's also an opportunity to profit. The invention of the car provided a need for oil, gas, rubber, and so forth. Changes in how, where, and when we work, at least in the United States, provide entertainment companies an opportunity to provide products to fill up leisure hours with fun things to hear, see, and do.

The Reason for the Perfect Storm

In the music and entertainment industry, this time it wasn't Columbus's crew or anyone even like them who caused things to change. Instead, it was a perfect storm of brilliant researchers, technicians, naïve, sticky-fingered factory workers, executives, creative businessmen and women who unknowingly developed the digital devices that led to events that once again rocked our perceptions. Entertainment, especially music, is important to most of us. So what's the big deal when we can acquire what we want free and online? It's the same music, video, movie, and it's free. We save time looking for it, and don't have to buy it on a piece of plastic, or even buy a bunch of other devices that allow us to enjoy it. What's wrong with that?

The phrase a "perfect storm" was popularized in the late 1990s by the book and movie of the same name. A perfect storm is a rare event, originally a natural phenomenon, now used to describe a paradigm shift caused by a convergence of multiple ideas or situations. The term is frequently related to anything political, social, or business-oriented, and the results of it are considered more often as negative, rather than positive. What got the perfect storm into widespread use as a business and cultural term was the book by Sebastian Junger, *The Perfect Storm: A True Story of Men Against the Sea.* Christopher Lehmann-Haupt (1997) provides a wonderful summary of this true story in his *New York Times* book review detailing how three storms merged off Nova Scotia. The rare natural event caused an 80-mile-per-hour storm and 100-foot waves, destroying the fishing boat *Andrea Gail* with the loss of all the crew.[4] No survivors! The reach of this perfect storm—that is, a rare event and outcome stemming from unusual circumstances—has been felt within and beyond just the entertainment and music industry in ways that I will show you as you read on in this book. Right now, I need you to stick with me to see how one innovation created a domino effect of changes across a wide spectrum of likely and unlikely places. You don't have to agree with how I describe it, and heck, maybe it is best you don't, but bear with me and see if, after you think about, it starts to make sense.

The Digital Storm

The industry's perfect storm—that is, what caused the destruction of the recorded music market—is beyond weird. Our early "perfect storm" people simply were trying to legally create products that would solve a digital storage problem and then sell and license the technology for a profit. The problem was simple: the solution was a mathematical nightmare. Digital technology disrupted the soaring profits of the music and entertainment industry when analog recordings and broadcasting were converted to a digital platform. It involves a cast of brilliant geeks (a term used positively in the industry) in audio engineering and digital technology, consumers, and employees. They, without knowing each other, lifted non-released CDs and DVDs and developed ways to use the Internet, personal computers, and devices that allow people to experience entertainment instantly and free. They created the perfect storm that spawned the creative destruction of the current industry's business model.

The Construction of Digital Signals

A digital signal is simply a row of numbers (called a binary code) of 0's and 1's or a magnetic pulse or no pulse that is converted by the computer chip into 0's and 1's, which is then converted into a "sound." The computer chip is actually creating the sound from the binary code that represents the original sound. If it's music that is encoded and converted into 0's and 1's, then it's the same music (sound) as what you'd hear live. It's being captured in a binary code, and then the computer uses the code to regenerate the music. We hear vibrations per second; thus 30–60 cycles per second or vibrations per second are "heard" by us as a bass sound. Twenty thousand per second is a very high-pitched sound that as we get older we usually can't hear. Dogs are able to hear much higher frequencies than humans. And the French fry cooker at most fast food restaurants is around 20,000 cycles, just enough to sound irritating.

Digital Devices

The key to the advent of small devices, and their popularity among consumers, is literally size, cost, and the quality of digital entertainment programs created from the digital signals' binary codes. That's a great advance in technology that now allows users to acquire any entertainment product they want, at any time, and in any place they happen to hear about it or see it. The benefits of such availability are enormous; but, there are potential problems associated with copyright violation, such that the traditional media and entertainment industry providers-movie studios, labels, and others- are experiencing varying forms of creative destruction. That is OK as long as the industry and the lawmakers figure ways to make sure the talents who create the products and the businesses who finance and market the entertainment products receive their fair share of the profits. That is where we are headed, just as the English Parliament did with the Statue of Anne with the copyright violation caused by the printing press. Managing changes and getting the most out of innovations just takes time!

The Decline of Record Sales

The "warn" sounds in analog and vinyl recordings are several different types of distortion, some of which are caused by the problem of trying to record on recording tape. Digital signals are also less expensive to use for telephone, television broadcast, and Internet transmissions. A few years ago, we had to use a converter box on old TV sets or throw them away because all TV broadcasts were converted to digital. Why? It saved bandwidth, had better quality, and was far less expensive to use. However, when the technology was converted into MP3 and then used by naive computer software geeks to upload entertainment products, the results were detestable to the music industry and crippling to the film, software, and computer game business. It cost the industry dearly, as *CNN Money* correspondent David Goldman reported. He notes U.S. sales of recorded music dropped from $14.6 billion to $6.3 billion in ten years.[5]

But back in the early 1990s, none of this stuff existed. Somebody had to invent a way to compress the digital signal to make it easy to make copies, broadcast, or transmit. At that point in time it couldn't be done, as the digital signal was just too much data. I was in Germany in the mid-1990s and visited a couple of studios where the big discussion was between the digital formats of MP2 and MP3. MP2 was the standard, but almost everyone thought (who had heard it) that MP3 was better. Both systems used compression to convert a single short recording into enough space to record it. But it was still too large to use. The trick was to eliminate some of the frequencies or sounds, which our ears could not hear anyway, into digital data to save space. Little did I know that at the time of my visit a graduate student named Karlheinz Brandenburg and his team of nerds were developing an MP3 software that would compress digital signals enough to quickly store digital copies of movies and recordings and even transmit them over the Internet.[6] They had no idea that would happen; they were just trying to solve a technologic audio problem. If they could solve the problem, they'd start an entrepreneurial company and license the process for a profit to the makers of devices, who would sell it to consumers. The consumers would then purchase the recordings they wanted and download them. Whoever thought the technology would be adapted for massive theft of property, the destruction of many careers, and the decline of an important segment of the entertainment industry?

The Storm Warning

I have written several books on the entertainment industry, and one of the ways I find information is to read the annual corporate reports. Polygram was at one time one of the big entertainment "majors," along with five others, who together controlled 80%–90% of the world's entertainment market. It changes each year but usually remains about an 85% average share of the market, with the remaining 15% coordinated by independents. I remember seeing one of the early reports in the 1990s that blew my mind and sent chills down my spine. Polygram had formed a "blue ribbon" panel of experts to best determine how to market digital entertainment products over the Internet for profit. Their conclusion was to sell the company, which they did as fast as possible, to Edgar Bronfman of the Seagram Liquor company in Canada. Why did they recommend selling the most profitable entertainment company at that time? Because they knew what was coming. Sadly, the software, computer gaming, and film businesses were about to have the same problems.

Seeders and Leechers

Somebody had to upload the new and in many cases prereleased CDs before hundreds of millions of people could steal billions of dollars' worth of entertainment products. The "seeders" and "leechers" are names for the persons doing the nasty deeds. A seeder is the person who illegally uploads a hard disk drive with the film, recording, or whatever. The leechers are people who use BitTorrent software to download a file from other computers using the same software. One provides the product; the other finds a way to distribute it. Sometimes, the leechers like to say the computer itself is the seeder, as if it were the computer doing the dirty deed all by itself. A tongue-in-cheek excuse for a crime when a crook gets caught is "the devil made me do it," which means they didn't know what they were doing. Right! Leechers use BitTorrent and other similar software because they are faster in grabbing parts of the data from any of the computers connected to the leechers' network. The more seeders in a system, the faster the download as more sources are available for the rip. Having BitTorrent is legal, but using it on illegal downloads isn't. When we write a song or a label creates a recording of that song, the owners of the song and recording have six property rights (discussed later).

BitTorrent

Steven Witt's book *How Music Got Free* describes how two North Carolina employees of a CD pressing plant stole unreleased CDs and used the MP3 technology to compress the music files. They then used BitTorrent to distribute those recordings to chat rooms used by thousands,[7] which in turn spread the unreleased recordings across the planet. How in the world was this happening? Nobody knew what was going on, but label executives started to get seriously concerned when the uploaded stolen albums started to outpace the sales figures on some of their newly released albums. Witt's account is an excellent read about all the crazy things that happened to facilitate the decline of the recorded music market, including the development of the technology, events, lack of aggressive legal action, and the naïve people who contributed to the "storms" that zapped much of the recorded sales market. At first, label executives turned a blind eye to the problem, hoping it would just go away. However, as new software was developed, it soon became clear that the front door of the candy store was left open and anybody who wanted to help themselves could and did. Screw ethics, the industry, the creative artists, and executives who had spent years developing their craft and knowledge, building the business that most of us use daily.

The Napster Storm

Shawn Fanning was a student at Northeastern University in Boston, John Fanning, his uncle, was an entrepreneur, investing in online software, and Sean Parker was a friend he'd met on a website message board.[8] Fanning was part of a computer security message service, using the handle w00w00. He asked the group for help, calling his new idea Music Net, which would allow music users to swap music files.[9] With the development of the software, venture capitalist Eileen Richardson was provided a copy to try. Her review for *Fortune* magazine's website states (2013),

So I went home that night and downloaded it. And holy shit. I remember at that time, folks started saying that people didn't want to download anything onto their computers anymore, because it was all getting to be web-based. But anybody would download that, easily. I just remember shaking, thinking: Oh my god, oh my god.[10]

The Perfect Storm's Destruction

The launch of Napster started the creative destruction process of the industry as 60 million users quickly learned to acquire millions of copyrighted recordings worth billions of dollars, resulting in a huge financial loss to both creative activity and to copyrights associated with the entertainment industry.[11] But Napster had a plan to market to the major labels its 60 million users. There was only one problem; they could not explain how anyone would make any money. Remember, the entertainment and music business is based on making profits, not giving away products in which the executives have invested millions to make available to the public. And look at how the labels perceived Fanning and Parker. They wanted to partner their company with the companies whose market they were destroying. Insane!

The Storm Chasers

The labels decided to use their professional organizations, the Recording Association of America (RIAA) in the United States and the International Federation of Phonographic Industry (IFPI), to fight Napster and all the other sites legally. Hillary Rosen, representing the RIAA, called Napster to explain the labels' position. According to Eileen Richardson at Napster,

Then I got to talk to Hilary . . . "You're letting people download . . . " I said, "You don't understand the technology. That is not what we're doing." She said, "You're infringing on copyrights." I said, "Well, I'd like to know which ones. Who, what, when, where, how?" And that's when she said, "Open up Billboard magazine! The top 200 are right there!"... And I said, "I'm sorry. I don't subscribe to Billboard magazine."[12]

Recording Industry Association of America (RIAA)

The RIAA represented the major record labels and filed lawsuits against Napster. Let the games begin! Napster lost case after case and quickly tried to change into a subscription business model. It tried to partner with the major labels and, at one point, offered the five major labels $1 billion to close the deal. At the time, the retail market for recordings was still worth about $35 billion a year, so their offer wasn't economically feasible compared to what they would lose.[13] Why accept $1 billion and lose most of your $35 billion market? Really insane! Napster burned a lot of bridges with the industry. After a brief merger with BMG (Bertelsmann Music Group), it was gone and so was BMG, as the parent company sold it to Sony.

Unfortunately, the hundreds of millions of people ripping off tracks forgot about the laws and ethics and clearly screwed the creative and business people who had invested their life's work and money into writing great songs, producing soul-wrenching masters, and of course, working hard marketing, promoting, and selling products. Several industry executives have commented to me, "I just do not understand how people who want a career in this industry could also destroy it." But did they destroy it? Or has it changed in a way that few are prepared to embrace?

Creative Destruction

Creative destruction is a business term that describes the process of innovation tied to creativity and business. Joseph Schumpeter first used the term in his book *Capitalism, Socialism, and Democracy* (1942), in which he described the process of "an industrial mutation that incessantly revolutionizes the economic structure from within incessantly destroys the old one, incessantly (creating) a new one."[14] In other words, change in society, business, and politics is usually caused by a discovery, invention, or theory. New products replace old ones, which provides consumers a way to improve or, at least, to seem to improve their quality of life. In politics, the process may lead to social or economic changes that, at times, may lead to tightened civil controls, economic downturns, job losses, destruction, and even war. Often the traditional establishment quickly adapts the new process or devices to solidify its political and economic control of the markets. The music and entertainment industry missed its chance. But it wasn't its fault entirely. The lawsuits against the pirates, thieves, and downloaders and the additional meetings the industry arranged with top congressmen and women and the U.S. Department of Justice have not taken the perfect storm off its path. A funny thing happened on the way to free; consumers found an alternative way to get free what once cost them money or time (watching advertisements) on the Internet. It's changed everything, including profits and losses, ethics, the market, consumer behavior, and the business structure of media and entertainment.

The creative destruction of the entertainment industry is not Napster's fault. It was just the bearer of the bad news. The

Figure 1.5 Technology brings change as consumers often use new devices and technology to their advantages and the traditional business models of broadcast and print media suffer. The decrease reflects the reduction of use in traditional media between 2010 and 2014 in 65 countries around the world. *Source*: "The Internet Is Gradually Replacing Traditional Media," by Statista Charts, with information provided by ZenithOptimedia (2015).

Increases in the Use of Digital Media Globally

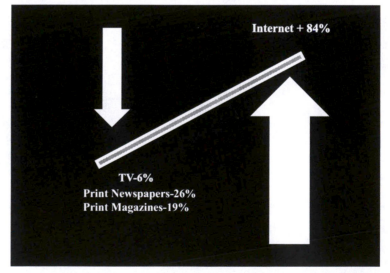

industry's gray-haired suits, creative insiders, and aging superstars feared this was coming and they never wanted to hear it, much less do something to prevent it. Clearly, something was going on and it was probably as earth-shattering as Columbus's trip, moveable print, capitalism, Marxism, the invention of the telephone, medicines, film, radio, TV, space travel, and maybe even gunpowder. Bang! It just causes a different type of explosion. Just as the telegraph replaced the Pony Express, the telephone wiped out the telegraph, the automobile zapped the horse and buggy, and the refrigerator melted the icebox industry, now it was iTunes, Amazon, and other digital providers that were wiping out retail stores and album sales. Digital downloads would sell lots of singles, but fewer albums. One comment often heard in the industry is how Steve Jobs "saved half of the industry" by destroying the other half. Fans didn't buy as many albums because now they could get the single they really wanted without having to purchase the entire album. Instead of the labels making a few dollars per album sold, they starting making just a few pennies on each digital single sold.

The Thrill of the Game

But the real problem was far worse, as over the next ten years additional illegal sites using BitTorrent and Cyber Locker generated over 50 million hits a month, many using the sites almost daily to kick the numbers over a billion. Pirate Bay, Kickass, and others facilitated massive theft of copyright entertainment products every day. Some of the sites added a social media side, allowing consumers to psychologically build a sense of community and false identity, and just as international spies and double agents pry information out of friends and associates, they were stealing something of value also. The thrill of the game was "don't get caught." Some of the users saw it as a conflict of good versus evil, with them being the "good guys," ripping off the labels and rich, arrogant acts that in their mind were "the bad guys." Silly as hell

Figure 1.6 The decrease in sales due to increases in streaming. *Source*: Nielsen, as displayed by Statista Charts (2014).

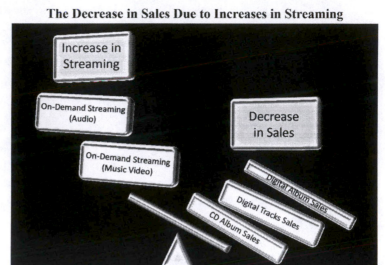

The Decrease in Sales Due to Increases in Streaming

of course, but entertaining and it didn't cost a dime. Meanwhile, the labels were starting to realize that it is damn hard to compete with free! Especially when consumers are now acquiring the products they spent millions creating. What a bummer!

Devastation

When the storm hits, there's no choice but to ride it out, but in the aftermath, what options do you have to respond? After a storm passes, there are varying degrees of devastation, which typically offer us two choices: to rebuild exactly what we lost, or to start over with something new or different that might better meet our current needs. When the digital supplier and Napster storms converged on the entertainment industry, it formed one of the most perfect storms and opportunities for creative destruction that any industry has ever faced. The traditional music and entertainment business is complex and the industry survivors have to pick up the pieces. Working in the business, I've seen the painful expressions on the faces of many industry executives, creative writers, producers, managers, professional musicians, and famous acts caused by the decline in product sales. There is a deep sense of loss not only to an industry they love but also to their financial existence.

Creating an entertainment product is not a one-person operation. Usually it takes money and many creative and business experts working together to create a song, recording, film, marketing plan, and so forth that may become a money-making machine with legs. It's risky, damn hard work, and the process is really a "perfect storm" of creativity and business, resulting sometimes in a hit and profits for everyone. But as good as industry professionals are in the music business, they'd make reasonable profits only on about 20%–40% of the acts signed. This means the majority of products fail. What we so easily call the entertainment industry is formed from a collection of artists, entrepreneurs, billion-dollar entertainment organizations, film production, record labels, the mass and social media (radio, television, print), cell phone networks, Internet portals, servers, technicians, producers, and others. It's an informal community of creative employees, lawyers, venture capitalists, and experts who create and market the entertainment products consumers purchase and listen to on the radio, Internet, and Spotify, and through films and other visual products, such as computer games.

Consequences of the Perfect Storm

Every product is usually the result of a combination of hard-working creative artists toiling away supported by businesses willing to spend millions on the chance to create some magic. In the old days, which seem like yesterday, if consumers wanted a new or older monster hit, as they discovered their need to hear it, they would buy a copy. Part of the royalties would be paid to the label to recoup its investments and part to the creators of the song to pay for the use of their copyrighted "property." However, the perfect storm nailed almost everyone in the business.

Success, if you want to call it that, is a mixture of oil and water that makes something out of nothing, with the results touching the hearts and souls of society. "It's not easy" is an understatement if you are an artist or know anyone who is. So, the pain felt by industry insiders is a rocket ride of gut feelings, financial risk, business sense, creativity, victory, and often failure. But the results of the perfect storm that hit them cost them almost everything they valued.

A perfect storm in business is a once-in-a-lifetime event, if it happens at all. But just as the print industry has reinvented itself, the entertainment and music industry is doing the same thing with streaming music and movies (e.g., Netflix and Spotify). It is still important and necessary to sell CDs and DVDs to consumers and digital downloads through pay services. But the profits just became a lot smaller part of the generated revenue since the perfect storm, and the industry can't survive on that alone. As more consumers switch to streaming the local radio and TV stations, the supporting networks of ABC, CBS, NBC, and Fox will continue to experience decreases in viewership in spite of the increase in the population. As technology continues to improve we may be able to completely cut the cable TV providers off since they are trending a little more like the Titanic, as consumers move toward viewing programing and listening to music on their computers connected to large-screen TVs.

Transitions of Revenue

Artists are usually angry about the way contracts are written once they discover that in the traditional recoupment system, they could have a gold record and not make a dime. The industry usually ended up on the profit side of that financial balance sheet. It takes money to make money in this game. Record labels invest about a $1 million or more in building a newly signed artist, who never has to pay it back (as we will see in Chapter 7). The industry recoups its "investments" out of the artist royalties from recording sales. Once the label's investment is paid off, the artists' share of the royalties generated would actually go to the artists. If (and this is the cool thing) the label never recouped all of its investment out of the artist royalties, the artists would still owe zero. The labels just made a bad investment and used the losses as a tax write-off. Of course, the perfect storm of technology and consumer defiance of the copyright laws has made it almost impossible for the industry to break even or profit on unit sales. (There will be more on this in Chapter 8—stay tuned.)

Equity Loss

Now, all that stuff is out there, available free on the Internet. It's almost as if 1 billion people in the world stole a few bucks out of five banks. There used to be five major labels; now the three that remain are struggling to build a new business model. So many people stealing a few bucks would empty the vaults of our mythical banks, driving them all out of business. Well, the vaults of the entertainment companies have also been emptied, as the inventory of classic film and music companies (most of the great copyrights in films, computer games, and recorded music) has been "lifted" and is available free online. That means if the labels and entertainment companies want to stay in business, they need to create new products. This means they have to be prepared to invest an average of $1 million into newly signed recording artists, and $10–$40 million into each new movie and TV series. And they know that new illegal consumer acquisitions will continue unless Congress and others feel some heat. See how difficult the business has become! Illegal downloads limit the potential size of the consumer base and market, financially crippling returns on the industriy's millions of dollars in investments, and they still have to balance their profit and loss statements. Try staying in the recorded music business when about 30 billion songs have been illegally downloaded and 95% of downloads weren't purchased. Time to embrace consumer innovation and the slide from sales to digital streaming.

Associated Negative Media Trends

The mudslide from the perfect storm is also starting to flow downhill toward the TV show providers, cable, and satellite networks, such as Comcast, Time Warner Cable, and Charter Communications. According to the Leichtman Research Group (2015), the largest pay-TV providers lost half a million subscribers in a three-month period (Q2–2015), which is a 400% increase in the loss of subscribers over the prior year. Ouch! The top cable networks lost 260,855 subscribers, satellite lost 214,000, and the Telcos such as AT&T and Verizon had less than half of a 1% increase.[15] Viewership is also down -36% to -14% for most of the TV programs on networks, such as Disney, the USA Network, A&E, TNT, and Nick at Nite.[16] It appears that 360,000 households switched or increased their broadband service after dropping the cable or satellite TV, sometimes with the same company.[17] Economics, timing, and convenience issues seem to be influencing many viewers switching to digital video-on-demand and Internet sources, as captured by Netflix, Amazon's Prime, and others, whose programming is available on large wall-mounted home computer display screens. Why pay all that money for cable TV and never watch most of the channels being offered when we can now find almost anything we want to watch on our computers? Instead of paying for two services, now only one is required. When revenue is decreased, it costs the industries that produce entertainment a range of related, direct, and indirect profits and creative people can lose their jobs or find they need to retrain to stay in an industry they love. Sadly, so many people have already illegally downloaded the best recordings and movies ever created, which is forcing the networks to improve existing or create new programming to draw in or try to retain viewers already satisfied with their stolen classics.

The Double Shuffle

A typical coin has two sides. That is so simple! The issues facing the entertainment industry have many sides. Politicians often move at the speed of molasses in winter, which means even now, about 25 years after this mess started, Congress and legislative entities around the world have failed to stop the illegal downloading. That really sucks for the creative folks and businesses that have lost billions of dollars to this storm. However, there is another side to the story and that one describes the results of the creative destruction process that most of us have experienced. When that process starts to happen money follows. The difference in quality between a manufactured CD and a downloaded one is impossible to hear. The revenue generated by the store purchase is about 100% greater than a pirated one, so the money stays in the pants of the thief and never reaches the industry. That's life and Congress may pass plenty of laws to stop it, but the horse is already out of the barn. So much has been downloaded that most people already have what the labels (and in some cases film production companies) created to sell. Why buy something you've already got? How does this crazy violation of laws happen in a western culture?

Capitalism

Over the years, attitudes about property rights have dramatically shifted toward a more socialistic or communistic theory. Capitalism is based in citizens' and companies' (assuming legal purchase) ownership of their own property. Here's an example: let's say you are William Colgate living in 1806 and you have developed toothpaste, candles, starch, and soap.[18] Consumers want the products, yet they cost money to make, market, and sell. We know his business model worked since we have many of his products for sale in pharmacies and grocery stores around the world. Currently, the company is still considered a huge success, with sales of $18 billion worldwide.[19]

We aren't selling toothpaste. And yet, songwriters and music publishers sell songs; labels sell recordings and the corresponding merchandise and tickets. Film production companies sell movies or the access to them. So, there must be a natural system of opportunity for creative product producers to generate the hope of profit and economic security for financial investors and others. In capitalistic countries, there is a natural balance between demand and supply. Private or public companies take the risks to develop products to fit consumer demand, and the price is usually based on whatever the consumer is willing to pay. However, nothing is really free. It stands to reason the companies developing and selling the products take the financial risk of failure or success. It's not a perfect system to say the least; yet, the products provided are usually of a much higher quality than products available under other political systems.

Communism/Socialism

Property rights are the opposite in communistic and some socialistic countries. In most cases, they follow some degree of Marxism, which sounds reasonable, yet just does not seem to work. All work is often assigned and workers are paid about the same salary by the government. Everyone is considered an equal part of the team of society. There are no property rights, as the government owns everything. Both systems claim that the "government" is a representation of the people, and yet under the dogma of the communist system, there is usually only one leader to vote for and you will do what you are told to do.

I visited the former Soviet Union several years ago, and in a meeting with a high government official we talked for a few minutes about music, entertainment, and property ownership, or copyrighted materials. He sincerely glanced over at me and the others I was with and said, "I just do not understand why music is not allowed to be used (heard) by everyone. It's such a wonderful thing—it should be free to everyone who wants to listen to it."[20] His kindness and sincerity caught me by surprise as I realized he just didn't get it! Why would anyone work so hard to create a product that improves the quality of life in a society if there's no personal reward for all the hard work and time spent creating it? Now, you understand why there are few to zero rich film, music, or rock stars in communist societies. There is no cultural sense of the way supply and demand drive the markets in consumer goods, products, and services. There's no profit in it and there may even be political blowback, so why even try? I've talked to several people who grew up and lived under a strict communist society and even under the socialistic political systems of Western Europe. Most of those countries are in financial trouble, just as we experience in capitalistic societies, yet many young people are hopeless when it comes to finding a job and career. We have entertainment products to enjoy during the difficult times in our lives, and now because many in Eastern Europe and Asia have lifted much of it, they may also enjoy it free, contributing to the perfect storm.

The Difference of Freedom

Think for a moment about something with which you might be more familiar: baseball players defecting from Cuba, or Russian gymnasts seeking asylum when the Olympic games were held in Western Europe. There are many stories of people who have left oppressive regimes to find the freedom to create and use their own voices and to personally profit from their hard work. Here is one you might not know—at least, I did not and I think it is very interesting. I found an article on the Autonomous Nonprofit Organization website (2010) that tells the story of the now largely forgotten composer Dimitri Tiomkin (1894–1979), who was born in Russia and immigrated to Europe after the Russian Revolution, which placed all citizens under communism. In the United States, he sold his jazz compositions to industry leaders and met major Hollywood film producers and directors, such as Frank Capra and Alfred Hitchcock. His success as a film composer won him several Oscars for songs he wrote for more than 100 film scores, including "Do Not Forsake Me, Oh My Darlin'" for the movie *High Noon*, and others for *Shadow of Doubt*, *Strangers on a Train*, *I Confess*, *Dial M for Murder*, *Red River*, *Rio Bravo*, *The Alamo*, and even *Mr. Smith Goes to Washington*.[21] Under capitalism, Tiomkin had a very successful artistic career and enjoyed all the freedoms associated with the capitalist free market economy. It's doubtful he could have enjoyed a similar record of success under communism. It is worth pausing to think about the great movie songs we might never have heard, as he likely would not have had the opportunity to write them under that system.

The ways of thinking and control issues are very different in non-free societies. In Moscow, at the time of our visit, citizens were not allowed to cross a street directly on foot. Instead, they were required to use the tunnels under the roads located at the end of each street. Late one night, the group I was with walked back to our hotel and had to get to the other side of the street, so we followed the rules and went down into the tunnel to get to the other side. We heard some fantastic classical music being played in the tunnel, echoing off the bright tile walls. We stopped and listened a few minutes, realizing how heartbreaking it must be for some of the best musicians in the world to have to perform in a damp tunnel for tips to survive.

Popular Culture

Popular culture is often born out of the innovative creative artists who see and perceive what we all can see, but they see it differently. That's often the first step in becoming an artist. Then the struggling artists may possess the passion and inner drive to explore what is unique about their visual and emotional perceptions and develop those into entertainment products that reflect their vision. Truly gifted artists normally have an amazing talent combined with the ability to express their creativity in a tangible form of artistic expression, such as a painting, song, script, movie, computer game, fashion design, or jewelry. Think of some of our history's great creative artists and read about their lives. As an example, take a quick look at Paris in the 1920s and then consider the accomplishments of innovators such as Claude Monet, F. Scott Fitzgerald, Josephine Baker, Ernest Hemingway, Juan Belmonte, Gertrude Stein, Pablo Picasso, Djuna Barnes, Salvador Dali, Man Ray, Luis Bunuel, T. S. Eliot, Henri Matisse, Leo Stein,

Henri de Toulouse-Lautrec, Paul Gauguin, Edgar Degas, and "Coco" Chanel. Race, gender, nationalities, sexual preference, and politics mattered to them as together they struggled with creativity, innovation, art, and culture. They all changed the world with their artistic "works," most of which consumers today find priceless. Many lived depressed lives, most were broke financially, some emotionally, and yet from their "works" consumers often find joy and beauty. In our modern world of "free" they may never have even tried to create what they did, and look at what humanity would have lost.

Shift in Consumer Behavior

So, if we start applying the "take whatever we want" mentality to entertainment and music products, where does that leave us morally and commercially? The illegal providers of the entertainment products didn't see themselves as bad guys or people designing systems meant to harm, cheat, or steal from others on such a large scale. Rather, perhaps, they saw themselves as building a virtual community of fellow pleasure seekers. In effect, they were the grandest of the cybercriminals who never grew up, who enjoyed never having to accept or face the results of their naive behavior.

I am not political, yet there are a couple of concepts about the types of political situations we live under that need to be mentioned. Not making any judgments but having traveled throughout the world, I believe the quality of one's life is usually happier and much better in a capitalistic society. I rarely hear anyone talking about the perfect storm's effects on an artist's notion of creativity as tied to different forms of government. However, this perfect storm has shown that many have forgotten the value of the entertainment products they have acquired freely, as most never consider that great art, medicine, and most of what we have in the West are not found in other types of societies. That is simply because there is no motivation or profit to do the work, take the risk, and spend years developing artistic talents in many societies. Remember that it takes a team of people, creative and business, working together to provide the high level of artistic products and live shows we have. Take the profit margin away and we end up with the quality of life in 1950s Cuba, but still we live in a modern society in the 21st century.

Copyrights (©)

Copyrights provided by the laws of the United States (Title 17 U.S. Code) and similar laws in other free countries provide protection—that is, ownership of their creative properties—to the authors of original works of authorship, including literary, dramatic, musical, artistic, and other intellectual works.[22] Copyrights encourage artistic and creative people often using their own inventions and discoveries to create, among other things, songs, musical recordings, paintings, movies, stage plays, books, photographs, choreographic works, computer-enhanced artwork, and films. I go into more depth on copyright laws in Chapter 3, but for now, let's use this nutshell summary to keep the conversation going.

Figure 1.7 When we author a "creative work" that meets the qualifiers, we own the "creative work entertainment product" (such as a song or a recording of a song), as our legal property. We also gain six "related rights" that must be paid if somebody else wants to make copies, distribute, use it in a public performance (a bar, as an example) or public performance by means of the Internet or digital transmission, or if they create a new copyright out of our original copyright. Penalties include jail time and as much as thousands in fines if the violation is caught and proven.

The Exclusive Rights

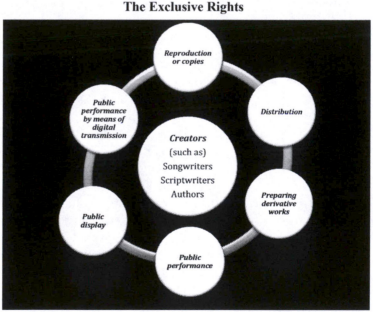

The perfect storm in the entertainment industry was clearly caused by many things, but this crazy change in attitude toward "free access" to others' property is a serious ethical and social problem that I believe will lead to a positive and fundamental change in how we acquire entertainment products. It will work only if, and when, the copyright owners who created the songs, recordings, films, books, computer games, and other products are paid fairly for their endeavors. You will have the opportunity to learn more about these exclusive rights in Chapter 3, and how they are in jeopardy today because of innovation and the widening use of technology in the entertainment and music industry.

What We Really Own

Are you starting to understand why there are very few, if any, entertainment superstars in countries without property rights? There's no money in it except maybe in tips and for musicians playing classical, public domain pieces, usually concerts approved by their governments. Even in our democracy, acts, songwriters, copyright owners, labels, and others are stuck with declining royalties in recorded music sale and movies, due not to our laws but to the advancement of technology and consumer behavior of acquiring products for free. How weird is that! Consider this: with many consumers downloading entertainment products without paying for them, what has happened to the legal protection artists should expect? Is copyright infringement merely a thing of the past? As recently as 2012, according to the Copyright Office Circular 1 (2012), the U.S. government reinforced, on paper at least, that "It is illegal for anyone to violate any of the rights provided by the copyright law to the owner of copyright."[23] Buying a CD or DVD does not give us any property ownership right of the entertainment products other than the plastic disc we bought that is full of "pits." Those holes or "pits" reflect back to a sensor in the device either the presence of a hole (computer analysis as the number one) or no reflection, which the computer analyzes as a zero. The string of ones and zeros becomes the binary code the computer chip in our device reproduces into the music or movie we want to see. We own the shining disc or digital download signal on a hard disk drive, but not the rights to the entertainment products that are digitally being reproduced.

Creator's Ownership Rights

Let's get creative and write a great song—at least we think it's great—about our boy- or girlfriend. She's so darn cute that let's also take a picture and video with our cell phone and then download them onto our computer. Let's really get crazy and record some tracks on our computer, using GarageBand or Pro Tools. Sounds good, looks even better. Man, I got talent and I can't wait (nor can you) to let her hear it. She'll be so impressed! But wait a minute. Who owns the song, picture, video, and music tracks we just created?

An *idea* for the song, video, or picture is not protected, but if it's written down, sung into an iPhone, recorded, or whatever, then it's yours, as long as it meets the legal definition of copyrightable material. In other words, songwriters who come up with an idea and turn it into a physical tangible form of expression (e.g., a song or video) stored as a binary magnetic recording now have a form of legal property and they own it because they created it. Further, it's original and has been placed it into a physical form. Cool, very cool. In this case, what they've created is now a legal piece of property that we own, similar to the ownership of our house, land, car, computer, first-born kid, blah, blah. Therefore, remember that an idea isn't good enough; you've got to write it down, record it, and get it into a physical form before it legally becomes property and yours. According to Chapter 300 of the Compendium of the U.S. Copyright Office Practices (2014),

> *A work of authorship may be deemed copyrightable, provided that it has been "fixed in any tangible medium of expression, now known or later developed, from which [it] can be perceived, reproduced, or*

Figure 1.8 Exclusive Rights Payments: Here is Figure 1.7 again. This time take a close look at the six rights from the perspective of how you are going to be paid for your product. If we rip off songs or a recording of the song, then the songwriter, music publisher and label will not be paid for their creative works. That's a violation of copyright law. Music publishers usually gain ownership of the song's copyright through a single song contract where they share 50% of the royalties with the songwriters. Labels pay for a mechanical license to gain the rights to record and distribute (for sale) the recordings of the song. When we rip the recording off, none of the copyright owners are paid. The arrows represent the licenses payment due to the copyright owners being ignored by some consumers and businesses that illegally use new digital technologies to bypass the law. The copyright owners of the song, book or script (as examples) © and the record label ℗, film production and book publishing companies ™ are not paid for their creative works.

Required Payment

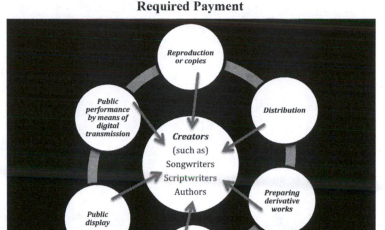

otherwise communicated, either directly or indirectly with the aid of a machine or device" . . . Specifically, the work must be fixed in a copy or phonorecord "by or under the authority of the author" and the work must be "sufficiently permanent or stable to permit it to be perceived, reproduced, or otherwise communicated for a period of more than transitory duration."[24]

The Useless Fine

If we create something we own, what property "rights" do we have? Unless we reside under a different form of government (a dictator, king, or communist or some socialist governments), we gain six powerful (at least, on paper) rights. Nobody is allowed to make copies, distribute, perform, create derivatives, digitally transmit, or publicly display what we've created without our permission! Do I need to say it again, place it into a larger font, scream, or pull more of my

Table 1.2 The Creative Circle. The Creative Circle represents the beginning of the magic of entertainment in the form of a three-minute story, such as those contained in a poem, song, fiction or nonfiction book manuscript, script of an original book, or even a concept turned into a computer software game. The authors who create these original works are the owners of the books, scripts, and songs they created. They are provided with property rights, defined as a copyright, granted by the United States and other free-world nations' governments and laws that respect copyrights. Entertainment-based companies, such as book publishers, record labels, and movie production houses, must pay for the rights to turn the authors' scripts, books, songs, and so forth into entertainment products consumers can buy, use, and enjoy.

Author/Copyright Owner	Agent	Acquiring the Rights/ Deal	Royalties or Payments	Collection of Royalties or Payments
Songwriter(s) (often sell copyright in exchange for a percentage of generated royalties)	Music publisher (usually acquire copyright and all six exclusive rights); usually acquires a deal by pitch or referral of material	Single-song contract	Royalties based on a percentage of generated royalties determined in the deal	1 **Mechanical:** Royalties from license for recording and sales and digital downloads 2 **Blanket licenses:** Public Performance Royalties from public performances 3 **Sync license:** From songs placed into movies, TV shows, and visual presentations 4 **Other sources**
Scriptwriter (retain copyright)	Literary agent and some talent agents (get a percentage of sales or paid a one-time fee)	Agency/union representation, such as the WGA or talent agent working independently (which you should be careful about)	Paid a percentage based on length and type of presentation, number of times viewed, or one-time fee	Payment is usually based on option price (give a producer the right to make a movie and the script is taken off the market) for 10% of the full price of the copyright of the script. If the producer exercises the option, the writer receives the other 90% plus additional fees for rewrites. Consider becoming a member of the WGA.

Table 1.2 Continued

Authors (retain copyright)	Literary agents (get a percentage of sales or paid a one-time fee), connect by providing a query letter and proposal	Agency / representation contract	Paid a percentage or one-time fee	Authors are usually paid a royalty based on each unit sold. However, be careful in the language of the contracts to make sure you will retain or be paid for e-books and associated rights, such as foreign, film, audiobook, and serials.
Computer game designer (often sell copyright in exchange for a percentage of generated royalties)	Direct to company or company agent representation	Agency / representation contract	Royalties based on a percentage of generated royalties determined in the deal	There are several types of "deals" available, such as collaboration agreements, licensing, development, and even employment contracts. Payments, lump sum or royalties, plus ownership depending on the type of deal and who owns the original copyright.

hair out? Wow! And the duration of time we own it is usually 70 years past the copyright holder's death, although it varies according to the type of distribution and country we reside in. In addition, if we illegally download just one recording, we've, in most cases, violated two copyrights, those of the songwriter and music publisher who own the song and that of the record company who owns the recording of the song. It's a federal offense in the United States that carries the maximum criminal penalty of as much as a $250,000 fine, plus five years in jail, per offense. Ouch! We could also get hit with civil penalties, which can total a minimum of $750, plus legal and court costs. But I'm not done yet. In some cases, the music publisher can also be penalized with a maximum fine of $150,000 in statutory damages if the publisher distributed the song, though we downloaded it illegally. You do the math. How much money do you owe and how many years should you serve if you were caught for all the recordings you've downloaded? Clearly, the fact that you have the software to download others' entertainment products does not make it right or legal. All the fines for all of the products illegally ripped off alone would probably be close to $1 trillion. But the industry knows that the vast majority of those who violated copyrights could never afford the fines or time in jail. So you got away with it! And if you are Steven Witt, you even wrote a book about it, for guess what? Profit!

Talent

Creativity is usually understood as the "talent" the artist presents in the products—that is, recordings, concerts, songs, videos, games, and movies. The amount of "talent" housed in an act, or presented in a recording, game, movie, and video, is determined by the business side of the industry as a profit or loss on its profit and loss (P&L) statements. Thus, talent in our industry has different meanings to various entities. The creative contributors (songwriters, musicians, engineers, producers, scriptwriters, and video and movie producers) analyze the quality of the "artistic value" of a final product, while the business side views financial returns or losses on its investments. Talent means creating entertainment products and showing the creative side of the industry and profits and losses to the business side. Having "talent" means you may be a great scriptwriter, songwriter, producer, act, or musician, and at the same time, it also means having the personality, knowledge, and ability to market, distribute, and sell creative products to the business side.

Valuing Entertainment

Think of the songs, films, media, and so forth that got us through falling in and out of love, divorce, wars, personal and national illnesses, and other events affecting our daily moods and life. What's being created, marketed, and sold are packages of the best and worst of humanity, and most everything in between usually being purchased, downloaded, listened to, or stolen is a song, book recording, or video game, or presented on stage or even a street corner. The "punch line," if you will, of a creative work varies by the type of media it is. In a film, it may be connections between the consumer and the creator around emotions such as courage, humor, or the intellectual pleasure of experiencing a good storyline. In a computer game, play, or book, it may be the unraveling of a suspenseful narrative. The potentially profitable packet can come in multiple forms (recordings, films, digital transmissions) that consumers may use to better connect to something they are feeling emotionally. In the film industry, it's the movie tied to the persona of the actors and storyline being sold to consumer emotions. It's the same thing for theater, books, computers, games, sports, politics, and any other form of entertainment. It's mental escapism, an imagination, hoot, and fantasy. And it's also one heck of a great business if you know what you are doing and have the talent to actually write, create, produce, market, and sell it to consumers who perceive the products as an innate value. It's often considered a very risky business, tougher than a slot machine in Las Vegas.

The entertainment industry is a unique combination of various forms of entertainment products that according to government sources and the accounting and research firm PricewaterhouseCoopers grosses about $546 billion in the United States and close to $1.8 trillion globally. I dig the M in million, love the B in billions, and get a little thunderstruck with the T in trillion. It's a business that grosses more revenue than some countries. The industry operates on the success of connecting their products to human emotions. We're not selling hot dogs or cars. We're selling the gas for the car, the mustard on the hot dogs. We're selling entertainment products consumers use to feel something about themselves— hopefully love, happiness, or joy. But it costs money, time, and really talented people to do it right. Consumer preferences are also a constantly moving target, changing, maturing, or seeing our individual worlds of mental reality.

Entertaining the Future

Great things are happening to change the filmed and recorded elements of the industry from a "purchase" model to an access / streaming model. Don't be surprised if it changes again in the future to something totally different, such as direct feed from the production companies or labels to the consumers. Don't be surprised if some type of phase-cancelling irritant tone is activated if you try to make a copy or upload your products to the Internet. That would take out all the middlemen and reduce the cost of distribution and maybe even the retail price. It takes guts, brilliance, and innovation to embrace new technology and create a new business model out of zip. We are in the middle of the shift from one model to the next one. Live theater, performances, festivals, and other entertainment forms have also enhanced their purchase models by expanding value-added opportunities. It's hard to sell CDs at live concerts anymore or even buy DVDs at a retail store, but people still desire to be entertained. That's not going away, but it's time to make sure the people (probably you in the future) who create it and businesses that invest in it profit from their hard work. My guess is that this new attitude of "if I can get it free online, it's okay" will not change quickly, so the industry will simply have to develop better access methods, such as streaming with access codes or, God forbid, government licenses and taxes.

The House Judiciary Committee

All of us are at fault for non-enforcement of the laws designed to protect artists and their works if we have ripped off the people who worked their butts off, investing time and money in creating, marketing, and selling the stuff we enjoy. It's not as if we could have anticipated this would happen, but now that we know, we have both an opportunity and obligation to do something about stopping the bleeding of profits from the artists and from the entertainment industry—that is, if you agree with me that is a problem in need of a solution. According to an interview by Sam York with Jacques Attali and printed as an article in the BBC News (2015),

> *a French polymath called Jacques Attali wrote a book that predicted this crisis with astonishing accuracy. It was called* Noise: The Political Economy of Music *and he called the coming turmoil the "crisis of proliferation" . . . Soon we would all have so much recorded music it would cease to have any value. . . . Music, money and power were all tightly interlinked, he wrote, and had a fractious relationship stretching back through history. Powerful people had often used music to try and control people. In the 9th century, for example, the emperor Charlemagne had imposed by force the practice of Gregorian chant "to forge the cultural and political unity of his kingdom". Much later, the arrival of capitalism and the pop charts gave moguls the chance to use music to extract large amounts of money from people . . . But at the same time, music can be used to subvert power, and undermine the status quo. Rock and roll in 1950s America, for example, helped to sweep away a raft of conservative social mores.*[25]

In his interview, Attali makes another prediction that might interest us (2015):

> *Attali also had another big idea. He said that music—and the music industry—forged a path which the rest of the economy would follow. What's happening in music can actually predict the future. . . . When musicians in the 18th century—like the composer Handel—started selling tickets for concerts, rather than seeking royal patronage, they were breaking new economic ground, Attali wrote. They were signalling the end of feudalism and the beginning of a new order of capitalism. . . . In every period of history, Attali said, musicians have been at the cutting edge of economic developments. Because music is very important to us but also highly adaptable it's one of the first places we can see new trends appearing.*[26]

Here we are in 2016 and the piracy figures continue to grow and increase. As an example, according to *Music Business Worldwide* (2016) there were 576 sites (many in third-world countries, I assume) that were "wholly dedicated to music piracy or contain[ed] significant music content."[27] They had more than 2 billion hits on the sites last year, which is 28%–30% of the 7 billion people in the world.[28] Congress has not been as vigilant as the entertainment industry has needed for the past 20 years and more, seeming to fail in proactively addressing the changes the new technology has brought to copyright laws, property transfers, and safe harbor restrictions. As a result, in some cases, absence of leadership at the national level has encouraged illegal transfers of property without payments with safe harbor restrictions that allow Google and YouTube to be seen as in compliance with existing laws. However, there may be a change in the density of the smog. In the fall of 2015, House Judiciary Committee Chairman Bob Goodlatte and Ranking Member John Conyers announced the next step in the House Judiciary Committee's ongoing review of U.S. copyright law. According to Chairman Goodlatte (2015),

> *Two years ago, the House Judiciary Committee began the first comprehensive review of our nation's copyright laws since the 1960s.*

The goal of our review is to ensure that our nation's copyright laws keep pace with the digital age. The next step in our review will be an opportunity for all interested parties, prior witnesses or others, to share their interests and concerns with Committee staff as we consider updates to our nation's copyright laws.[29]

And Ranking Member Conyers added (2015),

Copyright plays an increasingly important role in the American economy. The two-year copyright review highlighted several areas where the copyright community can potentially find common ground. I look forward to hearing from all stakeholders about their additional views and proposals that will strengthen the copyright system . . . America's copyright industries—movies, television programming, music, books, video games and computer soft-ware—and technology sector are vitally important to our national economy. The House Judiciary Committee's copyright review is focused on determining whether our copyright laws are still work-ing in the digital age to reward creativity and innovation in order to ensure these crucial industries can thrive . . . In the coming weeks the House Judiciary Committee will conduct several round-table discussions to hear directly from the creators and innova-tors about the challenges they face in their creative field and what changes are needed to ensure U.S. copyright law keeps pace with technological advances.[30]

Sad, and the money is gone, people and property owners have not been paid, but there is a light at the end of this depressing tunnel. It's taken 40 years to understand Attali's predictions about the destruction of personal property wealth. Use of the Internet, in particular, combined with naïve consumer behav-iors has allowed people to stack their hard disk drive full of the world's best entertainment products for free. However, our elected representatives have a full plate of political issues to deal with so copyright violations may not be very high on their list. It's got to be a thankless job and the responsibilities are certainly over the top, so I'm glad it's them and not me. But, eventually, it has to be you, as you will need to step up and provide our leaders with your views on such matters as fair use, copyright ownership, and digital as well as intellectual property rights, and you will want to do this when it is your creative product, process, invention, or innovative practice at stake.

In the long run, innovation and creativity will win, as people, you and I have to hope, will still need entertainment products to set their alarms to, answer the cell phone, celebrate their essence, cry when life hurts, and dance to the beat for joy. Let's all take a deep breath! After all, this is a business often defined as show business, priceless to the people who create it, work in the industry, and most of the consumers who enjoy it. As the famous song by Irving Berlin reminds us, there's no business like show business.[31]

Read On!

Learning *how* the industry works and about all of the corresponding business, legal, and creative opportunities (careers) will leapfrog you ahead of the millions of others who want to start an industry-related entrepreneurial venture, become a writer, musician, film guy, and so forth, or find a job and build a career in this business. There is much you need to know, and even as it goes through the creative destruction process of the natural business cycle caused by technology, innovation, and financial concerns of industry leaders, the industry is growing with new, exciting, better jobs. Opportunities in entrepreneurship are exploding and new concepts for entertainment products and their uses are being developed. Indeed, consumers want more entertainment products, not fewer. Getting consumers to pay for it, in some cases, may be a little tricky. But with your help, the industry will do it.

A Word About How This Book Is Designed

In trying to provide you with the tools you will need to succeed in the entertainment and music business, you will see

Figure 1.9 The value of your talents are a combination of knowledge (how business works and the uniqueness of the industry business models), your creative abilities (as an artist, writer, musician, sales, social media, promoter or marketer), your creation and staging of products or shows, plus the ability to network to meet industry executives in order to get your shot and to work successfully in the business.

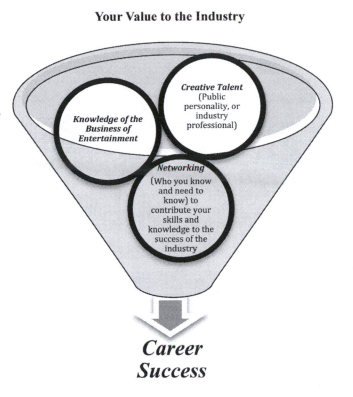

Your Value to the Industry

Knowledge of the Business of Entertainment

Creative Talent (Public personality, or industry professional)

Networking (Who you know and need to know) to contribute your skills and knowledge to the success of the industry

Career Success

I have packed a lot into these pages. Remember, it has taken me a lifetime, and my professional careers in the industry and as an educator, to lay all of this before you. Because the book is a handbook, it is meant for you to read now, in whatever context you have found it, and for you to return to as a guide and reference work. The pattern of the book is relatively simple, and I emphasize throughout how an insider sees the entertainment and music business, from the creation of products to legally protecting the product and the artists as their investment to getting the products to the consumers. You will sometimes see figures repeated across the chapters and some common topics introduced and revisited in different parts of the book. I did this intentionally to help you gain the knowledge you need and to understand how the industry is like a

3D puzzle that you can make into different shapes depending on what you want to accomplish with entertainment products. I started in this chapter to give you a big-picture overview of what it means and what it looks like to be involved in the entertainment and music industry today, which is continued into Chapter 2, on what I would say is your goal for being in and making products in this industry—how to find the "wow." Chapter 3 is focused on copyright law and how it functions in the entertainment and music industries.

This is a central concept in the book, and one you will see recurring throughout. Chapters 4–8 take you into the music business, starting with a clear explanation of how the business plan and business channel development are key to the kind of entrepreneurship required to be successful in the industry. In Chapter 5, I tell you the stories of two phenomenal songwriters whom I have had the pleasure to count as friends and family. There lies the basis for a "it's really a business" look at how the recording studio, the studio budgets, and the royalty processes all work from the monetization of the creative product standpoint.

With Chapter 9, we step inside the business of some of the remaining powerhouse record labels to see how they work (spoiler alert: the 360 business deal is really important) and from there, we move in Chapters 10 and 11 to looking at how you need to shape your career. By this point, as you read about how to make money with your brand and how to find and use an agent well, you should be on your way to fully realizing the intersections between creativity, money, the business, and the digital transformation of the industry. You may also be thinking as I have about solutions to this runaway marketplace that will allow us to keep making a living as artists and keep motivating authors to write and sell their works. Chapter 12 presents an overview of the ways we can still deliver our creative works to the market with an emphasis on the business of live touring. Just as you have the chance to build the recording budgets to see how much it will cost you to produce one song and one album, you will also learn how to manage the financial building blocks of live touring using the information on the business process you learned in Chapter 4. Finally, we will take a look at what I think the future of the entertainment and music industry might be and suggest to you how you can use the information in this book to pull together a coherent picture of the future of our industry.

Notes

1. Witt, Stephen. "Introduction." In *How Music Got Free: The End of an Industry, the Turn of the Century and the Patient Zero of Piracy*, 1. First ed., New York, NY: Viking/Penguin Random House, 2015.
2. Honeyman, Thomas. "How One Generation Was Single-Handedly Able to Kill the Music Industry." *Elite Daily*. June 6, 2014. Accessed May 30, 2015. http://elitedaily.com/music/how-one-generation-was-able-to-kill-the-music-industry/593411/.
3. Froman, Michael B.G. *2015 Special 301 Report*, 12. Washington, DC: Executive Office of the President of the United States, 2015.
4. Lehmann-Haupt, Christopher. "'The Perfect Storm': Shipwreck Story No One Survived to Tell." *NY Times: Books*. June 5, 1997. Accessed August 18, 2015. https://www.nytimes.com/books/97/06/01/daily/storm-book-review.html.
5. Goldman, David. "Music's Lost Decade: Sales Cut in Half." Money.cnn.com. February 3, 2010. Accessed August 21, 2015. http://money.cnn.com/2010/02/02/news/companies/napster_music_industry/.
6. Witt, Stephen. "Chapter 1." In *How Music Got Free: The End of an Industry, the Turn of the Century and the Patient Zero of Piracy*, 5–25. 1st ed., New York, NY: Viking/Penguin Random House LLC, 2015.
7. Ibid.
8. Nieva, Richard. "Ashes to Ashes, Peer to Peer: An Oral History of Napster." *Fortune*. September 5, 2013. Accessed June 15, 2015. http://fortune.com/2013/09/05/ashes-to-ashes-peer-to-peer-an-oral-history-of-napster/.
9. Ibid.
10. Ibid.
11. Ibid.
12. Ibid.
13. Ibid.
14. Schumpeter, Joseph A. "Can Capitalism Survive?" In *Capitalism, Socialism and Democracy*, 81. Sixth ed., New York, NY: Routledge, 2006.
15. "Major Pay-TV Providers Lost About 470,000 Subscribers in 2Q 2015." Leichtmanresearch.com. August 17, 2015. Accessed August 26, 2015. http://www.leichtmanresearch.com/press/081715release.html.

16. "US Consumers Continue to Turn Off Pay TV Subscriptions." *EMarketer Media & Entertainment*. August 26, 2015. Accessed August 26, 2015. http://www.emarketer.com/Article/US-Consumers-Continue-Turn-Off-Pay-TV-Subscriptions/1012906?ecid=NL1001.

17. "About 360,000 Added Broadband in 2Q 2015." Leichmanresearch.com. August 18, 2015. Accessed August 26, 2015. http://www.leichtmanresearch.com/press/081815release.html.

18. Cook, Ian. "Our Company History." Colgate World of Care. 2015. Accessed June 28, 2015. http://www.colgate.com/app/Colgate/US/Corp/History/1806.cvsp.

19. Cook, Ian, and Dennis Hickey. "Colgate-Palmolive 2013 Annual Report." Cogate.com. 2014. Accessed June 28, 2015. http://www.colgate.com/us/en/annual-reports/2013/financial/.

20. Russian Government Office (personal reminiscence by author). May 1, 1989.

21. El-Bakri, Diana. "Russian Born Composer Who Scored Hollywood." RT Question More. May 9, 2010. Accessed August 21, 2015. Autonomous Nonprofit Organization.

22. *Copyright Law of the United States and Related Laws Contained in Title 17 of the United States Code*. December 1, 2011. Accessed June 6, 2015. http://copyright.gov/title17/circ92.pdf.

23. "Copyright Basics." Copyright Basics, Circular 1. 2012. Accessed June 9, 2015. http://www.copyright.gov/circs/circ01.pdf.

24. "305 The Fixation Requirement." Copyright.gov. December 22, 2014. Accessed June 15, 2015. http://copyright.gov/comp3/chap300/ch300-copyrightable-authorship.pdf.

25. York, Sam, and Jacques Attali. "The Pop Star and the Prophet." *BBC News, Magazine*. September 17, 2015. Accessed September 17, 2015. http://www.bbc.com/news/magazine-34268474.

26. Ibid.

27. Ingham, Tim. "Global Music Piracy Downloads Grew by Almost a Fifth in 2015—Music Business Worldwide." *Music Business Worldwide*. January 21, 2016. Accessed January 21, 2016. http://www.musicbusinessworldwide.com/global-music-piracy-downloads-grew-by-almost-a-fifth-in-2015/.

28. Ibid.

29. Rexrode, Kathryn, and Jessica Collins. "Goodlatte & Conyers Announce Copyright Review Listening Tour." U.S. House of Representatives, Judiciary Committee, Chairman Bob Goodlatte. September 10, 2015. Accessed September 17, 2015. http://judiciary.house.gov/index.cfm/press-releases?ID=73277024–3AA3–464B-B4A3–2A907C9FEDB2.

30. Ibid.

31. "Irving Berlin—There's No Business Like Show Business Lyrics." 2015. Accessed August 18, 2015. http://www.metrolyrics.com/theres-no-business-like-show-business-lyrics-irving-berlin.html.

THE THREE WOWS   2

Good Enough

Are you good enough to compete with the best in the world already in place and rocking it as entertainers and industry leaders? The answer is yes, of course, but it's not easy to learn all you need to know at the entry level, let alone become the CEO of a major label. Do you get ahead with luck? With who you know? How about talent? Be yourself, whoever that is, and be nothing short of great. Be as weird as you want, just be great at it, and then hopefully, someone in the industry will take a shot at seeing if there is a viable, profitable market for it.

The Three Wows

When it comes to advice about how to "make it" in the entertainment industry, there are lots of opinions, and some advice is better than others. For instance, my long-time friend and successful music industry artist manager Louis O'Reilly has three things he considers when looking at signing an act. He calls them the three wows and you've got to have at least two of them. They are a *wow song*, *wow brand*, and *wow show*. Wow! Just say it to yourself out loud and you'll start to understand what Louis is talking about. In the theater and film business, it's a wow screenplay, script, or story, a wow brand in this case means a wow metaphor (*The Godfather*, *Harry Potter*, *Star Wars* as movies, *Cats*, *Chorus Line*, *The Music Man*, and *Rent* as stage play), and a wow show means the creative team of famous actors led by equally experienced producers and cutting-edge, creative-minded directors.

If you want to be a major recording artist, you need the three wows—song, image, and live show—to make enough money

Figure 2.1 If you're serious about an entertainment industry career in the creative side or as a business executive side you'll have to be able to either create or know when you see or hear a Wow song, image/brand, and live show.

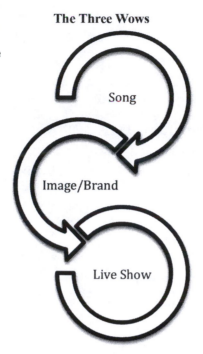

The Three Wows

Song

Image/Brand

Live Show

for you and those who work for you to become successful. One of the biggest mistakes you might make is to assume because you have talent you should become a big star. But as you can see from Figure 2.1, you need all three to become a sellable personality.

The Wow Song

Think about the wow songs from Lennon and McCartney, Roy Orbison's "Only the Lonely" and "Oh, Pretty Woman," Hank Williams's "Your Cheatin' Heart" and "I'm So Lonesome I Could Cry," The Beach Boys' "In My Room" and "Good Vibrations," Abba's "Dancing Queen," Freddie Mercury and Queen's "Bohemian Rhapsody," John Fogerty's "Proud Mary," Jagger and Richards's "(I Can't Get No) Satisfaction," and Dylan's "Just Like a Woman," "Knockin' on

Figure 2.2 The top-selling single songs of 2015 include "Uptown Funk," sung by Mark Ronson and Bruno Mars, with 1.76 million units sold (CDs, digital downloads, etc.). Second place is "Cheerleader," with 1.52 units million sold. "Take Me to Church" moved 1.25 million units, and "Love Me Like You Do" sold 1.19 million with the help of its being in the movie *Fifty Shades of Grey*. "See You Again" followed, with 1.17 million units sold, and Adele's "Hello" hit sixth place even with it being just released. *Source: The Official Top 40 Biggest Songs of 2015 Revealed*, Official Charts (2016).[1]

Top-Selling Wow Songs

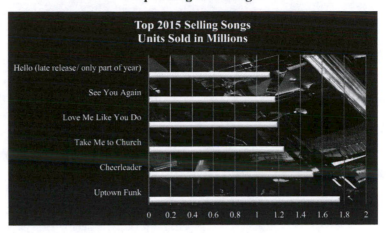

Heaven's Door," and "Like a Rolling Stone." All wows and, of course, each a cash machine for the writers and copyright owners. Figure 2.2 shows the top-selling songs of 2015.

The Wow Brand/Image

The second wow is about image and branding. Notice that Louis didn't say anything about talent, the ability to sing, write, act, and so forth, as they are ingredients required in building a consumer wow. Also notice that you need more than the ability to sing kind of in tune or to have a voice or sound. It takes more than hitting the notes to really deliver a believable live performance as an iconic hero, villain, good old boy, sexy dude, bad guy, rebel, or whatever. Think about an image as an identifier of an iconic image tied to an emotion and the "brand," as everything done to enhance it, connect it, and sell it as a very cool lifestyle. Think of the bee-to-honey attraction of women to Elvis and the Beatles. The sexy voices of Karen Carpenter and Patsy Cline, the way Roy Orbison describes his desire for a pretty woman. There's also Frankie Valli, Sam Cooke, Elvis, Bob Dylan, Whitney Houston, Bono, Stevie Wonder, Aretha

Franklin, Buddy Holly, Etta James, Johnny Cash, Gladys Knight, Chubby Checker, Janis Joplin, Alicia Keys, Donna Summer, The Police, George Jones, Billy Joel, Sinatra, Hank Williams, The Bee Gees, Adele, Al Green, Don Henley, Brian Wilson, Freddie Mercury, Tina Turner, Marvin Gaye, Ray Charles, and all of your favorite acts, too. I hear their voices in my head right now and I can even feel the emotion tied to each one. Every one of them a wow and industry cash machine!

Image

An image is a one- to two-word descriptor of the emotion people might feel when they hear the act's recordings and see the act live. A good old boy image with the youthful appearance may also be seen as sexy and fun, a villain or rebel might be emotionally felt also as a leader rebelling against the man, or as an outlaw playing Robin Hood. A funny but smart bad guy may be seen and felt by women as exciting and a challenge. The list goes on and on, and the most important thing for both the artist and industry executives is that it really connects with fans. That means buzz, fan base, and possible profits. As far as the acts are concerned, my belief is always be honest and portray your true passions. If you can't do that, then the act better be a great actor as it's difficult to fool the true fans.

A producer/engineer friend of mine told me a story about Johnny Cash, who was often seen as a rebel, and as a hero of the working class against the man or the system. Cash wanted to sing better, be more believable, real. So, he asked to be dropped off in the middle of the desert and left there. He felt he could sing about thirst, hunger, suffering, life, and the will to live if he really experienced a little of it. If he had not found a cave after a few days and had a spiritual revelation he later claimed he'd never have made it out alive. Listen to his voice over his career and listen for the difference. He also said he was really stupid to do it and would never try it again. So take his advice and don't try any death-defying adventures or drug-induced escapism in the name of "art" as there are many who do who are gone before their time.

Branding

Brands are a different and a deeper stage in the image game. The first connection to fans is usually an artist's "sound" connected to their image. If anything other than the music will get their attention, it's the image. Psychographic lifestyle analysis to determine the brand used in different fan lifestyles is then connected to the image and the act. Why do you like your favorite act or actress, recording artist, author, and others? In the music and entertainment industry the process of branding is the association of products or services to your fan base. The fans have in their heads their image of you (the artists) and the emotions (whatever they are) about you. Companies want to connect to your fans to expand their market, image, and profits. It often costs millions for them to make a deal with the people who control the act's image, which is usually record labels in 360 deals and personal managers in independent deals. Acts are regularly offered the same types of deals from different companies. So it falls to the labels, band members, and their management to determine which ones they should accept. It's not the act working for the company, who is trying to sell products and services, but the other way around. Now the automobile company, as an example, pays for the association with the act and their image through their brands of products with large sums of money.

Branding is usually tied to psychographic research based on lifestyles of the fans (see Figure 2.3). It's the fans' desire to enjoy the perceived lifestyles of the famous acts that connects the fans to the companies through the act's image and corresponding emotions. Sports clubs in the United States make about $12 billion a year from companies paying their leagues, teams, and players for the association. Bands, labels, and managers use branding as an additional income revenue stream to try to cover moneys lost to the decline of unit sales.[2]

Figure 2.3 Corporations often use branding to connect their products to consumers who like different types of music and acts. In some cases, the music can help shape the image of the company's products and services for a younger generation.[3]

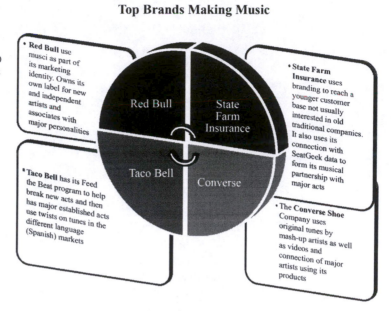

Top Brands Making Music

- **Red Bull** use musci as part of its marketing identity. Owns its own label for new and independent artists and associates with major personalities

- **State Farm Insurance** uses branding to reach a younger customer base not usually interested in old traditional companies. It also uses its connection with SeatGeek data to form its musical partnership with major acts

- **Taco Bell** has its Feed the Beat program to help break new acts and then has major established acts use twists on tunes in the different language (Spanish) markets

- The **Converse Shoe** Company uses original tunes by mash-up artists as well as videos and connection of major artists using its products

Major acts often tell newer artists to select companies their fans will find cool, trendsetting, and worth celebrating. The key element of success is pairing the act to the right product in exchange for tons of money. It's a win-win-win situation for everyone who embraces the acts: image and pocketbook, management and label revenues, and fans discovering products and services that fit their preferred lifestyle and world mental frame of reference. A second level of branding is to associate your company with the business of the band, such as a bank checking account, clothes, or whatever.

The Wow Show

Louis's last wow is the show. I've seen many famous artists performing wow songs live. Goosebumps! Most didn't need any glitz, fireworks, or stunts like riding a motorcycle over the audience's head as they're trying to sing a hit. The Canadian artist Gordon Lightfoot (look him up) once blew me away with just himself, guitar, and a bass player. Standing ovation. Here's the bottom line: the less talent, the

Figure 2.4 *Pollstar* (2015) reports Taylor Swift as the top touring act, with a gross total earning of $250.4 million, the average ticket price of $110.15, the average tickets sold 42,893, and the total tickets sold 2,273,328 from 83 shows in 53 cities.
Source: "2015 *Pollstar* year-end top 100 worldwide tours", accessed January 10, 2016. The annual *Pollstar* ranking of the concert industry's top performing artists is tabulated for all world-wide shows worked between January 1 and December 31, 2015. All ticket sales figures are calculated in U.S. dollars and are based on reported information and extensive research by *Pollstar*. Copyright 2016.[4]

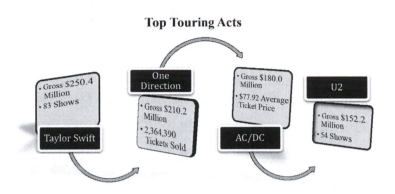

Top Touring Acts

more tricks, stunts, dancers, strobe light, and explosions are needed to stage the show. And in the end, it doesn't matter as long as the consumers are thrilled and pay for the thrill (see Figure 2.4).

The Event Business

Think about a concert as an experience or event instead of a time to sit and listen to the music. For the fans it's all about the excitement, fun, and being a part of the environment. For the promoter, venue, and acts it's all about making money. I was a very young guy when buddies of mine and I saw the Beatles. Of course, in a packed house of thousands it was just our luck we had some really good-looking girls right behind us in the cheap seats. Being the fool I am, I asked one of the girls directly behind me if she was going to lose it and start screaming when the Beatles hit the stage. She laughed and said, "No, I think that's just silly." The lights went out and we all pulled out our cigarette lighters (now it'd be a cell phone) and the Fab Four energetically hit the stage. The place went nuts. We heard the first part of the song drowned out by the screams. I felt the fist of the girl behind me hitting my head and shoulders as she uncontrollably screamed *Paul, Paul, Paul* . . . I learned a lesson in humility and kindness out of the deal. That's a wow show. What promoters want is a loud buzz from the audience as they leave the concert as that's an indication they want more and will come back next year for another experience or thrill. And how many billions of dollars did this wow show make and does still make for the individual Beatles and wannabe acts?

Artists Career Timeline—Beginnings

The biggest mistake you can make if you want to be an artist is to assume your gift of talent is special. It may be something you and your family enjoy, but in all honesty there are millions of people (out of billions on the planet) with the raw talent required to make it in the business. However, very few have the passion, drive, guts, and saneness of mind to give it a serious shot. In the beginning, consider your abilities in being a performer. Several steps or processes are usually required to determine if you're into performing for the sheer enjoyment of it or to perform for a career, making enough for a decent living, or to become a famous, glamorous, rich, world-renowned opera or rock star. We may discover a joy of music as a kid in school, church, choir, or even when singing with our favorite recording artist with our iPhone headphone buds, or whenever the idea of the joy of performing may come to mind. Take a look at the following steps (Table 2.1) to evaluate your ability to perform and think about what it will cost and your personal enjoyment.

The Do-It-Yourself Process

Digital products increasingly make almost everyone think that they can become a rock-and-roll star, glitzy movie star, or famous person. Good luck. The fact that we can make a singer sound talented with auto-tuning, create great music tracks without real musicians using Pro Tools, shoot a video or film with a cheap store-bought HD camera, and then upload the results to YouTube, SoundCloud, CD Baby, iTunes, or some other site does not mean that anyone will like it, enjoy it, or throw any money our way (see Table 2.2). Industry sources provide various numbers on the size of the web, ranging from about 250 million to more than

Table 2.1 Artist's Career Timeline—Beginnings. Playing a musical instrument, singing, painting, writing, and shooting film (as examples) are fun and very satisfying. Turning pleasurable creativity into a professional career is taking something we do for enjoyment and turning it into something very different . . . work!

Artist's Career Timeline: Beginnings
Discover musical preferences (genre)
Shape your abilities through private lessons and gigging
Discover performance abilities (instruments and vocals with others)
Develop musical style (based on your own creative personality)
Find other artists with similar styles, creativity, talent, and abilities)
Perform with others in choir or band to discover joy of performing with others
Learn to be the best you can be (what makes you unique as a vocalists, musician, or both)
Be a solo and a performing group/band
Evaluate crowd reaction of your performance (quality of talent, showmanship of yourself as a solo act, and member of performance group)
Evaluate crowd reaction of band or group (quality of talent, showmanship of act/show)
Evaluate performing for enjoyment vs. possibility of making it a business and career
Decide if you want to turn fun/hobbies into a business

759 million, with a quarter to half a billion active. In addition, various sources state between 8 million and 9 million servers appear to store 14.3 trillion to 40 trillion webpages. Each website has various costs, including hosting, planning, design, maintenance, and marketing expenses often between $5 and $20,000 annually in a small, solely owned page. Talk about a needle in a haystack—an illusion or naivete on a grand scale. The odds of anyone even finding a site are extremely low.

Table 2.2 Artist's Career Timeline—Turning Professional. The career time line: Turning professional is a different story than entertaining or creating artistic works for fun. Work is work and, to be honest, darn hard at times. Being a creative artist for a living raises consumers' expectations of the act to be professional as now they are paying for the experience.

Artist's Career Timeline: Turning Professional	
Decide if you want to turn fun/hobbies into a business	
Form the professional act either as a solo or band	
Decide on your legal form of business by making the following decisions and then filing the paperwork with the local, state, and national authorities	*Who owns the business of the band?*
	Who owns the name of the band?
	Who is the boss or is it shared (which will not work in the long run)?
	Who will control the books and pay the bills?
	How will you pay bills?
Develop the unique qualities of the band (what will make it sellable to night club owners, promoters, and booking agents)	*Sound*
	Visual performance
	Material to be performed
	Songs to be written
	Vocal blend and performance
Time to get serious about the band as a career and profession. All members of the band should sign a partnership, LLC, or corporation agreement that sets in writing the decisions made earlier about your legal form of business.	

Anyone can use an inexpensive Mac or PC and GarageBand, Pro Tools, or Logic software to record a *great-sounding* song. Almost anyone can use ChordBank, Tabletop, Animoog, iMaschine, Traktor DJ, and Beat Maker to create various chords and sounds.[5] If you don't know how to operate the software, simply register at Lynda.com and follow the step-by-step instructional videos. You can also buy or rent an inexpensive 1080 to 4K high-definition camera to shoot a music video or movie to dump on YouTube or any of the other sites. The industry already uses RED cameras (high-quality digital cameras) and Final Cut Pro software to shoot and edit movies. So can you! You can even record your latest song on your iPhone.

The DIY Infrastructure

Once you've created a recording, won't you need some seed money to get started? Register a request for funding with PledgeMusic, BandPage, Patreon, See.me, and of course, Kickstarter. Then upload the recording to iTunes, CD Baby, TuneCore, DistroKid, The Orchard, and Ditto Music to distribute it for sale as a CD or digital download. For a better understanding of your potential costumers (demo and psychographics) consider using Next Big Sound, Musicmetric, Buzzdeck, Google Analytics, or Prizm (Claritas). For promotion, catalog your tracks at BandPage, SoundCloud, Bandcamp, and other providers. Use the same sites to "brand" your image and sound to potential consumers. Register with Bandsintown, Sonicbids, ArtistData, Stageit, and Concertsinyourhome.com to book a live appearance or event. You may have to pay Eventbrite, Brown Paper Tickets, Splash, or Ticketmaster (if you become big enough) to sell tickets to the event. Download AmpKit, iLectric, and Audiobus to enhance the sound of your live performances. Provide and sell merchandise through Chirpify, Limited Run, Square, and Thehub.fm. Finally, join Artist Growth, Topspin, and other software providers to learn how the industry works, make industry connections, and control your financial events and books.[6]

Many Internet start-ups offer a viable business model by providing a free platform where artists can (in effect) give away all their rights and recordings in exchange for the hope of being discovered. The sites make money from the joining fees, upload cost, or, if they get popular enough, advertisement money. However, most labels are not interested in an act that has already given away its product to be discovered. What are they going to sell? Which is why labels are now music companies marketing acts and products to fans for money. They want to sign acts that have developed a fan base willing to pay or at least steal the recordings, and then purchase the corresponding merchandise, touring, and so forth, marketed worldwide. In other words, do not be in a rush to give your recordings away as it may destroy any serious possible future market and major label deals.

Money, Money, Money!

Now that we have computer-generated music production software (e.g., Pro Tools, Q, Logic Pro), and computers used in the daily production of movies, has the value of a recording studio engineer increased or decreased because of the use of computers? Strange to think that one's worth as a professional could be devalued by technology, isn't it? But today worth is measured by whether technology changes what you do. Some skills and knowledge prized today may be worthless in the future. The future for people who want a successful career in the production of movies, video, and music is therefore tied to . . . computers! The key is lifelong learning and keeping up with the technology, economics, and social changes, growing with the times and creating experiences to keep yourselves ahead of the herd of people who'd like your job. That's true in any business at any time. And by the way, great audio engineers and producers are considered almost priceless in the eyes of the artists whose careers depend on their knowledge and skills to help them become rich and famous. Value and worth are part of the economics of the qualitative side of entertainment; money and profits are what make the business tick and are a key measure of success of the wow.

So, what is money and how is it viewed? We use it every day, yet do we really understand what money is and its functions in our society? Thousands of years ago, bartering was used instead of money, because money didn't exist. Cows, grains, leather, gold, silver, diamonds, and many other things were exchanged for goods and services. How many cows are you worth? Will you dig this ditch for an old cowhide? As an audio engineer what is your value or worth? If audio engineers were around thousands of years ago, you'd have to decide how many cows, grains of wheat, pieces of bronze, or whatever your "work" was worth. A record producer would be worth more "cows" and so forth than the audio guy, just as they are worth more green paper today. Actually, the audio engineers and record producers were "worthless" thousands of years ago because electricity and the technology didn't exist. Yet time changes things and so it goes . . . wonder what is next?

Let's dig deeper into the revenue streams of entertainment: How is money made? Who makes it? How do we establish value for what we create? Does it matter if we go it alone or sign with a label?

The Value of Knowledge

People got tired of trying to figure out the value of "cows" compared to labor or skills so in about 550 BC, King Croecus of Lydia (now part of Turkey) pressed the first official gold and silver coins (currency).[7] One last thing to consider: different countries use different types of currency that often have "floating" value or worth. Who cares? Probably you, if you're an artist, label, songwriter, publisher, and so forth, as the amount of money paid (royalties) for use of your copyright in the United States may not be the same as in other countries, due to their laws or lack of them, and as the exchange rate between the dollar and other currencies changes daily. Starting to see how crazy it gets? How much money (cows) will you agree to exchange your recordings, movie script, or songs for? Well, what are they worth? That's another great question, as in this business we've got to make it and distribute it (usually with promotion) before we know if anyone is going to find enough "value" in it to even steal it (see Table 2.3)! Okay, let's hope they buy it or at least pay for streaming it!

Table 2.3 Artist's Turning Professional as a Performing Act. The Artist's Career timeline: turning professional as a performer is difficult, as we have to combine the oil and water of creativity and business. First develop a product or service that consumers will find valuable. That means better than what exists in the popular culture or at least as good as (but not the same as) the famous acts, artists, and authors who thrill us now.

Artist's Career Timeline: Turning Professional as a Performer	
Use your unique sound, visual performance, and musicianship to determine these elements	*Image*
	Brand
	Show
Practice, practice, and do it again, again, and again	
Put on a small show and evaluate the response to each song performed by observing the fans' attention and response to each song. Have a person nobody knows mingle with people attending and listen for their comments. If possible, have that person ask people enjoying the act what they like most about the act. Note the responses and gender, age, and other possible demographics of the respondents. If this is working, use the information to determine the act's potential market. Then once again, review your band agreement and these elements.	*Who owns the business of the band?*
	Who owns the name of the band?
	Who is the boss or is it shared (which will not work in the long run)?
	Who will control the books and pay the bills?
	How will you pay bills?
Develop the unique qualities of the band (what will make it sellable to nightclub owners, promoters, and booking agents based on the audience responses and comments).	*Sound*
	Visual performance
	Material to be performed
	Songs to be written
	Vocal blend and performance
Develop a business plan to determine the feasibility of turning the band into a profitable venture. Complete these steps.	*Describe your company (what it offers to others).*
	Products (shows, recordings, merchandise available).
	Market analysis (research the market place and the local industry to determine your potential market and income)

(Continued)

Table 2.3 Continued

Artist's Career Timeline: Turning Professional as a Performer	
	Build your business team model on the management structure that fits your band's business.
	Determine your marketing plan, including an electronic press kit, promotion, publicity, website, aggregators, distribution, and partners.
	Complete your financial projections to run the business, and profit and loss statement. Write the executive summary.

Value of the Wannabes

The traditional industry has suffered financial direct hits in the marketplace due to consumer access to entertainment products without payment. Still, the recreation, leisure, and entertainment global market is estimated to be $1.8 trillion and growing at a rate beyond 6%. How large is the independent industry that is not part of the established industry? Probably a few million, but if we really "stretch the rat," cross our fingers and legs, and throw in everything including a kitchen sink, it's probably between $100 million and $1 billion. That is about .5% to 1% of what the established industry RLE global market generates. Want to be a songwriter? What's the value of your songs used in recordings or movies to the public if they can get them free?

The Net Zero Market

What is your independently released upload (songs, videos, films) worth to consumers who don't even pay for the established industry's products? With the Internet and computer software many have created their own computer-based recordings and videos and thrown them up on YouTube or another site. Sadly, very few indie products show the quality of innovation and creativity it takes to get people very excited.

Figure 2.5 The persons who create the song or script own what they have created. If it is good enough an agent may represent you and pitch your creative work to a production company, which may then license its use and turn it into a recording, movie, computer game, stage play, and so forth. This acts as a natural filtering system that often keeps non-quality productions from appearing as consumer products. However, now with the Internet, anyone can leap over the industry by uploading his or her creative works online. Very few wannabes' creative products can compete with the professional industries. However, now instead of having a few poor-quality entertainment products available (whatever that may mean) we have millions, or should I say billions of amateur products online, taking up server and bandwidth space. It is wonderful that anyone can create and jump in the game of entertainment with our home computers and the Internet; sadly most of what is uploaded can't be sold or even stolen.

The Industry "Filter" System

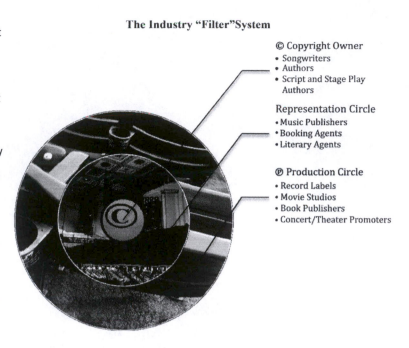

© Copyright Owner
- Songwriters
- Authors
- Script and Stage Play Authors

Representation Circle
- Music Publishers
- Booking Agents
- Literary Agents

℗ Production Circle
- Record Labels
- Movie Studios
- Book Publishers
- Concert/Theater Promoters

The fact that we can *now* use technology to make a recording, video, or film that sounds and looks good doesn't mean they're good enough to make any money, let alone launch a career. Millions have tried and very, very few have succeeded. They simply do not have any of the three wows that are usually produced only by the pros who over years have gained the knowledge, skills, and professional talent required to create wow products and shows consumers dig. The quality of the indie lyrics, musicians, camera people, directing, and singing (without auto-tune) simply can't compete with the pros. Why do cover bands sing only others' hit songs and recording? Because they don't have the talent, experience, and knowledge to create it themselves and in most cases if they do have the talent, the songs or their "sound" has never been promoted or marketed. And they rarely sound or look like the real thing! People who hit the clubs get only "part of the emotional experience." All they may see is a knockoff band that sounds a little like the real thing. It's rare for a band to play their own material or someone make their own film that blows up into a wow hit and serious moneymaker. Why?

Drumroll, please—because there is a brand-based message in a song, storyline in the video, plot, acting, and about a million other things rarely found in a self-made recording or movie. Many have known this for generations. The fact that we want to write a book doesn't mean we have the skill, persistence, or even talent to finish a novel and sell it. Even great writers have agents. And to make it more difficult the amateur stuff is competing with the best in the world, created, financed, and marketed by the preeminent folks in the established industry. But what the heck—take a shot at it anyway! You will learn a ton and get nailed with a dose of reality.

Just Do It Anyway

I love a challenge! It tends to make life a little more interesting. Wisdom comes from experience, success, and failure. But as most of us know, failure is a sure thing if we don't take a shot. So, take it and get the education, experience, knowledge, contacts, and everything else you need to be or create the best you can be. That said, because you can create an entertainment product does not mean you should post it and give it away online. Why would the industry want to hire you or sign you as an artist or songwriter if everything you've created has been given away? And the industry is really going to be interested in you or your products if they perceive that hiring you (for you knowledge, talent, and skills) or signing you or your creative works—songs, recordings, and so forth—can help them make a profit. Remember, it's a business after all.

The Phony Success of Virtual Media

It's very cool to have a song, recording, or video posted for billions to catch. Want to give it a shot? Buy a computer and software and get creative. To get better at what we do, hopefully it will not cost us 10,000 hours to get good enough to

Figure 2.6 Many take a shot at stardom by writing their own song and then use either a home studio or recording software, such as Pro Tools or Logic Pro, to record the song. What do your own? The song is yours and copyrighted at the time it is completed (if original and in a tangible form) and the recording is also owned by you as you created it. hat most forget is the business side of registering the copyright, forming a legal business, and then registering the business with local, state, and sometimes national government agencies. The © symbol is often used (but not required since the Berne Convention) for the song you've created. The ℗ symbol is the copyright symbol for the actual recording of the song that we created in the studio or on your computer. The ® symbol gives notice that you've registered the name of your company (your own label) with the Library of Congress Patent and Trademark office. The ™ symbol is used to indicate the same thing but it is not registered with the government.

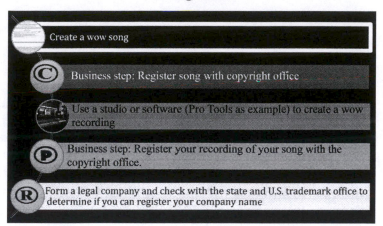

Taking a Shot

create something good enough that consumers want it. As only one or a few people are involved in the ownership, the costs for creating the recording, uploading, and collecting any royalties are far less than in the traditional industry. If you understand the industry's business principles, the profit margin is lowered to almost zero. That means the odds are really in the independent industry's favor. Instead of paying thousands of employees, as an indie only you and the corresponding companies, such as CD Baby or SoundCloud, are involved. If you can get just a few thousand to buy an album or streams, you can make a living—maybe.

Levels of Success

Success on YouTube, Beats, Spotify, and other sites is laughable until you hit about 20 million or more hits. Let me explain. How valuable are the songs and recordings you've uploaded to SoundCloud, Soundcast Music, ReverbNation, MySpace, #music, Earbits, Last.fm, Beats, or even Spotify? The answer is zero or darn close to it. Some of the sites pay nothing or next to zip, and others pay legal rights royalties that are sadly significantly *less than a penny* per stream or video. So in

Figure 2.7 Start with a song, then record it on Pro Tools or in a studio, and follow up by the mastering process. Once you've got the "product" you may place it into retail for sales or streaming royalties by uploading the master recording to many websites, such as CD Baby, TuneCore, INgrooves, The Orchard, and many others. Different sites accomplish various tasks for you, such as registering with ASCAP, BMI, and SESAC, and others do not. Some will also provide marketable CDs and others won't. Don't forget to register your companies (songwriting, music publishing, and label) with local, state, and national organizations to receive any royalties and to pay taxes.

Beginning Steps of Creation Tied to Business

reality, loading your song as a recording or video (property you own) to an Internet site is about the same as giving it away. If you are lucky you'll probably make enough money to buy a cup of stale coffee at McDonald's, and only if you get a few hundred hits. Sadly, even a *million stream*s on Spotify will gross less than $5,500 for the master right (copyright of the recording) and less than $550 paid to the songwriter(s) and publisher(s), usually split between the two. That turns out to be only $0.00521 per play on Spotify, $0.00012 per play on Amazon Cloud, and $0.01122 on Rhapsody.

YouTube videos pay between $0.00175 per play for all rights, including video, master, and publishing, and only $0.00032 per play or view for only the master rights. Get a million streams on Spotify and you'll receive a whopping mailbox check for $5,210 if you're the label and about $260.50 if the sole songwriter. Add a cowriter and that amount is halved.[8] And to make it even worse, the majority of videos on YouTube (which controls most of the market) are not monetized! Which means no royalties are paid and you've given the products of your creative efforts away. At that point, they are basically worthless to the established industry. Be careful about signing away your copyrights (songwriter, publishing, and recordings) to some of these "Internet suppliers" as you'll need to have them if you become popular enough to seek a major label or publishing deal. Labels have been squeezed by the economics of scale of illegal downloading and need all the revenue streams possible when making a decision about which acts to sign. If your publishing is already gone, they may pass on signing you as an act, as without the music publishing royalties their risk for making a profit increases.

YouTube Royalties

The different levels of royalty payments for audio streams vary as the rates are based on profits, *or* the amount of passive to interactive use, *and* the year the provider was established. The rates for audio streaming are so low they might just as well be free. Videos are the same situation, as YouTube controls most of the market and it pays on the type of channel deal, the number of plays, and third-party involvement, plus or minus advertisements; but you have to hit about 100,000 views or more before advertisements are added and the majority of videos receive less than 500 views. Oh, yes, one other thing: the more plays or views on YouTube the less each is paid, meaning the more popular the video the less its rights holders are paid per stream, view, or play.[9] Confused yet?

The amateur independents get the double shaft. If they control all their property rights and receive a million streams, mailbox money will still be only about $6,000. Hard to live on that! If the song is ripped a million times the pay is zero, which is not a heck of a lot different than getting legally paid. Now if we could just sell our recordings on iTunes we'd receive 70%

of the purchase price or $7.00 per $10 album and a little over 90 cents on each single. But sales at iTunes are fading quickly (down 13% in 2014) and less than 5% of cataloged albums sold more than 2,000 copies.[10] Indeed, why buy when the cost of one album lets us subscribe to Spotify or one of the other sites and we can listen to anything we want, anytime we want it, anywhere. Or we can wait for the advertisements to finish and pay zip. Who is making the money off the amateur creative individuals using their computers and a little sleight of hand to "distribute" their songs and recordings to the billions on the planet? Not the people who create and own their entertainment products but the Internet companies that claim that for just a few bucks, they can provide global distribution, international Internet radio, and so forth. Most indie acts that follow this path end up learning a tough street lesson: without massive publicity, promotion, and marketing, the chances of financial success are about as secure as winning a lottery. The actual odds are even worse.

Aggregators

Let's see now, one song divided into about $1.8 trillion? How much of that pie is yours? The answer is probably zero, if you forget to connect to the right business organizations in the industry that represent you and collect license fee royalties, *and* pay you. If you want to be paid, then join the industry agencies or pay an Internet aggregator, such as CD Baby or Ditto Nation. You pay them and then they'll collect the money and take their share out, reducing the amount of royalties you'd receive, if any. In the States, if you're a songwriter or music publisher, join ASCAP, BMI, or SESAC, as they collect "performance royalties." A record label must pay for a mechanical license fee, usually to the Harry Fox Organization or directly with a publisher to legally record a song they do not own. Labels also join SoundExchange if they want royalties collected for various types of digital transmissions of the recordings they own. Songwriters and music publishers must remember to notify the performance rights organization (PRO) (ASCAP, BMI, or SESAC) of which they are a member if they want royalties paid directly to them. Or you can a deal with a middleman company, such as SoundCloud, CD Baby, or many others that may need a registration fee to upload your album to a streaming service, iTunes, or whatever. They take a percentage of the royalties from the collection agency (Harry Fox, ASCAP, BMI, SESAC, SoundExchange, etc.), which also took out its percentage first. Therefore you're being paid the remaining amount after two commissions, plus a registration fee you could have done yourself. But they're actually a bigger problem.

The Money Changers

There is little doubt the independent industry middlemen are making the majority of their money off of the amateur creative marketplace. Clearly from the streaming rates where a million streams is worth about $6,000, if CD Baby posts your recording for streaming, it will also take a piece of the pie for administering your rights and royalties. Most of the deals, connections, and streaming could be done by yourself, but for a few bucks it will connect the dots and put your material into its system, plus take a percentage of whatever royalties roll in, if any. CD Baby is a company out of Portland that merges the manufacturing process for CDs or vinyl if needed, the iTunes, Amazon, Google, Rdio, Spotify, and other digital registrations, and a distribution link to the Orchard and other distribution to 15,000 retail stores (usually if your CD is ordered), and provides promotional distribution through SoundCloud, PledgeMusic, and others for Internet radio airplay. It will keep you legal if you use a cover song in your own recording (which is a good thing); however, this is something you can also do yourself by checking out the agency that industry uses, harryfox.com.

CD Baby or some other website may also have its own store to sell your product and pay royalties on each unit sold. It may cost you about $10 for a single and $50 for an album to get the process started. It also registers you as a writer/publisher with ASCAP or BMI and other agencies that collect money in foreign territories. That costs about $35 for a single and $90 for an album. It fulfills a valuable niche for the independent artists who want to take a shot at uploading their songs and recordings to the distribution and streaming companies that use and sell them *if* there is a demand for them. And that's where the problem is! Most of the streaming and retail markets will not stock, play, stream, and so forth anything from these types of middlemen distributors unless they can profit from it. You'll probably see advertisements that your CD will be available in thousands of stores worldwide, which is true, but they'll rarely have it in stock and then will tell the customer "we can order it for you." So what—customers can buy it themselves or rip it if it's popular enough, so why would anyone buy a copy at a higher price with a retail or online store that has to sell it at a higher price to survive? Just a word of caution: most of the sites offering you streaming to the planet are making their profits off of your registration and/or subscription fees. And in most cases, you could have joined the professional industry organizations yourself. But it takes much more than uploading a song, recording, video, or homemade movie to the Internet to become successful. Let's face it, who even knows it exists? Maybe there's a different way.

The Foundation

What does it take to create a profitable book, poem, film, game, video, or recording that catches consumers' imaginations? The answer is the talent of creativity and the business of entertainment. The foundation of a house or building is what everything else in a structure is built on. Same thing in the entertainment business, and this time the foundation is a wow story presented as a song, script, book, or whatever. With a foundation of sand, instead of solid rock, the first real storm that hits it will probably destroy it. In the entertainment industry the business is built on a solid foundation of a wow story, fairytale, book, or song that becomes a *communicative link* between the actors, artists, and consumers. That's where it begins, looking for a great story to tell, even if it's only three minutes long. What does it have to say, how can it provide a humorous, courageous, or scary emotion created by the actors, artists, singers, musicians, and others, who usually have to work together to create *magic*? The actors, recording artists, and lighting and sound guys are not the story, but they are responsible for how well the story is told through their body movements, speech, facial expressions, singing voice, and musical talents! It takes time to learn how to do it well and money to make it, usually lots of it. Unfortunately, a bad song, script, or story is just like building your million-dollar beach house on sand. Bad investment! How can we gain the knowledge and experiences required to write, create, and sell foundational creative works to the industry?

The Traditional Process

The best directors understand how to motivate great performances out of the actors, producers get it out of recording artists, and even agents have to sometimes bug the heck out of writers to get what they need to sell. They have to know who's looking and what they're looking for before they can slide into their glass-walled executive offices. And the industry moguls and movers and shakers need to know who the agents or you are before you'll get an appointment to get past the front door. If they don't know you, forget it, unless they've heard of your work or you've got a credible insider referral. The technicians from lighting to sound to the cameramen and women are usually passionate workaholics, proud of the quality of their work and reputation, respectively of all others including coworkers and executives, as they are often asked to accomplish the impossible. It's a process to create and almost always a financial risk to stand up a Broadway play, wrap a movie, or close a recording project. Computer games may start with the spark of an idea, but it's the storyline in the game often detailing the battle between good and evil, win or lose, and we get to play the hero. Think of the types of skills and knowledge required to develop just one new game or software, let alone sell millions of copies to consumers.

The Profit Motive

Thus, the business side of the industry is built on the shoulders of a great story positioned into a sellable product for the purpose of making a profit. Once completed, experts in sales, promotion, and marketing have to spin it out in a way consumers might "discover" it. Sometimes anything goes, but the names of the iconic actors, directors, and performing artists are branded and pushed out to fans and new potential consumers on a global scale. That also costs money! Media—traditional, mass, or social—is not free. The problem is how to get people to notice the product, connect emotionally with it, and then to buy or use it. Over the years, the business side of the industry has developed many tricks to help that happen. The industry discovered in its beginning that making some of the actors and singers famous personalities actually increased profits. That was a business decision that continues today.

The Deal—Get It in Writing

The legal side of the industry represents the laws of the countries we live in, which allow for ownership of created products (called copyrights), and lawyers who make sure that everyone plays by the rules (laws) of the game. Even the people who are creating and financing the products are locked into paperless contracts. Sadly, we already know what a disaster parts of the music and film industry have experienced because of the naive and discouraging lack of enforcement by our representatives. But attorneys are even more important as they provide the legal glue that often holds the players in the creative process and businesses to their respective tasks. Guess how this trillion-dollar business operates? As a business, with contracts being offered and signed daily and decisions about investments and careers that are difficult to make every day. But never forget that it's a business first, right after the importance of the art, without which there is not a business.

Changing Markets

Film production companies are realizing that many fans have moved to Netflix and other streaming sites. The labels have had to change from a "purchase the CD model" to market an act's image as a "brand" in order to recoup investments and profit. Labels brand the act with the song, image, look, or "feel" of live performances, to perceived consumer emotions that will (hopefully) make them want to buy it, creating sales for the label, leading to merchandise, corporate sponsorships, concert tickets, and anything else you can think of, but mostly profits! The film industry is the same way and has been for a long time. George Lucas reportedly made more money off *Star Wars* merchandise than as the producer of the actual film. First and always, this is a business, maybe glamorous and crazy, yet the successful products customers fall in love with are mostly made by extremely professional techniques, producers, directors, creative performers, and others working with executives and administrators to make money. Simple, isn't it?

The Industry Perspective

Far worse is what the industry thinks if you posted your creative works online and failed to get millions of sales, streams, or hits. Why would anyone want to employ or sign a writer or act if "the best stuff they've got" is already available free online? Why would anyone want to employ someone stupid enough to give copyrights away? Starting to see the problem? Now, prospective acts and writers need a fan base or previous success to prove to the labels they are profitable. Failure to monetize creative products at a significant level is an indication you're not ready, don't know enough about the business of entertainment, or are just not good enough. There's nothing wrong with trying and making a few bucks, however, even when a million streams or rips are seen as a failure by the industry. At least they'll take notice. But consider this: it may be smarter to perfect your knowledge and creative craft before trying to get noticed by giving it away. Failure has always been my greatest teacher—passion, confidence, and faith my foundation, which have always helped me not to give up and to be a little smarter next time.

The Professional Advantage

Great films, recordings, computer games, software, and other entertainment products take a significant financial investment, usually a great song or story to tell, lots of very expensive creative technicians, a real producer and directors to make, let alone additional millions to promote, distribute, and sell. Roll the dice. People who make this stuff have a passion and love for the art and craft. They are the very best at what they do. People who financially invest in the process and products also have a passion for it, but they invest to make profits and are smart and wise about the risks and opportunities. Consumers who acquire the products free or rip them are seen as %@*&#^ (I better not say it) by almost everyone in the industry. But their kids do it too. Young people who want a career in the industry need to understand it, know it, have a passion for it, and start protecting it.

Talent and Knowledge

You can't wait around to be "discovered" because, honestly, nobody's looking for you unless you give them a reason! What are the laws that affect this industry? Who do we need to know? What is unique about the business? How can you connect with the important key people and businesses? By the way, that's called "networking." Where can we find the money to learn and maybe start a company? What do we need to know about copyrights, laws, and contracts? What qualities and quantities are required to market, promote, financially break even, and make a profit?

Why would another company want to acquire my creative products or me? The two most common abilities required are talent and knowledge. Consider the *who, what, where, when,* and *how* questions about what you want in the industry to help you start making a plan for career success. *What* do you want to do? *How* are you going to do it? *What* do I need to know before I jump in? *Who* do you need to know? *What's* the industry looking for? *How* can you contribute? *Where* can I find the right connections? *When* are opportunities available? *How* will I pitch my creative works or self, if seeking a job? *Who* to? *Who* are my ideal mentors, customers, bosses, markets, and coworkers? Ask yourself, *how* and *where* I can improve my abilities? *When* can I gain networking or interning experience, or work daily in the industry? One way is to start your own industry-related business and find, meet, and gain what you need to know to make yourself a valuable asset. Daydreaming, wishing, and throwing up your song, recording, or film on the Internet isn't enough unless you make a great buzz in the millions.

Professional Opportunities

The key is to use the business of the industry to your advantage. Contribute a creative work someone in the industry decides to finance, and you are on your way if you do not sign a foolish contract. Learn and breathe how the business works and figure out what you're good at. Grab your chance at a wonderful lifelong career working as a technician or businessperson making and selling products that entertain almost everyone on the planet. Jump in; don't be afraid to get wet or make mistakes. However, if you never try, you've just made a fatal mistake. It is not easy, but great things never are, so the decision is yours. According to 2014 U.S. government employment data sources there are 46,600 people employed as independent workers in the industry, which should increase by 12.5% over the next eight years.

Creating Value

Instead of waiting to be discovered, proactively pursue your career passion by perfecting your talents and knowledge of the arts, culture, humanities, and the business of entertainment. Network to meet industry insiders and have something of value to show them in the manner of talent, knowledge, product, or skill. It's not easy, but it's worth it and then some. Don't be a fan, asking for an autograph or "selfie," for as much as this industry loves its fans, very few are considered seriously to be given a shot at a career. Additionally, there are many opportunities for careers in the business according to our government's data. Take a look at the following summarized reports (Table 2.4) and think about what you would like to do and where you fit in. These are the latest figures available (2012–2014) showing the number of businesses, jobs, and the average wage. Notice the total wages of American workers in the following information: $5,927,513,234,040.00. Check please! Then notice the numbers of businesses in the industry you may want to work for (if you enjoy it) or seek a gig from and the total number of employees. The total number of employees into the gross revenues generated provides the last figure, which is average salary, based on the value of your occupation. Notice that it ranges from $10,801 if you want to work at a drive-in theater (I didn't know they still existed) to $141,541 if you become a software publisher or (as another example) $130,000 if you would rather be a film or video distributor. Maybe you should consider making sure the person at the drive-in gets the film, instead of just working at the drive-in.

Independent Artists, Writers, and Performers

While change is natural, developing or adapting a new business model is hard, and managing creative content is even more difficult. Almost everyone in the industry has changed the way they accomplish business by adapting traditional skills to the new paradigm. And yet, at the same time many of

Table 2.4 The North American Industry Classification System (NAICS). The North American Industry Classification System (NAICS) is the latest federal government statistical data on the number of jobs in the entertainment and related industries. Note each type of job has a unique number and that many have subcodes related to similar jobs. Take a look and you'll find the number of businesses, the average number of employees, the total annual wages, and the annual wages per employee. Have fun!

NAICS Code	Types of Industry or NAICS Sub Code	Sub Industry	Annual Establishments	Annual Average Employment	Total Annual Wages	Annual Wages per Employee
10	Total, all industries		9075876	115557595	$5,927,513,234,040	$51,295
	1026	Leisure and hospitality	793468	14625406	$307,030,768,985	$20,993
334	NAICS 334 Computer and electronic product manufacturing		18992	1047250	$110,975,227,966	$105,968
	3343	Video equipment manufacturing	691	19764	$1,777,226,456	$89,923
443	Electronics and appliance stores	Retail	48854	488418	$21,883,743,168	$44,805
451	Sporting goods, hobby, book, and music stores	Retail	53041	614563	$12,038,486,442	$19,589
	45114	Musical instrument and supplies stores	3858	31408	$870,021,275	$27,701
	4512	Book, periodical, and music stores	8285	89050	$1,753,846,505	$19,695
	451211	Book stores	7229	81601	$1,547,692,857	$18,967
	45122	Recorded tape, CD, and record stores	No data available	No data available	No data available	No data available
	45392	Art dealers	5228	17104	$863,487,829	$50,483
511	Publishing industries, except Internet		37116	721441	$70,516,770,871	$97,744
	51113	Book publishers	3790	64228	$5,046,029,724	$78,564
	5112	Software publishers	14931	310895	$44,004,443,247	$141,541

Table 2.4 Continued

512	Motion picture and sound recording industries	29164	384839	$25,732,739,262	$66,866
	5121	Motion picture and video industries	25553	368763	$24,512,755,283	$66,473
	51211	Motion picture and video production	17916	214321	$20,351,259,680	$94,957
	51212	Motion picture and video distribution	535	7398	$961,729,997	$130,000
	512131	Motion picture theaters, except drive-ins	4289	127489	$1,678,289,191	$13,164
	512132	Drive-in motion picture theaters	230	1950	$21,065,077	$10,801
	51219	Postproduction and other related industries	2583	17605	$1,500,411,338	$85,225
515	Broadcasting, except Internet	9744	282264	$23,094,579,171	$81,819
	5151	Radio and television broadcasting	8334	219765	$16,685,933,164	$75,926
	51511	Radio broadcasting	5616	90136	$4,935,153,650	$54,752
	5152	Cable and other subscription programming	1410	62500	$6,408,646,007	$102,539
516	Internet publishing and broadcasting	No data available	No data available	No data available	No data available
517	Telecommunications	39772	850936	$68,613,325,877	$80,633
	5172	Wireless and communication carriers	8630	155470	$11,588,339,241	$74,537

(Continued)

Table 2.4 Continued

NAICS Code	Types of Industry or NAICS Sub Code	Sub Industry	Annual Establishments	Annual Average Employment	Total Annual Wages	Annual Wages per Employee
	5174	Satellite telecommunications	777	9077	$917,657,647	$101,092
518	Data processing, hosting, and related services		16418	276773	$25,591,467,578	$92,464
	54181	Advertising agencies	19057	189809	$17,844,822,786	$94,015
	54192	Photographic services	13780	56767	$1,615,495,060	$28,458
	61161	Fine arts schools	12094	83313	$1,557,985,472	$18,700
71	Entertainment and recreation		132631	2094689	$73,013,251,195	$34,856
	711	Performing arts and spectator sports	51724	444794	$36,387,475,968	$81,808
	7112	Spectator sports	6309	135853	$17,558,490,642	$129,247
	7111	Performing arts companies	9844	114631	$5,171,624,756	$45,115
	71111	Theater companies and dinner theaters	3332	62607	$2,128,557,135	$33,999
	71112	Dance companies	898	10933	$414,846,261	$37,945
	71113	Musical groups and artists	4961	34208	$2,316,709,316	$67,725
	7113	Promoters of performing arts and sports	7174	120052	$4,461,483,891	$37,163
	713	Amusements, gambling, and recreation	74734	1503654	$31,884,622,282	$21,205
	7114	Agents and managers for public figures	4548	22139	$2,509,519,307	$113,353
	7115	Independent artists, writers, and performers	23849	52119	$6,686,357,372	$128,290

Source: NAICS www.census.gov (2015).

the same "talented gifts," creative abilities, networking, communication skills, courage, and business practices that made musical recordings such a pleasure (should I say treasure) in the past are still absolutely required to make great new music and entertainment products in the future. About 15%–20% of the total market is independent. It fluctuates annually. The figures are based on revenue, not on uploading or projected sales by individuals.

There's another list of creative career facts and figures provided by government sources that gives us additional information on the entrepreneurial side of the industry (Table 2.5). These are the latest figures available (2012–2014) showing the number and occupations of small entrepreneurs, work-for-hire, and independent contractors, such as producers, musicians, and camera operators, and the average wages. Notice the total number of employees, the percentage of jobs in the indie side of the industry, and the projected increase in employment over the next few years. The projected increase for creative, innovative, skilled, bright, educated artists, technician assistants, and bookkeepers is a shockingly high 12.5%. Far ahead of almost every other industry. Sounds like a heck of an opportunity to me! So what do you need to have and know to fit into the industry's bright future?

The Industry Team

You need ideas, a team of creative individuals and business executives who together make stuff we want because when we use it we seem to help ourselves feel . . . something. You need to understand the basic business processes that are used in the entertainment industry and what questions to ask to determine if you have a shot at success. What's your "value" to the people who create entertainment products (as we saw in Fig. 1.9)? Excellent creative final products they can sell. What's the value of your talents to yourself? A great career! Should the people who create the entertainment we enjoy be paid anything? If the money stops, you can bet so will the classic creative products. What is the value of creative hard work to consumers?

Table 2.5 NAICS Jobs for Creative Artist, Technician Involved in Production. NAICS jobs for creative artists and technicians involved in production are important as they are the professionals who make the recordings, movies, books, and computer games we all enjoy.

Occupation Title	Code	2012 Employment (in thousands)	2012 Percentage of Industry	Projected 2022 Employment (in thousands)	Projected 2022 Percentage of Industry	Employment Change 2012–2022 Number (in thousands)	Employment Percentage of Change 2012–2022
Total, all occupations	00-0000	46.6	100.0	52.4	100.0	5.8	12.5
Fine artists, including painters, sculptors, and illustrators	27-1013	3.9	8.3	4.3	8.3	0.5	12.5
Actors	27-2011	3.9	8.4	4.4	8.4	0.5	12.5
Producers and directors	27-2012	3.0	6.4	3.3	6.4	0.4	12.5
Writers and authors	27-3043	2.9	6.2	3.3	6.2	0.4	12.5
Secretaries and administrative assistants, except legal, medical, and executive	43-6014	2.1	4.5	2.5	4.7	0.4	18.3
Audio and video equipment technicians	27-4011	2.0	4.3	2.4	4.5	0.4	18.1
Public address system and other announcers	27-3012	1.9	4.0	2.1	4.0	0.2	12.5

Table 2.5 Continued

Craft artists	27-1012	1.8	3.9	2.1	3.9	0.2	12.5
Office clerks, general	43-9061	1.8	3.9	1.9	3.7	0.1	6.9
Musicians and singers	27-2042	1.7	3.7	1.9	3.7	0.2	12.5
General and operations managers	11-1021	1.1	2.3	1.2	2.3	0.1	12.5
Laborers and freight, stock, and material movers, hand	53-7062	1.1	2.4	1.3	2.4	0.1	12.5
Executive secretaries and executive administrative assistants	43-6011	1.0	2.1	1.0	1.9	0.0	0.3
Multimedia artists and animators	27-1014	0.9	2.0	1.1	2.0	0.1	12.5
Agents and business managers of artists, performers, and athletes	13-1011	0.8	1.8	0.9	1.8	0.1	12.5
Entertainers and performers, sports and related workers, all other	27-2099	0.8	1.7	0.9	1.7	0.1	12.5

(Continued)

Table 2.5 Continued

Occupation Title	Code	2012 Employment (in thousands)	2012 Percentage of Industry	Projected 2022 Employment (in thousands)	Projected 2022 Percentage of Industry	Employment Change 2012–2022 Number (in thousands)	Employment Percentage of Change 2012–2022
Bookkeeping, accounting, and auditing clerks	43-3031	0.8	1.7	0.9	1.7	0.1	12.5
Business operations specialists, all other	13-1199	0.7	1.4	0.7	1.4	0.1	12.5
Photographers	27-4021	0.7	1.5	0.8	1.5	0.1	12.5
Self-enrichment education teachers	25-3021	0.5	1.1	0.6	1.1	0.1	12.5
Public relations specialists	27-3031	0.5	1.0	0.5	1.0	0.1	12.5
Technical writers	27-3042	0.5	1.1	0.6	1.1	0.1	12.5
Film and video editors	27-4032	0.5	1.0	0.5	1.0	0.1	12.5
Sales representatives, services, all other	41-3099	0.5	1.2	0.6	1.2	0.1	12.5
Accountants and auditors	13-2011	0.4	0.8	0.4	0.8	0.0	12.5
Graphic designers	27-1024	0.4	0.8	0.4	0.8	0.0	12.5
Set and exhibit designers	27-1027	0.4	0.8	0.4	0.8	0.0	12.5

Table 2.5 Continued

Sound engineering technicians	27-4014	0.4	0.9	0.5	0.9	0.1	12.5
Security guards	33-9032	0.4	0.8	0.4	0.8	0.0	12.5
Museum technicians and conservators	25-4013	0.3	0.6	0.3	0.6	0.0	12.5
Dancers	27-2031	0.3	0.7	0.3	0.7	0.0	12.5
Music directors and composers	27-2041	0.3	0.7	0.3	0.7	0.0	12.5

Source: NAICS www.census.gov (2015).

Damn good question! Stick with me now to find out how to protect, sell, and sustain what you want to do, which is why you are reading this book right now!

Notes

1. Copsey, Rob. "The Official Top 40 Biggest Songs of 2015 Revealed." January 4, 2016. Accessed January 11, 2016. http://www.official charts.com/chart-news/the-official-top-40-biggest-songs-of-2015-revealed__13270/.
2. Hampp, Andrew. "Are Brands Music's New Bank? Lars Ulrich and Top Execs on the $1.3 Billion Up for Grabs." Billboard.com. September 26, 2014. Accessed January 14, 2016. http://www.billboard.com/articles/6266322/billboard-brand-roundtable-lars-ulrich-top-execs-billion.
3. Diaz, Ann-Christine, and Shareen Pathak. "10 Brands That Made Music Part of Their Marketing DNA." Advertising Age Special Report Music and Marketing RSS. September 30, 2013. Accessed January 14, 2016. http://adage.com/article/special-report-music-and-marketing/licensing-10-brands-innovating-music/244336/.
4. "Pollstar Year End Top 100 Worldwide Tours." Pollstarpro.com. 2016. Accessed January 11, 2016. http://www.pollstarpro.com/files/charts2015/2015YearEndTop100WorldwideTours.pdf.
5. Nagy, Evie, Glenn Peoples, Alex Pham, and Nick Williams. "Your Career in Your Hands." www.billboard.biz. July 6, 2013. Accessed May 29, 2015.
6. Ibid.
7. The History Files. "Maeonia & Lydia C. 13th Century—547 BC." The History Files: Middle East Kingdoms, Ancient Anatolia. 2015. Accessed June 29, 2015. http://www.historyfiles.co.uk/KingLists MiddEast/AnatoliaLydia.htm.
8. "The Streaming Price Bible—Spotify, YouTube and What 1 Million Plays Means to You!" The Trichordist. November 12, 2014. Accessed August 31, 2015. http://thetrichordist.com/2014/11/12/the-streaming-price-bible-spotify-youtube-and-what-1-million-plays-means-to-you/.
9. Ibid.
10. Ibid.

THE RULES
OF THE GAME

3

Many young artists and creative people just do not understand the essential elements of creativity and the rules of the game of business. We often think that "rules" get in the way of creation and imagination, and stifle the free flow of ideas. Ironically, when it comes to "creative works," the "rules" of the copyright law provide authorship and ownership to you, because you created it and it's original in a tangible, physical form (see Figure 3.1). Because of the copyright law, you own what you've created, and, shazaam, you can usually profit from your efforts if somebody will pay for it or for the use of it.

In its own way, the stability of the law makes it possible for artists to be more confident they'll be paid for their creativity

Figure 3.1 If we have a great idea, who owns it? Nobody, unless it is original, complete, or finished, and placed into a physical, tangible form.

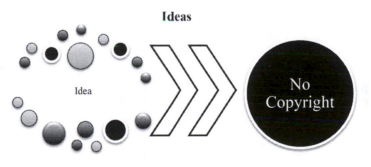

Ideas

Idea

No Copyright

Sometimes we get weird ideas about many things that can be a creative product (such as a story, poem, movie script idea) that might be very cool.

Ideas can't be owned (they are just thoughts in our minds).

and efforts. At least that's what it is supposed to do, so let's start there. You'll need to understand some of the basic copyright laws in order to work in the industry. Accordingly, you'll see segments of the copyright law in this chapter to illustrate the main reason we have and care about the copyright laws. Reading the laws themselves can be really boring, yet it's also extremely important you know the essentials if you want to succeed in the industry (see Table 3.1).

Table 3.1 First Unique Copyright Registrations. As technology has changed over the last couple of hundred years the copyright law has been able to expand its definition of what is copyrightable. As examples, copyrights went from covering only maps and simple book publication in 1790 to sound recordings in 1972.[1]

Registration	Year	Creative Work
First federal registration of a work	1790	John Barry's book *The Philadelphia Spelling Book*
Registration of the Statue of Liberty	1876	Henry de Stuckle and Auguste F. Bartholdi secured registration number 9939-G for the "Statue of American Independence"
First motion picture registrations	1894	The Edison Kinetoscopic Record of a Sneeze, also known as Fred Ott's Sneeze
First television show registration	1947	"Unexpected Guest" by Hopalong Cassidy
First computer program registration	1964	John F. Banzhaf's computer program to compute automobile distances
First sound recording registration	1972	Bob and Dorothy Roberts's "Color Photo Processing Cassette, An Accurate Sound Signal and Oral Instruction System for Processing"

Not Quite 20 Questions, But Let's Explore

Think about this for a moment: When more than one song-writer composes a new song, who owns it? Who is the legal author? What rights does the owner have and how can they be used to monetize your song, film, or whatever? How do you get your song placed with a major recording artist or film? How does a film company get the rights to use the song or the recording? Why would a recording artist want to record your song, and a label spend the money to produce a recording of it, and then spend more money for the marketing, promotion, publicity, and sale of each unit on different platforms? If you write a script or film a movie or video, do you own it? Who owns the song or the script, and the song when it is on a recording? How many copyrights are involved in a recording, video, or film? How can you get paid for the movie you shot? Why is a script important to a movie production company? Why would an actor want to be in a movie of a script you've written?

The industry is the "establishment" with the power, money, distribution, and connections that can mass-produce your script and so forth into an entertainment product consumers *might* want to buy or use. The song, script, play, or whatever is yours, right? What you've created you own because it's your property and that's the law. Now, how can you get it to the right people and the industry-related companies interested in using it? Why is it important for us to know the answers to legal and industry questions before trying to place whatever we've created with industry executives? Here we go again with more questions that might be asked once a contract has been offered. What rights should you sell? What rights are *yours* to sell? What's an option? Let's assume the industry executives find the song or script, movie, recording, or whatever is a wow! Now what? What's the wow worth? What does that mean or, even better, why do they consider it a wow? Why would Brad Pitt want to be in a movie script you've written? How does a theater company acquire the rights to stage (they say "stand it up") the play *The Music Man*? How much is it going to cost and to whom do you send the check? And the question you never want to ask, why can't they just do it and pay me what I am owed?

A Brief History of Copyrights

Our copyright law is related to the invention of the printing press in 15th-century England. Moveable print was the technology of the time that allowed mass distribution of printed information just as the Internet and digital downloading are transforming the entertainment industry.[2] The invention of the printing press freaked out the politicians and religious leaders, who controlled citizens politically by restricting the flow of information to the public. They thought the technology of the printing press would allow citizens to share ideas and communicate with each other. Of course, just as politicians in some countries today shut down the Internet and cell phones when they're having political problems, they wanted to do the same thing about 260 years ago in merry old England.

Instead of banning the printing press technology, the English government decided to allow the use of the printing press technology *but* to censor what was printed. Control freaks, don't you just love them? Their stated reason was to manage the distribution of "Protestant religious heresy and political upheaval." In effect, the government forced authors and book publishers to obtain a *license to publish* from the Stationers' Company, which was established and controlled by the government in 1557.[3] Of course, this still allowed the top dogs to control what people knew by controlling the distribution of information.

When the licensing requirement was about to end in 1710, the English Parliament did something very unexpected. With courage and honor they passed the Statute of Anne, which is often considered the first copyright law. It changed things to clearly state that a book manuscript is legally defined as property, owned by the author. Second, it empowered authors with several ownership property rights, including control of copies and distribution. The statute's design was to "encourage learning by vesting the copies of printed books in the author or purchasers of such copies."[4] And as a consumer, if you didn't pay for the book, you'd be fined and the book illegally copied would be destroyed. According to the Avalon Project of the Yale Law School, Lillian Goldman Law Library (2008), the statute established the following fines or penalties:

> *Whereas printers, booksellers, and other persons have of late frequently taken the liberty of printing, reprinting, and publishing, or causing to be printed, reprinted, and published, books and other writings, without the consent of the authors or proprietors of such books and writings, to their very great detriment, and too often to the ruin of them and their families: for preventing therefore such practices for the future, and for the encouragement of learned men to compose and write useful books; may it please your Majesty, that it may be enacted . . . That the author of any book or books already composed, and not printed and published, or that shall hereafter be composed, and his assignee or assigns, shall have the sole liberty of printing and reprinting such book and books for the term of fourteen years . . . and That if any other bookseller, printer or other person whatsoever . . . print, reprint, or import, or cause to be printed, reprinted, or imported, . . . the proprietor or proprietors . . . shall forfeit such book or books, and all and every sheet or sheets, to the proprietor or proprietors (and) . . . That every such offender or offenders shall forfeit one penny for every sheet.* [5]

The Statute of Anne supported the methods of financial livelihood of the authors and publishers and required a fine and destruction of illegal copies. The one cent per page fine would be about 38 cents per page today and as an example, a $3.00 fine for a 300-page book in 1710 would be a fine of $114.79 now. Plus, the book you ripped off would be destroyed.

Article 1, Section 8 of the U.S. Constitution

The authors of the U.S. Constitution did not miss the Statute of Anne's inference in Article 1, Section 8 (2015), which

> *gives Congress the right to legislate copyright statute to promote the Progress of Science and Useful Arts, by securing for Limited Times to Authors and Inventors, the Exclusive Right to their respective Writings and Discoveries.*[6]

Let's break it down.

"Promoting the progress of science and useful arts" is accomplished by providing legal ownership and, therefore, potential financial rewards to the individuals who invest their time and money into scientific discoveries (e.g., inventions and medicines) and useful arts (e.g., performing and creative arts), including songs, scripts, recordings, movies, and other entertainment products. Interestingly, the word "useful" has not been challenged as removing it would lead to censorship. The Compendium states (2014),

> *the Copyright Act does not explain what level of creativity is necessary for a work to qualify as a "work of authorship" under the Act. Section 102(a) of the Act states—without further elaboration—that "[c] copyright protection subsists . . . in original works of authorship fixed in any tangible medium of expression, now known or later developed, from which they can be perceived, reproduced, or otherwise communicated, either directly or with the aid of a machine or device."*[7]

Copyright protection exists from the time the work is created in a fixed form. The copyright immediately becomes the property of the author who created the work. The copyright is instantaneous, and in reality, when you register your claim of a copyright all you are doing is registering your "claim"

Figure 3.2 The six exclusive rights of owning a creative "work" include reproduction (someone who wants to make a copy), distribution, use in a public performance (e.g., a bar) or public performance by means of the Internet or digital transmission, and derivative (used when someone makes a new copyrightable work based on our copyrighted work).

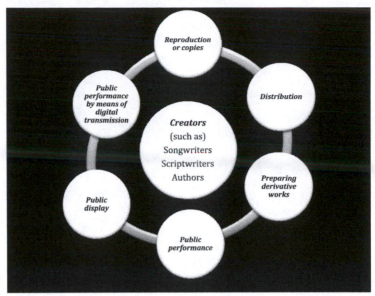

that you own it. The government registers the time and date of your claim.

The "exclusive rights" bestowed are as follows (see Figure 3.2):

1. Reproduction or copies
2. Distribution
3. Preparing derivative works
4. Public performance
5. Public display
6. Public performance by means of digital transmission

The phrase "by securing for limited times" means ownership rights do not last forever—only for a stated period of time before the copyright (ownership) falls into public domain. Then it's owned by the public and can be used by anyone without payment. This "duration of ownership" has changed over the years from 14 years (renewable) to the life of the author plus 70 years. Indeed, almost everything about copyrights, the Copyright Office, and the registration process has changed over time. The Compendium, Chapter 100 of the U.S.

Figure 3.3 The length of a copyright is for "limited times." The duration was 14 year renewal with another 14 years, for a total of 28 years in 1710. Currently, the duration of a copyright after 1978 is life of the creator of the work (owner), plus 70 years. There are some exceptions, such as a work made by two or more authors, where the work is protected for life of the authors and then 70 years after the death of the last surviving author. A work-for-hire is protected for 95 years from the first date of publication or 120 years from the creation, whichever is shorter. In addition, for works created but not published or registered before January 1, 1978, the protection is provided for life plus 70, 120, or 95 years, depending on the type of authorship. In some cases the copyright may extend until 2047. For copyrights already protected before January 1, 1978 (transitional copyrights), they are protected for 28 plus 28, under the 1710 act, and later that was changed to 28 plus 67 under the 1976 act, plus the Sonny Bono Act, for a total of 95 years. However, all works published in the United States before January 1, 1923, are now in the public domain. Confused yet?

Copyright Duration

Copyright Office and Copyright Law: General Background, provides us with the following (2014):

> *In May 1790, when Congress ended the first federal copyright law, the U.S. Copyright Office did not yet exist. Instead, authors and publishers recorded their claims with federal district courts and submitted copies of their works (in those days, book, maps, and charts) in support of their applications. These works, known as deposits, were stored in a variety of places, including in the U.S. Department of State and the U.S. Department of the Interior. As of 1846, the Smithsonian Institution and the Library of Congress shared them. This meant that records of copyright ownership were scattered among different government offices, and despite the federal scheme of protection, there was neither a consolidated tracking system nor centralized plan for preserving or using deposited works.*[8]

Major Copyright Acts

You can see the legal underpinnings that are the foundation of the entertainment industry. Like the industry itself, the copyright laws are changing and adapting as new modes of production, innovation, technology, and consumer experiences evolve. The law remains a constant to which artists and creators should be able turn to for protection of what they have created. So, as Sue Greenberg (2011) reminds you, "If you are performing, selling artwork or otherwise making money as an artist, you're engaged in business."[9] We need the law to protect our business as the creative works are our investments at several levels and, certainly, they are investments for studios and labels. We need to advocate for creative works to be protected in a timely manner as the industry changes around us. Look at Table 3.2 to see how the legal system has responded to changes in creative production. Can you say it is doing enough?

The Digital Marketplace

Now that we understand the concept of copyrights, the trick is trying to find someone who'd want to buy or use (listen

Table 3.2 Major Copyright Acts. The majority of the copyright acts (laws) pertain to the duration of the copyright, registration, and what is copyrightable.[10]

Date	Purpose	Time	Entertainment Industry
1790	This law provided a term of 14 years with the option of renewing the registration for another 14-year term.	28 years	Map, chart, and book ownership to authors and proprietors.
1831	Music added to works protected against unauthorized printing and vending.	42 years	The business of music and entertainment is now legal.
1865	Congress enacts "An Act to Amend the Several Acts respecting Copyright."	42 years	Added protections for photographs and photographic negatives.
1870	The law added "works of art" to the list of protected works and reserved to authors the right to create certain derivative works, including translations and dramatizations.	42 years	Allowed derivative works to be created and owned.
1895	Congress mandates that U.S. government works be not subject to copyright protection.	Zero	All government documents are non-copyrightable.
1897	Congress enacts a law to protect music against unauthorized public performance.	N/A	Legal foundation for collection of performance royalties for songwriters and music publishers.
1909	Certain class of unpublished works now eligible for registration. Term of statutory protection for a work copyrighted in published form measured from the date of publication of the work. Renewal term extended from 14 to 28 years.	56 years	Widened the definition of unpublished works that could be registered. Duration for "published works" started with the date of publication not creation. Extended the duration of a copyright.
1912	Motion pictures, previously allowed to be registered only as a series of still photographs, now protected works.	56 years	Motion pictures, film, and so forth are copyrightable.

Table 3.2 Continued

Date	Purpose	Time	Entertainment Industry
1914	U.S. adherence to the Buenos Aires Copyright Convention of 1910.	56 years	Established copyright protection between the United States and certain Latin American nations.
1953	Recording and performing rights extended to nondramatic literary works.	56 years	
1955	United States becomes party to the 1952 Universal Copyright Convention as revised in Geneva, Switzerland.	56 years	U.S. copyrights extended to other agreeing nations and theirs to our country.
1972	Effective date of the act extending limited copyright protection to sound recordings fixed and first published on or after this date.	56 years	Sound recordings provided limited copyright protection.
1974	United States becomes party to the 1971 revision of the Universal Copyright Convention at Paris, France.	56 years	Extended world copyright protection in agreeing nations.
1976	The 1976 Act replaced the 1909 Copyright Act and changed much of how copyright law operates, including as follows:		
	Duration of most works	Life, plus 50 years	Replaced the 56-year duration with duration of life of the author plus 70 years.
	Anonymous, pseudonymous, and works made for hire	95 years from publication	
	Copyright at the moment of fixation	N/A	Copyright instantaneous at time creative work is placed into a tangible medium of expression. Duration no longer based on date of registration or publication.

(*Continued*)

Table 3.2 Continued

Date	Purpose	Time	Entertainment Industry
	Author may terminate previous grants of copyright	N/A	Author may request in writing "recapture of copyright" within a 35–40-year window.
	Added protection for additional types of work	Life plus 50 to 95 years based on type of registration	Pantomime, choreography, architectural, mass works, and vessel designs
	Fair use rights	N/A	Congress added exceptions and limitations to the exclusive rights.
	Date of activity January 1, 1978	N/A	Listed as the 1976 Act but date of rule was January 1, 1978.
1980	Copyright law amended to include computer programs.	Life plus 50 to 95 years based on type of registration	Software copyrightable
1982	Willful infringement of copyrights for commercial/private financial gain subject to criminal laws.	N/A	Intentionally selling or using someone else's creative property (recordings, etc.) for money is a crime. Consider sampling as example.
1984	Record Rental Amendment of 1984 grants copyright in a sound recording the right to authorize or prohibit the rental, lease, or lending of phonorecords for direct or indirect commercial purposes.	N/A	As an example, the use of streaming (rental, lease, or lending) of recordings is not legal without the license from the copyright owner. Spotify and others have to pay labels licensing fees to use their recordings.
1989	The effective date of United States' adherence to the Berne Convention for the Protection of Literary and Artistic Works, as revised in Paris, France, in 1971.	N/A	United States joins agreement at Berne (Switzerland) Convention to expend copyright protections in world territories. U.S. required to drop mandatory use of copyright notice (©).

Table 3.2 Continued

Date	Purpose	Time	Entertainment Industry
1990	The Visual Artists Rights Act allows authors of certain types of visual works of art certain moral rights of attribution and integrity.	Life of author	The authors of visual works have the right to not have their name or demand to have their name associated with their works. Also, the right to prevent the distortion, mutilation, destruction, or modification of the visual work.
1990	Effective date of the Computer Software Rental Amendments Act	N/A	Grants the owner of copyright in computer programs the exclusive right to authorize or prohibit the rental, lease, or lending of a program for direct or indirect commercial purposes.
1992	Renewal registration becomes optional on a prospective basis. Any work in its twenty-eighth year of copyright protection no longer requires a renewal application with the U.S. Copyright Office in order for the copyright to extend into and through the renewal term.	Additional duration to 75 years	As such, all works initially copyrighted between January 1, 1964, and December 31, 1977, were renewed automatically, even if the party entitled to claim the renewal copyright failed to file a timely renewal with the Office.
1992	Effective date of the Audio Home Recording Act	N/A	The Act requires the placement of serial copy management systems in digital audio recorders and imposes royalties (DART royalties) on the sale of digital audio recording devices and media that are distributed to the copyright owners.
1993	Copyright Royalty Tribunal Reform Act of 1993	N/A	Eliminates the existing Copyright Royalty Tribunal and replaces it with the Copyright Arbitration Royalty Panel (CARP) to recommend to Congress statutory license rates.

(Continued)

Table 3.2 Continued

Date	Purpose	Time	Entertainment Industry
1997	The No Electronic Theft (NET) Act	N/A	Defines "financial gain" in relation to copyright infringement, sets penalties for willfully infringing a copyright either for the purposes of commercial advantage or private financial gain or by reproducing or distributing (including by electronic means) phonorecords of a certain value.
1998	The Sonny Bono Copyright Term Extension Act	Life plus 70 years for 1978 and later copyrights and 95 years for current valid pre-1978 copyrights	Extends the term of copyright for most works by 20 years.
1928, 1998	The Digital Millennium Copyright Act of 1998 (DMCA) adds provisions to the Copyright Act, including the following:	At least 50 years of duration for any kind of work	World digital copyright rights and protections
	It provides for the implementation of the World Intellectual Property Organization (WIPO) performances and the World Copyright Treaty (WCT).		The right of distribution and rental, and a broader right of communication to the public. The WCT provides a "three-step" test to determine limitations and exceptions.
	Phonograms Treaty (WPPT)		Prohibitions against circumventing technological measures protecting copyrighted works

Table 3.2 Continued

Date	Purpose	Time	Entertainment Industry
	Section 512 safe harbors from liability for Internet service providers		Safe harbors allow digital providers to exist as long as they follow certain guidelines of operation. Basically, it is an attempt by governments to strike a balance between innovation and copyright owners' rights. What it really does is provide a way for individuals who upload copyrighted songs and recordings to do it illegally and then not get a lawsuit filed against them, as they don't have enough money to make it worth hiring the attorneys. In the long term, the safe harbor foundations could be removed as subscription providers catch on.
	An exemption in Section 117 of the Act		Permits the temporary reproduction of computer programs made in the course of maintenance or repair; clarifies the policy role of the U.S. Copyright Office; and creates a new form of protection for vessel designs.
2004	The Copyright Royalty and Distribution Reform Act	N/A	Phases out the Copyright Arbitration Royalty Panel (CARP) and replaces it with the Copyright Royalty Board.
2008	The Prioritizing Resources and Organization for Intellectual Property Act	N/A	Creates the new government position of the Intellectual Property Enforcement Coordinator and prohibits the export and import of infringing copies.

to) what you've created instead of ripping it off. Consumers of entertainment products may understand the meaning of copyrights better by rearranging the roots, from "copyright" to "right to copy." Sadly, we are learning it's more difficult now to make money due to the advancement in digital technology and the appalling naive consumer behavior regarding copyrights and property right laws. Let's assume you turn "your great idea" into a song, a recording, and then a video. There may even be a movie in it!

Do you own an idea? No! According to Circular 92 of the Copyright Law of the United States (2011),

> *In no case does copyright protection for an original work of authorship extend to any idea, procedure, process, system, method of operation, concept, principle, or discovery, regardless of the form in which it is described, explained, illustrated, or embodied in such work.*[11]

Nobody owns an idea (no matter how original), unless it is converted into what the industry calls *intellectual property*. What in the world is that? If you want to make a living in the industry as a creative writer, musician, artist, or producer, you better find that out fast. The truth is that almost everything from recordings to sports teams' logos are considered intellectual property *owned by* an individual or a company. The key words are *owned by*, which relates to entrepreneurship and *business*. Forget about the brain rush you had when you came up with the great idea and start thinking about "How can I turn my idea into a piece of legal property?" *Property*, something you can own and sell! What a concept! After all, that's what this industry is really about, selling and licensing products (allowing companies and consumers to use/enjoy creative products) for a profit. So, in the eyes of the law, even though your idea comes from inspiration, it is as real as your car, or the mortgage on a house *once* it's placed into a tangible form as defined by copyright law. As you will see, once the idea has a concrete delivery mode involving teams of people working to make the idea a reality, it can be bought and sold for profit. Different types of "properties" may be protected through the process of intellectual properties, of which copyrights are one.

Converting "Ideas" Into Intellectual Property

How do you turn a great idea into a form of intellectual property you own? I know you will remember from Chapter 1 that if it's written down, sung into an iPhone, or recorded, then it's yours as long as it meets the legal definition. In other words, the songwriter who came up with an idea and turned it into a physical, tangible form of expression (magnetic or tangible) now owns the song he or she created and the recording stored in some type of device, called a phonorecord (see Figure 3.4).

Fixation Requirements

To be considered as a copyrightable creative work what you've created must be "fixed" into a tangible medium of

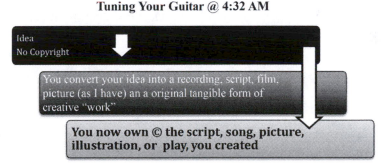

Figure 3.4 We do not own an original idea until we convert it into a physical, useable form, such as a song, movie script, or stage play. It must be original, complete, or "finished" and placed into a physical storage (phonorecord), such as a written document or a digital copy stored on a flash or hard drive. Talking about it to another person does not give us ownership or authorship of the idea. It has to be in a physical form before we can claim authorship (I thought of that) and ownership.

expression. According Chapter 300 of the Compendium of the U.S. Copyright Office Practices (2014),

> *A work of authorship may be deemed copyrightable, provided that it has been "fixed in any tangible medium of expression, now known or later developed, from which [it] can be perceived, reproduced, or otherwise communicated, either directly or indirectly with the aid of a machine or device"* . . . *Specifically, the work must be fixed in a copy or phonorecord "by or under the authority of the author" and the work must be "sufficiently permanent or stable to permit it to be perceived, reproduced, or otherwise communicated for a period of more than transitory duration."*[12]

Now you know why, how, and when a creative work is copyrighted. The instant it is "fixed" into the physical form, you're protected. *Not when it's registered*, but when it's created. Most will say it's the paper received from the Library of Congress when registering your creative work. *Wrong*. The U.S. government does not issue a "copyright" to you. It's actually your *claim of ownership* of what you've created, such as a song, recording, photograph, film, or video, as long as it fits the legal definition. The certificate of registration (and the registration number on the certification) represents your claim of creation backed by the government.

The writers who wrote the song now equally own it, as it's a legal piece of property. How can you prove you own the copyright (song)? Easy—register a claim that you created it with the U.S. Copyright Office.

If you create a song or recording, then the song in most of the free world is legally yours and defined as property. Is it worth anything? That depends on how many people want to buy it or listen to it. In the free world, you have the right to do it, so read on to discover how easy it is to register your claim of ownership on songs and other creative products you create (see Figure 3.5).

Figure 3.5 There are three easy steps in the registration process. After creating your "work" in an original, tangible form you may register your claim of copyright with the Library of Congress at www.loc.gov.

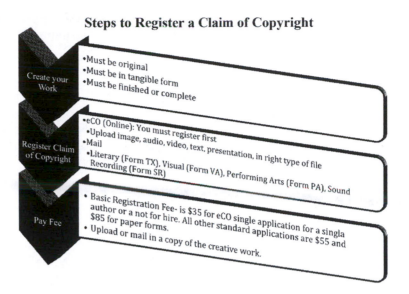

Steps to Register a Claim of Copyright

Create your Work
- Must be original
- Must be in tangible form
- Must be finished or complete

Register Claim of Copyright
- eCO (Online): You must register first
- Upload image, audio, video, text, presentation, in right type of file
- Mail
- Literary (Form TX), Visual (Form VA), Performing Arts (Form PA), Sound Recording (Form SR)

Pay Fee
- Basic Registration Fee- is $35 for eCO single application for a single author or a not for hire. All other standard applications are $55 and $85 for paper forms.
- Upload or mail in a copy of the creative work.

Legal Foundation

In the music and entertainment business, copyrights are the legal foundation of the industry. Why? Because entertainment products are property owned by the individuals and companies that create or buy them, and that can be worth a few pennies to billions, depending on how much consumers enjoy them. Yeah! And it's a great life, making money doing something you love. According to the Compendium of the U.S. Copyright Office Practices, third edition, the purpose of registration is as follows (2014):

> Under the current copyright law, a work of authorship is protected by copyright from the moment it is created, provided that the work is original and has been fixed in a tangible medium of expression. See 17 U.S.C. §§ 102(a), 408(a). Although registration is not required for a work to be protected by copyright, it does provide several important benefits:
>
> 1. A registration creates a public record that includes key facts relating to the authorship and ownership of the claimed work,

105

as well as information about the work, such as title, year of creation, date of publication (if any), and the type of authorship that the work contains (e.g., photographs, text, sound recordings).

2. Registration (or a refusal to register) is a prerequisite to filing a lawsuit for copyright infringement involving a U.S. work.

3. To claim statutory damages or attorney's fees in a copyright infringement lawsuit, a work must be registered before the infringement began or within three months after the first publication of the work.

4. A registration constitutes prima facie evidence of the validity of the copyright and the facts stated in the certificate of registration, but only if the work is registered before or within five years after the work is first published.

5. A registration provides information to prospective licensees, such as the name and address for obtaining permission to use the work.

6. A document that has been recorded with the U.S. Copyright Office may provide constructive notice of the facts stated therein, but only if the document specifically identifies a work of authorship and only if that work has been registered.

7. The deposit copy(ies) submitted with an application for registration of a published work may satisfy the mandatory deposit requirement, provided that the applicant submitted the best edition of the work.

8. A registration is necessary to secure the full benefits of a preregistration that has been issued by the U.S. Copyright Office.

9. The U.S. Customs and Border Protection Service may seize foreign pirated copies of a copyright owner's work, provided that the work has been registered with the U.S. Copyright Office and the certificate of registration has been recorded with the U.S. Customs and Border Protection Service.

10. A registration is required to claim royalties under the compulsory license for making and distributing phonorecords.

The purpose of the copyright law has been challenged in courts; however, according to the Compendium (2014),

> U.S. courts have analyzed the purpose of the Copyright Clause in a number of cases. The Supreme Court has interpreted the Copyright Clause to mean that copyright laws should promote both the creation and dissemination of creative works. See, e.g., Golan, 132

S. Ct. at 888–89. Thus, "[t]he Framers intended copyright itself to be the engine of free expression." Harper & Row Publishers, Inc. v. Nation Enterprises et al., 471 U.S. 539, 558 (1985); see also Golan, 132 S. Ct. at 890.

("By establishing a marketable right to the use of one's expression, copyright supplies the economic incentive to create and disseminate ideas.")[13]

Property

Let's dive in a little deeper. What is property? The definition of personal property depends on where you were born and where you now live. There are different laws in different countries, depending on their form of government. As an example, property rights in the United States and most of the free or western world are defined as ownership. Ownership is connected to a perceived value tied to economics. According to BusinessDirectory.com (2015) property is a

Quality or thing owned or possessed . . . Article, item, or thing owned with the rights of possession, use, and enjoyment, and which the owner can bestow, collateralize, encumber, mortgage, sell, or transfer, and can exclude everyone else from it. Two basic kinds of property are (1) Real (land), involving a degree of geographical fixity, and (2) Personal (anything other than real property) which does not involve geographical fixity. Personal property is subdivided into tangible property (any physical animate or inanimate object) and intangible property (intellectual property).[14]

Types of Intellectual Properties

Intellectual property rights (ideas turned into products and devices) become something of "value" as determined by consumers' "wants and needs" to be entertained. And the way we value something is by negotiating how many cows (only kidding) or how much money we'll pay for the experience. Ideas, remember, are not protected. However, the corresponding products "created" from an "expression of an idea in a physical form" may be copyrightable if they meet certain legal definitions. The recordings, films, TV shows, videos, computer games, performances, and other products sold, performed, or used by businesses and the public in exchange for money, licenses, and royalties are paid to the copyright owners. In most capitalist countries these industry processes and products usually fall into three legal areas—copyrights, trademarks, and patents. All three protect different types of intellectual property ownership rights.

Copyrights (©)

Copyrights provided by the laws of the United States (Title 17 U.S. Code) and similar laws in other countries provide protection (ownership of their creative properties) to the authors of original works of authorship, including literary, dramatic, musical, artistic, and other intellectual works.[15] Copyrights encourage artistic and creative people (often using the inventions and discoveries) to make, among other things, musical recordings, paintings, photographs, literary works, choreographic works, computer-enhanced artwork, and films. So what is a "copyright"?

Trademarks (TM, SM, and ®)

Trademarks protect the name and iconic symbols of products of their manufactory company marketed to the consumer market. Examples include business icons and logos, such as Coke, Coca-Cola, and Pepsi for soft drinks, and artist and band names, labels, and even the Rolling Stones red licking tongue logo to the corresponding acts, merchandise, branding, endorsement, corporate sponsorships, and other revenue-generating products and services. According to the U.S. Patent and Trademark Office release, "Protecting Your Trademark: Enhancing Your Rights Through Federal Registration" (2015),

> *A trademark is generally a word, phrase, symbol, or design, or a combination thereof, that identifies and distinguishes the source of the goods of one party from those of others. . . . A service mark is*

the same as a trademark, except that it identifies and distinguishes the source of a service rather than goods.[16]

Patents

Patents are issued by the government to protect ownership (property) in an original invention by *preventing* others from making, using, or selling the invention. The U.S. Patent and Trademark Office (2014) defines a patent as

> *"the right to exclude others from making, using, offering for sale, or selling" the invention in the United States or "importing" the invention into the United States. What is granted is not the right to make, use, offer for sale, sell or import, but the right to exclude others from making, using, offering for sale, selling or importing the invention. Once a patent is issued, the patentee must enforce the patent without aid of the USPTO.*

1. *Utility patents may be granted to anyone who invents or discovers any new and useful process, machine, article of manufacture, or composition of matter, or any new and useful improvement thereof;*
2. *Design patents may be granted to anyone who invents a new, original, and ornamental design for an article of manufacture; and*
3. *Plant patents may be granted to anyone who invents or discovers and asexually reproduces any distinct and new variety of plant.*[17]

The Value of Creativity

Enough! In spite of the laws we probably should understand better (most of us are not lawyers, including me), at least we can start to understand the mind-set of the creators and business executives who have invested their careers and money into creative products and now see them melting away because of technology and consumer behavior. Frustrating to say the least, but that ship has left the port and will probably not return. So what about you? The value of entertainment products has always been tied to the ownership laws of copyrights and consumer demand. Now things may be changing to a newer business model for generating profits. Nevertheless, according to the current laws whenever someone wants to use your property (e.g., songs, film, scripts, books) without notice and payment they are probably violating laws. That's the purpose of licenses used in the professional industry to keep everything legal and to make sure the copyright owners are receiving royalty payments (see Figure 3.6).

Is it becoming clear to you why there are very few entertainment superstars in countries without property rights? Now that you've read some of the laws, can you see the financial situation the industry is facing as consumers download entertainment products without paying for them? Remember, according to the Copyright Office Circular 1 (2012), "It is illegal for anyone to violate any of the rights provided by the copyright law to the owner of copyright."[18]

Figure 3.6 We have seen this before in Chapter 1, but now it may make more sense, as you understand creative works are property and must be paid for when downloading, copying, or using. As an example, the exclusive rights payments include license fees collected and paid to the copyright creators and owners.

Copyright Exclusive Rights Licensing

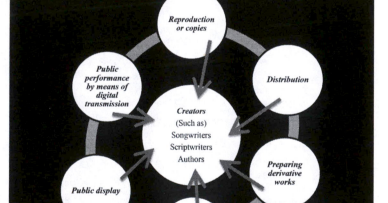

A Snake in the Grass

Sadly, among those in entertainment professions, recording artists are often the least informed because they don't learn how the business actually works. Egos are important but they seem to sometimes clutter the clarity of a superstar's understanding of the business of entertainment. Similarly, many of the creative people who act, direct, promote, fund, and market entertainment products do not understand the fundamental elements of the business as a system. They are the creators of products that generate billions of dollars, yet due to the complexity of the business they often trust others to take care of what they see as "details." As a result, you need to know how all the parts and roles associated with creative productions and business administration operate or you may become a victim of an unscrupulous manager, label, or entertainment company desperate for profits at your expense. Be careful also with the entertainment accountants and lawyers who might mismanage your contracts and royalties for the same reason. Most of the industry is highly respected and honest, but where there is money, you may also find a few rats. You want a career in the entertainment industry, not a life of mistakes that lead to lawsuits and bankruptcy. Let's not go there! By the way, most consumers don't care what happens to their favorite artists or their careers; consumers just enjoy the creative products and often want them to be free. The invention of the printing press scared the heck out of people who controlled the government and writings in the 15th century. The results of today's digital technology and the Internet have flipped the heck out of the entertainment industry's old business models. However, very smart and talented people in the industry are making more money than ever. How are they doing it?

Don't Be a Starving Artist

The industry and laws will change with the times. The entertainment industry, which is based on capitalism, creative efforts, hard work, financial investments, and profits, now has to compete with the idea that these very things should be free. The fact that we, as consumers, can acquire recordings and other entertainment products without paying for them does not make it right or legal. If you figured out how to steal a car or other property and get caught, you know what happens. But not in this business—now free is often expected. That's why Congress is looking at the laws and labels and other industry-related businesses are morphing into new profitable business models that provide a reasonable financial return on their investment.

After all, listening to music on the radio has been free since the invention of radio; you just had to wait for the DJ to play the song/recording you wanted to hear. But a key difference here is no consumers could control when or how much of a song they might hear again. Now, thanks to technology, you and all consumers have that access and power!

Cloud Servers

That's why the industry is headed quickly toward the streaming of entertainment model where consumers can enjoy any product they want, when they want it, in a millisecond by pushing a button or clicking a mouse. Recordings and other entertainment products are also available free as in the past or provided through new forms of distribution, such as Netflix, Amazon, and Spotify. With a subscription service you're not buying a copy of the product you want to listen to, use, or watch.

Selling Access

All you are receiving is *access* to it digitally through a monthly subscription payment. Some of the services are still free, and you still have to watch or listen to the commercials, just as we've done since the invention of media. Even if you buy a CD or DVD you do not own the entertainment embedded in it—you own just the plastic disc. Same for a book because you are not legally allowed to make copies of an entire book, recording, movie, or most other forms of entertainment. Why? You know the reason! Besides, you can find just about everything on the Internet, including Google books. However, lawsuits have been filed against them in many countries and their services have been banned in some for violations of privacy and copyright issues. The authors of the thousands of books scanned by Google have not been paid a dime, so it's restricted to show only short views of content, and they are not allowed to show entire publications. They may use the search to sell data about you to companies, but the original authors have received zip.

Perceptions of Free

Even with the new models of distribution, many creative people and businesses are not being paid much money. The problem has been to find an economic solution that will infuse the use of digital technologies into consumers' perception of value and expectations and still reward the entertainment product creators fairly. In spite of this weird situation, the music and entertainment industry is quickly changing its business models to adapt to the situation. Finding revenues in a free product market is difficult. By understanding: creativity, business principles, entrepreneurship, intellectual properties, consumer behavior, demographics, psychographics, metadata analysis and how the industry works, we can better understand how and why the industry is developing its new business models based on interactive streaming, branding, management of digital content, digital access distribution, rights management, and other sources of revenue. Just when we thought we knew it, here's another wrench in the system to consider.

Safe Harbor

Question: If we understand the law, how can YouTube and other services exist? A couple of reasons: first, they are "allowed" to operate as long as they pay a major percentage of their income to copyright holders, such as labels.

Second, they are currently considered a legal safe harbor. Think of it this way. If you are out on the water in your boat and a huge storm starts whipping the waves out of control, you'd probably want to find a "safe harbor" to anchor until after the storm. The Internet, Napster, and most of the other services are the "storm" that opened innovative distribution of entertainment products for sale or purchase, except most users wanted it free. Why are safe harbors allowed?

What's legal and illegal regarding the distribution of digital copies of creative products through businesses modeled on unknown innovation and technology is difficult to define. Most of the time, these types of issues are decided in court. That's what has happened in this situation as the courts have tried to strike a balance between copyright holders' legal rights and the development of business using the new technologies, and that is why we have some website providers legally designated as "safe harbors." As an example, a court case in Scotland between the *Shetland Times* and the *Shetland News* (1997) left the court trying to define the balance between copyright holders and new technology innovators:

> *A key question which remains under-explored in worldwide legislation (although it is addressed in the DMCA) is whether an intermediary, which provides a hyperlink to a site where illicit content is available, is liable for that content. The point is a vital one, not just because hyperlinking is the lifeblood of the Net, but because hyperlinks to unknown sites of unknown content are generated automatically by locational tools such as search engines every time a user requests a search . . . it would seem important to avoid placing unreasonable burdens of potential liability on search engine providers in respect of content to which they link. Current legal advice for those building commercial websites is now not only to disclaim liability for content linked to, but also frequently to seek to avoid any risk by requesting permission to link, on the grounds that linking might somehow equate to making, or authorizing the making of, an illegal copy; . . . this could prove to be an unfortunate practice in policy terms as it restricts Internet growth and connectivity, and may encourage extortionate demands from the site to which the request is made.[19]*

The World Intellectual Treaty and the WCT

As you probably guessed by now, copyright holders are trying to figure out a way to monetize their entertainment products, especially in light of the changing landscape of consumer technology devices. What's legal and what's not? Because of the technology, consumers now have control over the distribution and want any and all kinds of entertainment free. And at the same time, the providers were granted a legal "safe harbor," claiming the people stealing the creative works through their sites were the copyright infringers and not they. As we know, most people are very decent and do not understand that free downloads of entertainment products shaft the copyright owners. That is a serious problem! How can we solve that problem? Over the years, the World Intellectual Property Organization (WIPO) and the World Copyright Treaty (WCT) have provided some of the answers. In a short nine-page document many free countries of the world have together demanded copyright protection under 25 short articles related to the Berne Convention. Cory Janssen in *Techopedia* (2015) outlines the WCT as

> a special agreement enacted by a consensus of over 100 member states . . . for the Protection of Literary and Artistic Works (Berne Convention) and the International Convention for the Protection of Performers, Producers of Phonograms and Broadcasting Organizations (Rome Convention) . . . WCT was created to address changes in digital technology and communications, particularly the distribution of digitally protected works . . . WCT provides several key updates, including:
>
> - Protects computer programs as literary works per Article 2 of the Berne Convention.
> - Protects data compilations created as intellectual property in any form. This protection does not extend to actual data.
> - Stipulates that licensees must provide adequate legal remedies against anyone who knowingly enables or facilitates any type of copyright infringement related to unauthorized

electronic rights management, distribution, broadcasting or communication.[20]

Fair Use Rights

Fair use rights allow for free use of copyrighted works in criticism or commentary, news reporting, teaching, academic scholarship, and research. No payment for the use of the materials is required. Specific standards to determine what, how, where, and when copyright works may be used as a fair use right are based on factors including purpose, character, and the nature of the work, plus the amount used and its effect on the value of the original work. According to the U.S. Copyright Office Fair Use Index (2015),

> *Fair use is a legal doctrine that promotes freedom of expression by permitting the unlicensed use of copyright-protected works in certain circumstances. . . . provides the statutory framework for determining whether something is a fair use and identifies certain types of uses—such as criticism, comment, news reporting, teaching, scholarship, and research—as examples of activities that may qualify as fair use. Section 107 calls for consideration of the following four factors in evaluating a question of fair use:*
>
> 1. *Purpose and character of the use, including whether the use is of a commercial nature or is for nonprofit educational purposes.*
> 2. *Nature of the copyrighted work.*
> 3. *Amount and substantiality of the portion used in relation to the copyrighted work as a whole.*
> 4. *Effect of the use upon the potential market for or value of the copyrighted work.*[21]

Of course this is great as it provides a valuable service and source for people to learn stuff. That's part of the rules of the game: you create it, you own it (for limited times), but people can use small sections of it for research, criticism, and education. Sadly, some musicians, songwriters, or producers who like to use "samples" don't understand that using the part of one recorded musical note is a violation. Just get permission by acquiring a license from the copyright holders and you're fine. However, we're back to asking questions and we need to know what to ask to receive a correct answer. As an example, what types of licenses are required? Where can I file for one, and how much will it cost? I hope you're starting to understand why you need a basic knowledge of copyrights to work at almost any level in the entertainment and music industry. So if you're a song, script, or book author, a movie producer, label, or huge entertainment company, how can you register your products with the Copyright Office in Washington, DC?

Types of Registration

There are several ways of registering creative works of authorship for a claim of copyright. The word *authorship* is now associated with the word *ownership* (of what you created), which is defined as your *personal property*. How *valuable* is it? That is determined later by public demand and usage if you can get the industry or potential consumers to just discover it. However, at this point it is important to remember what you're claiming. If you created it yourself, you claim authorship. *And* since you're the only one who created it, you should also claim ownership of the copyright. If two people wrote a script or song, then both are considered authors and share ownership. Chapter 2 of the Compendium of the U.S. Copyright Office Practices, third edition, describes the types of registration (2014):

1. *Basic registration is an application for a basic registration that is used to register a copyright claim in a work created or first published on or after January 1, 1978, and covers the full term of the copyright. This type of registration may be obtained with a Standard Application or a Single Application (provided certain eligibility requirements have been met).*
2. *Group registrations are an application for a group registration that is used to register a claim to copyright in a group of related works that qualify for a single registration.*

117

3. *Supplementary registration is an application for a supplementary registration that may be used to correct or amplify the information in a basic or renewal registration.*
4. *GATT registration is an application for a GATT registration that is used to register a copyright claim in a work in which U.S. copyright was restored by the 1994 Uruguay Round Agreements Act (URAA).*
5. *Renewal registration is an application for a renewal registration that is used to cover the renewal term for works copyrighted before January 1, 1978.*[22]

Types of Registration

The two most popular ways to register your claim of copyright with the Library of Congress Copyright are the electronic (eCO) upload and the old snail mail paper forms, including TX, VA, PA, SR, SE, and CON.[23]

Table 3.3 details the snail mail registration process.[24] Claiming a copyright of a creative work is really a two-stage process, listing the author(s) of the work (section 1 in the attached SR form) and who owns the work (section 4 under CERTIFICATION of the attached SR form). According to 202.3 of the Compendium (2015),

> *Registration and recordation are two separate procedures: claims to copyright are registered, while documents related to copyright claims, such as agreements to transfer or grant a mortgage in copyrights, are recorded.*[25]

Certificate of Registration

At the end of an evaluation process, the Copyright Office either refuses or provides a *certificate of registration* containing a registration number. The process is described in the Compendium as follows (2014):

> *The U.S. Copyright Office examines applications for registering claims to copyright and any accompanying deposit copy(ies) to determine whether they satisfy the statutory requirements for*

Table 3.3 Forms. The old snail mail registration process is still used today by the Copyright Office. You may want to consider the eCO process, which is quicker and costs less money.[26]

Type	Use of Paper Forms	Link to Website
TX Literary Works	Registration of published or unpublished nondramatic literary works, excluding periodicals or serial issues. This class includes a wide variety of works: fiction, nonfiction, poetry, textbooks, reference works, directories, catalogs, advertising copy, compilations of information, and computer programs.	http://copyright.gov/forms/formtx.pdf
CON Continuation Application	The Continuation Sheet is used in conjunction with Forms CA, PA, SE, SR, TX, and VA when you need to continue the registration process.	http://copyright.gov/forms/formcon.pdf
VA Visual Arts	Registration of published or unpublished works of the visual arts. This category consists of "pictorial, graphic, or sculptural works," including two-dimensional and three-dimensional works of fine, graphic, and applied art, photographs, prints and art reproductions, maps, globes, charts, technical drawings, diagrams, and models.	http://copyright.gov/forms/formva.pdf
PA Performance Arts	Registration of published or unpublished works of the performing arts. This class includes works prepared for the purpose of being "performed" directly before an audience or indirectly "by means of any device or process." Works of the performing arts include: (1) musical works, including any accompanying words; (2) dramatic works, including any accompanying music; (3) pantomimes and choreographic works; and (4) motion pictures and other audiovisual works.	http://copyright.gov/forms/formpa.pdf
SR Sound Recordings	Registration of published or unpublished sound recordings. Form SR should be used when the copyright claim is limited to the sound recording itself, and it may also be used where the same copyright claimant is seeking simultaneous registration of the underlying musical, dramatic, or literary work embodied in the phonorecord. With one exception, "sound recordings" are works that result from the fixation of a series of musical, spoken, or other sounds. The exception is for the audio portions of audiovisual works, such as a motion picture soundtrack or an audiocassette accompanying a filmstrip. These are considered a part of the audiovisual work as a whole.	http://copyright.gov/forms/formsr.pdf

(*Continued*)

Table 3.3 Continued

Type	Use of Paper Forms	Link to Website
SE Serial	Registration of each individual issue of a serial. A serial is defined as a work issued or intended to be issued in successive parts bearing numerical or chronological designations and intended to be continued indefinitely. This class includes a variety of works, such as periodicals; newspapers; annuals; and the journals, proceedings, and transactions of societies. Do not use Form SE to register an individual contribution to a serial. Go to www.copyright.gov to register such contributions.	http://copyright.gov/forms/formse.pdf

registerability, including copyright ability, and otherwise comply with the Office's regulations. Based on its findings, the Office then either registers or refuses to register the claims.[27]

The eCO filing takes up to 8 months and paper forms about 13 months for processing. For the mail-in forms select the type of work you'd like to register a claim on and then follow the links to download the instructions and forms. Complete the form fully and then add a check and a copy of the creative work.

Certificate of Recordation

What happens when you sell your copyright (ownership) in the entertainment product (work) you've created? Usually, the new owners request a change of the name of the "owners" of the copyright registration at the Copyright Office, and they will then receive back under the same registration number a *certificate of recordation* (section 4 under TRANS-FER of the attached SR sample). The Compendium states (2014),

any transfer of copyright ownership or other document relating to copyright may be recorded in the U.S. Copyright Office, subject to certain conditions. The recordation of documents pertaining to transfers or other ownership matters is voluntary, but recommended because: (i) it provides constructive notice of the facts stated in the recorded document if certain conditions have been met; (ii) when a transfer of copyright is timely recorded (within one month of its execution in the United States or two months of its execution outside of the United States, or any time before a conflicting transfer is recorded), the recorded transfer prevails over a later executed transfer; and (iii) a complete public record may mitigate problems related to orphan works.[28]

Deposit Copies

To complete the registration process, a copy or copies of your work must also be submitted at the time of application. Also, detailed in the Compendium (2015) is the following:

As a general rule, the applicant must submit a complete copy or copies of the work to register a claim to copyright . . . In specific instances, the deposit copy(ies) may be submitted in digital or physical format. The deposit copy(ies) must conform to certain requirements depending on the type of work, the deposit requirements, and whether the work is published or unpublished . . . Once the Office receives the registration materials, a registration specialist will examine the deposit copy(ies) to determine if the work is eligible for registration. The Office will not return the deposit copy(ies) or the identifying material to the applicant.[29]

Electronic Registration

The eCO registration process is quicker and less expensive for you to file. It allows for digital uploading of files. You have to register once, and then every time after you'll be greeted with a friendly "hello" and a simple step-by-step menu to follow. Easy. Get started by completing the application and following all of the screen prompts.[30]

According to the Copyright Office official website (2015), the eCO electronic process may be used to register the following types of claims: *literary works, works of the visual arts, sound recordings, work of the performing arts, motion picture/audiovisual work,* and *single-issue serials.*[31]

The Business of Entertainment

Once we own a creative product then we have the right to sell it, which is the purpose of the entertainment industry. You can't stop at writing the clever books, songs, and scripts, producing the music, playing the instruments, producing the shows, and lighting the sets. Your creative work has to have quality, interest, and market value to reach the level of a true entertainment product. However, every so often God gives a few of us the talent and ability to create (and, therefore, own) wow books, scripts, musical recordings, and so forth, and then the industry steps in to finance the creative individuals making the recordings, movies, books, and computer games, and the corresponding businesses that distribute, promote, and market the products to consumers. And somewhere in that process, we hope the people (you) who create the products are paid for the brilliant work and the businesses that invest millions make a profit on their investments.

The Mob Pirates

In the history of book publishing, printers and booksellers who reprinted copyrighted works without securing the manuscript rights for new printings were called pirates and the illegal publications were called pirated copies. In the 18th century in England, the copyright laws, derived from the Statute of Anne, as seen earlier, were strictly enforced. Offending printers and booksellers, anyone they used to sell the publications, called street hawkers or mercuries, and any members of all those people's families could be thrown in jail for the illegal printing, and they all would have to pay large fines to be released. Nothing like Johnny Depp jumping around in a Disney movie! This kind of syndicate was organized crime, and we have the same system operating today, though the look of the products and how we access them have changed.

In the mid-1950s, using the same equipment and processes as the label, pirates would record the album into a new matrix and then make a mother (that's what it was called) and press a few hundred thousand new copies. Then they'd also use the same printing company equipment as the real guys, slide the new illegal album off the presses and into the jacket, seal it with shrink-wrap, and bingo, free albums they'd sell at discounted prices to bar owners, clubs, and even legal retail stores. All that money went directly into the mob's pockets and the labels, writers, publishers, and artists got zip. The same situation took place with cassettes and CDs in the record business and in the film business with videotapes and later DVDs. Ouch! The FBI was chasing a few bad guys ripping off the labels and movie companies for millions. And don't forget the markets overseas where the local officials and cops were sometimes ignoring the problem and the FBI was not allowed to operate. It was a constant battle as the mob just kept moving the equipment to different sites, yet every so often someone would get caught. There were also the small operators or pirates, but they had the problem of not only the FBI but also the mob if they got too big.

Today's Pirates

As we've already stated the pirates of today are the people sitting all around us. They are not the Blackbeards or even Johnny Depps of yesterday, pointing their swords toward the movie companies and labels, laughing with a patch over an eye, making hundreds of millions of illegal copies and then laundering the money in a foreign country. Today, we have the tools necessary to commit the crime ourselves. Of course, when any of us downloads a favorite movie, computer game, or recording we're not trying to make copies and sell them for a buck. Most of the billion or so modern pirates simply want to use the entertainment products free or become the local "hero against the establishment" by uploading the stuff on a bit.net site. However, as you can now see, there's not any need for the mafia as consumers simply do it themselves. But the damage is far more devastating than the mob could ever accomplish. Wow, and that really is a *wow*.

Why should the industry stay in the game, when consumers can obtain entertainment products free or at a cost that is too low to support the creators and copyright owners? Why would anyone want to learn how to become a great writer, singer, actor, production expert, investor, and so forth if the odds of making a living and having a great career are zero? That's what is starting to take place in the business. If you spend millions making a great recording, developing an act, or producing a wow film, as it is released and in some cases before it's distributed for sale or use, it's already available online free through bit.net. Sucks @+!@! So, if we want some form of entertaining product that we'd want to take the time to enjoy, then let's figure out how to make the systems founded in capitalism, currently struggling with the realities of wonderful technological advances, profitable. Let's explore some of the game players and their positions by looking at the copyright law.

The Decision Makers

Congress is stuck in the middle between the laws it has created by the authority provided in the Constitution and the potential new products and services innovative technology may provide in the future. This is a difficult marriage with historical ties to the adaptive uses of any new "thing" created, as we have already learned about with the Statute of Anne and the invention and use of the printing press in merry old England in the 1700s. As the late President Harry S. Truman used to say, "the buck stops here," and that's what is now happening in Congress. As most of us already know Article 1, Section 8, of the U.S. Constitution states very clearly that Congress shall have the right *"To promote the Progress of Science and useful Arts*, by securing for limited Times to Authors and Inventors *the exclusive Right* to their respective Writings and Discoveries." Here are a few other interesting parts of Section 8 that we may want to address. Notice the differences between what Congress, the president, and the Supreme Court are allowed or not allowed to do. Thus, when the president does something as an executive act in an area that Congress is supposed to control, it is questionable if it is legal. Another tough job and position to be in. Welcome to politics; the questions about intellectual property rights fall squarely into the political discussion. There are also conflicts between creators, owners, users, and requests for new laws that may uphold the Founding Fathers' writings. Here is Section 8 of the U.S. Copyright Law:

Article 1, Section 8

- *The Congress shall have Power To lay and collect Taxes, Duties, Imposts and Excises, to pay the Debts and provide for the common Defense and general Welfare of the United States; but all Duties, Imposts and Excises shall be uniform throughout the United States;*
- *To borrow Money on the credit of the United States;*
- *To regulate Commerce with foreign Nations, and among the several States, and with the Indian Tribes;*
- *To establish an uniform Rule of Naturalization, and uniform Laws on the subject of Bankruptcies throughout the United States;*
- *To coin Money, regulate the Value thereof, and of foreign Coin, and fix the Standard of Weights and Measures;*
- *To provide for the Punishment of counterfeiting the Securities and current Coin of the United States;*
- *To establish Post Offices and post Roads;*
- *To promote the Progress of Science and useful Arts, by securing for limited Times to Authors and Inventors the exclusive Right to their respective Writings and Discoveries;*
- *To constitute Tribunals inferior to the Supreme Court;*
- *To define and punish Piracies and Felonies committed on the high Seas, and Offences against the Law of Nations;*
- *To declare War, grant Letters of Marque and Reprisal, and make Rules concerning Captures on Land and Water;*
- *To raise and support Armies, but no Appropriation of Money to that Use shall be for a longer Term than two Years;*
- *To provide and maintain a Navy;*
- *To make Rules for the Government and Regulation of the land and naval Forces;*
- *To provide for calling forth the Militia to execute the Laws of the Union, suppress Insurrections and repel Invasions;*

- *To provide for organizing, arming, and disciplining, the Militia, and for governing such Part of them as may be employed in the Service of the United States, reserving to the States respectively, the Appointment of the Officers, and the Authority of training the Militia according to the discipline prescribed by Congress;*
- *To exercise exclusive Legislation in all Cases whatsoever, over such District (not exceeding ten Miles square) as may, by Cession of particular States, and the Acceptance of Congress, become the Seat of the Government of the United States, and to exercise like Authority over all Places purchased by the Consent of the Legislature of the State in which the Same shall be, for the Erection of Forts, Magazines, Arsenals, dock-Yards, and other needful Buildings;—And*
- *To make all Laws which shall be necessary and proper for carrying into Execution the foregoing Powers, and all other Powers vested by this Constitution in the Government of the United States, or in any Department or Officer thereof.*[32]

The Fork in the Road

Do we have property rights or don't we? That is the question. It would be a different story if we lived in Castro's Cuba, where everyone is supposed to share equally, yet we know that system is a joke unless you want to live a very restricted life. There is very little motive to create or do much of anything, as there's no reward, so what you get is what the government allows you to have. Boring! I have more faith in the individual and believe if given the freedom, protection of rights, and motivation many of us will work hard to contribute to the quality of life of everyone. In a free society, we have the right to profit from our hard work and intellectual gifts *if* we want to pursue, create, and develop something that may positively contribute to our society. And if we actually do it, it's the people who decide if they want or need it through the commercial free market system. The more consumers want something, the more money the owners make, if they are smart, understand the law, and don't give it away, as many are doing with their music on the Internet. In other words, a free market does not mean free products. It means the freedom to develop and create goods and services, and sell them for something of value, such as cows (remember this from our discussion of money?) or greenbacks. Where does Congress draw the line? That is a very difficult question to answer; however, let's take a peek at the possibilities.

"Property Owners" and the Google Books Lawsuit

The way I read Article 1, Section 8 and the whole law makes me want to acknowledge that the people who create copyrightable products (art/entertainment) own them as long as they meet the legal definition of useful arts. Songwriters own their songs, novelists own their manuscripts, scriptwriters own their scripts, and arrangers own their musical arrangements until they fall into public domain. Record labels own the recordings they finance by great artists, and movie companies own the films, TV shows, and movies they financed, marketed, and sold.

Let's take a look at the Google lawsuit to which I allude in some other chapters. Here is the summary of the lawsuit and its major parties:

> *The suit alleges that . . . universities allowed Google to scan millions of books from their respective libraries, and then used those works to create a shared digital repository known as the HathiTrust Digital Library. The complaint alleges that the repository contains nearly 12 million works, that roughly 73% of those works are protected by copyright, and that the universities allow faculty, students, and patrons to view, print, and download full text copies of those works without permission.*[33]

It will come as no surprise to you that it takes very hard work, talent, and usually significant capital to create books, movies, and computer games. The outcome of the Google lawsuit, which was handed down in October 2015, has set a legal foundation for the end of all creative arts and the entertainment industry, as we know it, as the presiding judge has also overruled the foundation of intellectual property rights.

A Legal Argument

Let's take a closer look at this real problem caused by the technical advances in digital communication and software. To understand the potential significance of the Google decision, let's recap what the U.S. Constitution provides us as protections for our physical, personal, and intellectual properties,

> *We the People of the United States, in Order to form a more perfect Union, establish Justice, insure domestic Tranquility, provide for the common defense, promote the general Welfare, and secure the Blessings of Liberty to ourselves and our Posterity, do ordain and establish this Constitution for the United States of America.*

Notice the words *we the people, establish justice, promote the general welfare*, and *posterity*.

"We the people" means "by the authority of the people"—yes, citizens of the United States—that members of Congress (as our elected representatives) have the authority and responsibility to protect our fundamental rights.

> *The Constitution of the United States of America is the supreme law of the United States. Empowered with the sovereign authority of the people by the framers and the consent of the legislatures of the states, it is the source of all government powers, and also provides important limitations on the government that protect the fundamental rights of United States citizens. [Emphasis added]*

Also notice the U.S. Constitution is the source of all government powers, with important limitations on the government. Article 1, Section 8, also states that Congress shall have the right "To promote the Progress of Science and *useful Arts*, by *securing* for limited Times to *Authors* and Inventors the *exclusive Right to their respective Writings and Discoveries*" (emphasis added). Okay, so Congress gave us the right to own what we've created for a limited time, which is currently the creator's life plus 70 years. In addition, the Constitution also defined our ownership rights as exclusive rights. In the context of creative works, the word "exclusive" provides the copyright creator or owner 100% *control* of the rights granted by the Constitution, currently defined as six exclusive rights:

> *Subject to sections 107 through 122, the owner of copyright under this title has the exclusive rights to do and to authorize any of the following:*
>
> *(1)* To reproduce the copyrighted work in copies or phonorecords;
> *(2)* *To prepare derivative works based upon the copyrighted work;*
> *(3)* To distribute copies or phonorecords of the copyrighted work to the public by sale or other transfer of ownership, or by rental, lease, or lending;
> *(4)* *In the case of literary, musical, dramatic, and choreographic works, pantomimes, and motion pictures and other audiovisual works, to perform the copyrighted work publicly;*
> *(5)* *In the case of literary, musical, dramatic, and choreographic works, pantomimes, and pictorial, graphic, or sculptural works, including the individual images of a motion picture or other audiovisual work, to display the copyrighted work publicly; and*
> *(6)* *In the case of sound recordings, to perform the copyrighted work publicly by means of a digital audio transmission.* [Emphasis added]

Fair Use Right

We are not finished yet as there are some exceptions to copyright owner exclusive rights, titled "fair use" rights. They are usually very positive for the education of society and a meaningful tool to be used to improve the public welfare, knowledge, and quality of life. The fair use exceptions as defined by Congress and by the ways stated in the USCO Mass Digitization Letter from the Office of the Register of Copyrights (2011) are as follows:

> *When analyzing the fair use defense, courts review the four non-exhaustive factors set forth in Section 107 of the Copyright Act: (1) the purpose and character of the use, including whether such use is* of a commercial nature *or is* for nonprofit educational purposes; *(2) the nature of the copyrighted work; (3) the* amount and substantiality of the portion used in relation to the copyrighted work as a whole; *and (4)* the effect of the use upon the potential market for or value of the copyrighted work. *[Emphasis added]*

Why?

The key questions the Google lawsuit appears to raise and the questions that should matter to you are as follows:

- Does Google have the right to scan full books and offer segments to help citizens?
- Does it have the right to do it for free without licenses from the copyright holders? Google grossed $55 billion in 2013 with all the other revenue sources and $66 billion in 2014.
- Why not just buy the books?

Misunderstanding the © and ℗

Google gained the right to copy "branding features," such as logos and illustrations, without the copyright holder's permission. Money, under the ruling, will be exchanging hands between the two parties even though the copyright holders are not receiving any monies or royalties. The authors usually give up their copyright in exchange for a percentage of each unit sold; thus the book publishers are not receiving any money, so the authors who wrote the books will not be paid as their deals are with the book publishers. If the publisher isn't paid, neither will the author be paid. In the music business, there is a different situation as Congress has approved ASCAP, BMI, and SESAC to collect public performance royalties (©) for music publishers and songwriters and SoundExchange to collect digital transmission royalties (℗) for the labels, and digital royalties for the royalty artists and recording session musicians on non-interactive digital streams. Why are royalties owed? Because radio, media, digital streaming, nightclub owners, and businesses who use the songs to set an atmosphere to increase business are making money off someone else's property—namely, the songwriters and publishers who own the rights to the songs (©). ASCAP, BMI, and SESAC in the United States and hundreds of PROs in other countries collect these royalties for songwriters and music publishers residing in their home countries. The labels are paid digital streaming royalties, as they own the recording of the songs (℗) used by businesses. Many recording artists are concerned as the rates for streaming are low $.0002 per play and as the acts' deals are signed directly with the label, which may or may not pay them any royalties.

The Google lawsuit and digital innovations are forks in the road. The transformations they bring should not be seen as a deterrent to the entertainment and music industry. You will be part of the decisions that will shape the industry, which starts with a carefully crafted business plan and understanding of not only your target markets but also how the business of entertainment is monetized. What follows will enable you to develop your entrepreneurial and leadership skills in ways that may contribute to the overall growth and transformation of the industry.

Notes

1. Rudd, Benjamin W. "Notable Dates in American Copyright 1783–1969." Copyright.gov. Accessed February 20, 2016. http://copyright.gov/history/dates.pdf.

2. Feather, John. "The Book Trade in Politics: The Making of the Copyright Act of 1710, 'Publishing History'" The Statute of Anne, 1710 (1/6). 1980. Accessed June 8, 2015.

3. Deazley, R., L. Bently, and M. Kretschmer. "Commentary on the Statute of Anne 1710, in Primary Sources on Copyright (1450–1900)." 2008. Accessed June 8, 2015. http://www.copyrighthistory.org/cam/commentary/uk_1710/uk_1710_com_272007105424.html.

4. Ibid.

5. "The Statute of Anne April 10, 1710." *The Avalon Project, Document in Law, History and Diplomacy.* 2008. Accessed June 8, 2015. http://avalon.law.yale.edu/18th_century/anne_1710.asp.

6. *Copyright Law of the United States and Related Laws Contained in Title 17 of the United States Code.* December 1, 2011. Accessed June 6, 2015. http://copyright.gov/title17/circ92.pdf.

7. "COMPENDIUM: Chapter 200 Overview of the Registration Process." Copyright.gov, U.S. Copyright Office a Department of the Library of Congress. December 22, 2014. Accessed June 10, 2015. http://copyright.gov/comp3/chapter200.html.

8. "COMPENDIUM: Chapter 100 U.S. Copyright Office and the Copyright Law: General Background." Copyright.gov, U.S. Copyright Office, a Department of the Library of Congress. December 22, 2014. Accessed June 10, 2015. http://copyright.gov/comp3/chap100/ch100-general-background.pdf.

9. Greenberg, Sue. *Artist as a Bookkeeper*, 15. 2nd ed., St. Louis: St. Louis Volunteer Lawyers and Accountants for the Arts, 2011. Print.

10. "U.S. Copyright Office—Information Circular." Accessed February 20, 2016. http://copyright.gov/circs/circ1a.html.

11. *Copyright Law of the United States and Related Laws Contained in Title 17 of the United States Code.* December 1, 2011. Accessed June 6, 2015. http://copyright.gov/title17/circ92.pdf.

12. "305 The Fixation Requirement." Copyright.gov. December 22, 2014. Accessed June 15, 2015. http://copyright.gov/comp3/chap300/ch300-copyrightable-authorship.pdf.

13. "COMPENDIUM: Chapter 100 U.S. Copyright Office and the Copyright Law: General Background."

14. "Property." BusinessDictionary.com. 2015. Accessed June 4, 2015. http://www.businessdictionary.com/definition/property.html.

15. *Copyright Law of the United States and Related Laws Contained in Title 17 of the United States Code.*

16. "Protecting Your Trademarks: Enhancing Your Rights Through Federal Registration." USPTO.gov. Basic Facts About Trademarks. 2015. Accessed June 6, 2015. http://www.uspto.gov/sites/default/files/BasicFacts.pdf.

17. "Circular 40: Copyright Registration for Pictorial, Graphic, and Sculptural Works." *Copyright Registration for Pictorial, Graphic and Sculptural Works.* 2013. Accessed June 7, 2015. http://copyright.gov/circs/circ40.pdf.

18. "Copyright Basics." Copyright Basics, Circular 1. 2012. Accessed June 9, 2015. http://www.copyright.gov/circs/circ01.pdf.

19. L. Edwards and C.E. Waelde. Quoted in Arpi Abovyan, "Challenges of Copyright in the Digital Age: Comparison of the Implementation of the EU Legislation in Germany and Armenia," 95. Unpublished PhD thesis, Munich: Utz, 2014.

20. Janssen, Cory. "WIPO Copyright Treaty (WCT)." Techopedia. Accessed June 15, 2015. http://www.techopedia.com/definition/26952/wipo-copyright-treaty-wct.

21. "More Information on Fair Use." Copyright.gov. U.S. Copyright Office, a Department of the Library of Congress. April 1, 2015. Accessed June 10, 2015. http://copyright.gov/fair-use/more-info.html.

22. "COMPENDIUM: Chapter 200 Overview of the Registration Process."

23. "ECO Registration System." Copyright.gov. U.S. Copyright Office, a Department of the Library of Congress. 2015. Accessed June 7, 2015. http://copyright.gov/eco/.

24. "Form SR." Copyright.gov. May 1, 2005. Accessed June 10, 2015. http://www.copyright.gov/forms/formsr.pdf.

25. "COMPENDIUM: Chapter 200 Overview of the Registration Process."

26. "Forms U.S. Copyright Office." Accessed February 20, 2016. http://copyright.gov/forms/.

27. "COMPENDIUM: Chapter 100 U.S. Copyright Office and the Copyright Law: General Background."

28. "COMPENDIUM: Chapter 200 Overview of the Registration Process."

29. Ibid.

30. "Welcome to the ECO (Electronic Copyright Office) Standard Application Tutorial: A Guide for Completing Your Electronic Copyright Registration." *Welcome to the ECO, Standard Application Tutorial.* Accessed June 8, 2015. http://copyright.gov/eco/eco-tutorial-standard.pdf.

31. Ibid.

32. "The Constitution of the United States: A Transcription." National Archives and Records Administration. Accessed October 1, 2015. http://www.archives.gov/exhibits/charters/constitution_transcript.html.

33. (S.D.N.Y. Oct. 5, 2011) ("HathiTrust First Amended Complaint") "Legal Issues in Mass Digitization: A Preliminary Analysis and Discussion Document." Copyright.gov. October 1, 2011. Accessed October 5, 2015. http://copyright.gov/docs/massdigitization/USCOMassDigitization_October2011.pdf.

THE ENTERTAINMENT BUSINESS 4

Introduction

The *business of business* is to make a profit! The business of the entertainment business is to use creativity for profit. One of the biggest mistakes we can make is to try to be in the entertainment and music business *without* understanding how business works. Another key mistake is downplaying how important desire and passion are in making a career out of creativity. To make it even more fun, this industry has some unique barriers to entrance built around what you need to understand *before* trying to pitch a new song, recording, film, book, script, game, or even yourself, to secure a deal or employment. Let's face it, if you're trying to sell property you've created (e.g., a song, script, recording) or yourself as the talent, you're in business, even if you don't know it (see Figure 4.1)!

Figure 4.1 People with amazing talents may forget that to be successful they often need to be an entrepreneur, essentially a businessperson with connections and knowledge of how the business within the entertainment business works.

Career Success

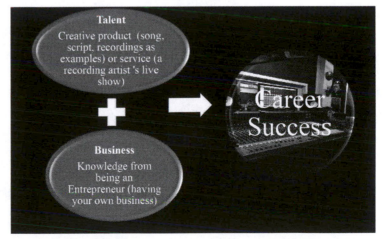

135

Entrepreneurship

Now, let's really dive into the concept of "business" by exploring the process of building an actual business model related to what you may want to achieve in the industry. I am providing some basic concepts to spur thinking and to get you to try to come up with helpful answers to the questions you may have.

Take a few minutes to review the illustration (Fig. 4.2 below) adapted from the book *Business Model Generation* by Osterwalder and Pigneur (2010).[1] It provides a step-by-step process to help us understand how a start-up business venture is constructed. Once you have looked it over, consider how to apply each step to the types of opportunities in the entertainment and music business you find interesting. Osterwalder and Pigneur (2010) claim there are nine steps or "building blocks" to consider before starting a business.[2]

The Language of Start-Up Ventures

How can you become successful in the entertainment and music business? Find a problem consumers are experiencing, develop a service or product that is the solution to the problem, and then build your own venture to sell to consumers. Businesses in the entertainment industry sign talent, scripts, and so forth, and buying, licensing, and signing "talent" are a business-to-business deal. If you're the talent then you're also the "business" executives in the industry will "deal" with if you've got something they want. What a concept! So, if you're a creative talent, you're also going to be a business and more than likely the owner of the business also! If you are a songwriter, you're in business! If you write scripts, act, sing, or play music, believe it or not, you're in business. Here's the big surprise: to be a successful creative person, you are usually required to also be a businessperson who owns your own writing, acting, singing, performance, or other related business. And your product or service is usually the selling of your own songs, scripts, and so forth to other businesses that desire to license your "creative works" as books, recordings, movies, or computer games. So what is the name of your business? What type of legal business is your company? What's it going to cost you to run your business? And the questions go on and on, but let's start with building a business model.

Business Model

A business model is required to determine the type of business you'll need or need to be to sell your products or services. There are many different types of business models; yet, no matter which style you choose, there are nine commonly required steps in developing your business model. Remember the business model is simply a plan rooted in how you are going to sell your songs, recordings, or live performances as a product or service. Figure 4.2 shows you how the business model flows with it fully completed. Note that each step has a related prompt that links the product or service to the consumer and, ultimately, to the profit the product or service will generate.

The Artist as Businessperson

Often, successful creative people begin their careers locally as a sole proprietorship, frequently choosing an engaging business name, not their own, allowing them to do business as (DBA) or within a framework that is designed to attract customers. At first, many creative people do not even know they need to be a business or understand the purpose and processes of commerce. Are you a creative artist and you want to get hired by the local club, promoter, or film festival, or sell your paintings at a street festival? Lesson number one is to realize the club owner, festival promoter, gallery manager, or anyone who is hiring talent, placing the art on a wall, or showing a film short is making the money, not you.

Figure 4.2 According to Osterwalder and Pigneur (2010), the first step in the building-block process is to determine your customer. Who they are, age, gender, income, total number in the United States, and location of zip plus four that you can often find through government sources and subscriptions to research companies, such as Prizm. With the customer segment understood, you can make assumptions about how consumers will value your product or service, and how to deliver it to them. Step 4 considers the business's relationship with customers, such as a one-time sale or creating sustained interest to develop repeat customers by providing strong customer product support. Steps 5–9 require the new business owners to consider financial assumptions about revenue, costs, expenses, key partnerships, and activities that will make the company run smoothly, finishing with the cost structure of the business, and the related administration and operations in support of the business.[3]

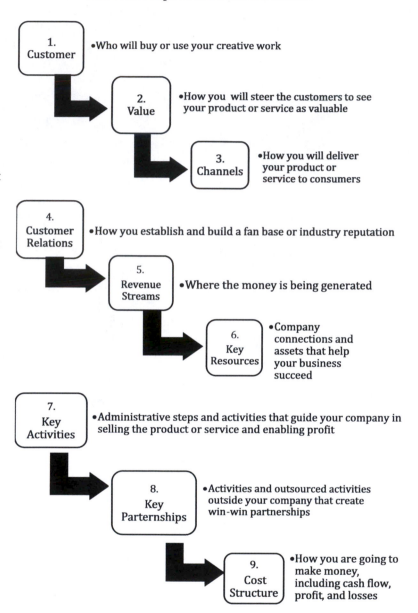

The Nine Steps of the Business Model

1. **Customer** •Who will buy or use your creative work

2. **Value** •How you will steer the customers to see your product or service as valuable

3. **Channels** •How you will deliver your product or service to consumers

4. **Customer Relations** •How you establish and build a fan base or industry reputation

5. **Revenue Streams** •Where the money is being generated

6. **Key Resources** •Company connections and assets that help your business succeed

7. **Key Activities** •Administrative steps and activities that guide your company in selling the product or service and enabling profit

8. **Key Parternships** •Activities and outsourced activities outside your company that create win-win partnerships

9. **Cost Structure** •How you are going to make money, including cash flow, profit, and losses

They're betting whatever you've got or can perform will increase the number of customers who buy a beer, meals, paintings, or tickets and to them, that's a profit that grows their business. Whatever you've got is a means or a way for owners of businesses to bring people in who will spend money, and the way they do that is to hire your business to provide whatever you artistically do in order for their business to make money. This is a business-to-business industry even when you're just getting started. When you're getting started as an artist or someone who's helping artists sell their creative "works," you need to ask yourself the following building-block questions about how to develop the business model (Table 4.1).

Table 4.1 The Building Block Questions. Answer the question(s) for each building block to help you understand the type of business model that will best help you sell yourself, creative works, or services (e.g., artist management) to other businesses that need them to make a successful profit.[4]

Order and Name of Building Block	Definition	Examples of Questions You Might Ask
1. Customer segments	Defining your customer	Who is your customer? Fans, labels, music publishers, and so forth.
2. Value propositions	The "perceived value" your product or services provide to your customers	What problem are you solving for your ideal customers?
3. Channels	The ways and methods of distribution of your products and services	How can you make your customers aware of your products and services and get them to them?

(*Continued*)

Table 4.1 Continued

Order and Name of Building Block	Definition	Examples of Questions You Might Ask
4. Customer relationship	The types of professional relationships between the company (you) and your customer	How do you find and retain customers to upsell your products and service? Is it automated, or is there personal assistance available if needed?
5. Revenue streams (note that it is often more than one stream of money)	Represents the gross income from sales minus expenses, which is the net profit or loss	How can you monetize different types of customers or sell to established customers with different or new products? In the music and entertainment industry, think about subscription fees and free products driven by advertisements, such as what Amazon and Spotify do.
6. Key resources	The physical, financial, and intellectual property or human beings who make a company or product valuable	What is the value of your songs, videos, and other creative products or what is the value of your knowledge and contacts to an employer? Why would someone want to hire you and not others?

Order and Name of Building Block	Definition	Examples of Questions You Might Ask
7. Key activities	The most important thing you or your company does to make a profit or to become more valuable to any employer	What are the key things your company does to make it more profitable in the marketing, production, reduction, and so forth, or for an individual? This could include networking, knowledge, or increasing creativity.
8. Key partners	The suppliers of goods and services that will help your company or you achieve more value to the industry	What people, places, things, and knowledge do you need to acquire to be as successful as possible?
9. Cost structure	The total cost and expenses of doing business	How much will it cost you to gain the knowledge, experience, products, and services that will make your company successful?

The Establishment

The need for a sound business model is the same for major films, labels, touring production companies, and so forth, where your value is based on how much fans, movie ticket buyers, and radio and TV listeners and viewers spend on whatever you've got to offer. In more ways than you would imagine, any stardom you might achieve and your measure of success are tied to the value of the money you (your image and brand) generate—usually for someone else. The trick is for your artistic efforts and products to emotionally move the customers enough for them to pull out the credit card or lay down the cash. Before you get upset that I'm being so honest about the money angle, never forget that it's really about the art, passion, emotions, message, experience, and about a billion other things art provides emotionally to consumers. Money, as we learned in Chapter 2, is simply the language of the deal.

Small Business Administration Questions

Even if you don't want to own a business it's smart to either start a venture or work for one related to your passion. The experience usually confirms your desire to work in the industry, gives you a chance to apply the basics of knowledge, develop contacts, and experience business-related realities needed later to help your employer solve problems and make a profit. As a college student, typically, you may have enrolled in an internship program to experience the opportunity of using what you learned in class to inform what you need to be able to accomplish to be successful in the industry. The quicker you learn, the more valuable your talents or skills are to your success and to the industry. Of course, when you're starting to launch a creative career, the catch-22 is that nobody important in the industry is likely to know who you are or what you've created. Why would they listen to your songs, films, and so forth, or hire you? The surprise is that you'll get your shot!

Your Shot

Club owners to industry executives understand that their future is tied to new ideas, talents, products, and innovation, which only new generations can contribute. It's still not unusual for new bands to pay to play or confirm a number of fans will appear before a club owner gives them a shot. It's a business after all! Word of mouth is always helpful as long as the buzz is positive. Negative comments or referrals from others who gave you a shot where you did poorly cause a slow, painful artistic death. So plan ahead, understand others' needs and agendas (as promoters or club or gallery owners), and you'll quickly find opportunities for work.

Starting a business or evaluating a business opportunity can't be happenstance. The questions provided by the Small Business Administration (2015) are great places to get some guidance about what you should be thinking before you start doing. Here are a few of those questions and some of my comments, which I hope will make these questions relevant to your particular needs and interests in having your own business.

1. Why Am I Starting a Business?

Good question—the answer is simple, to make money for the other guy and yourself. Many beginning positions in entertainment are entrepreneurial in nature. After all "To be or not to be?" is the question Shakespeare famously wrote about big choices. If you want to be an act, songwriter, manager, or scriptwriter, do it now and get better at it every day. However, from day one *be yourself*; be creative when you need to be and a businessperson at the same time. Be your own business and sell it as you sell yourself because that's what you're really doing every day if you're an executive or even a local act playing a gig. Gain success by having others experience your art, act, and gigs, and if they dig it, hopefully they will talk, talk, talk about it.

Start a Business by Creating Market Demand

Now if your product or service is getting exposure, is heard, seen, experienced then other club and gallery owners and fans will hear about it and your price goes . . . up. Business

143

and product demand is the language executives in most businesses use to make their important decisions. Do you know the language? For example, what are these: a controlled composition clause, WIPO, SoundExchange, the formula, recoupment, advances, options, a sync or master license? To create a demand for yourself as an artist or a product you may want to sell, you've got to know the language used in the industry. That's the way people talk to others about business opportunities and the potential profits and losses tied to what you're offering. How does everyone make money? Forget about ego for a minute and focus on the business as a business if you want to really succeed in the entertainment industry. Art is art, business is business, and when you can tie the two together (if you have something consumers want) you may never be able to spend all of the money it may generate.

Try working for a small, established industry-related entrepreneurial company, such as an independent label, promotion, agency, or production company. More success, knowledge, and contacts lead to better job offers and eventually the possibility of starting your own company. But this time instead of starting on the bottom, trying to get noticed, you're a major player and maybe even on top, controlling a valuable niche in the film, music, art, or touring industry. Rock on! Everyone needs the experience and contacts that bring wisdom, and the industry businesses will evaluate you on "who you are and what you can do." An internship connected to a college that offers related industry majors is a great way to discover how successful businesses are designed. So once again, why should you start a business? To learn what you need to know to make yourself valuable to the industry. Unless you're filthy rich, you'll probably have to work to survive this cruel, mean world (only kidding). Actually a job can be a great experience, lots of fun, and an opportunity to meet others and contribute some type of kindness to the world. If you're in the entertainment or music business, ditto that and add hard work, and you might have an amazing, blessed life.

The Deal Within the Deal That Starts Your Business

Nothing beats the fun of performing or excitement of selling the book you've written! But that great feeling starts with the deal based on green paper. I've mentioned before the "don't give it away concept," hoping you'll realize the power you have as a writer or act, when someone in the industry likes what you've got and it hasn't been distributed yet. Why *not* do something fun and get paid for it? Most of us have seen our parents labor at gigs they really do not enjoy, so consider a career that may make less money, but provide you with a rich lifetime of experiences. Further, with a business of your own, you can give back, as you'll probably be helping to create or sell products and acts that may improve the quality of people's lives. That's a good thing! So when you're offered a contract put the ego aside and consider what's really important. In many cases, you'll find it may not be the amount of green paper you'll get in exchange for your time and efforts, but the opportunities for experiences and emotions it provides.

2. What Kind of Business Do I Want?

What do you want to do and can you actually do it well? Let's take a quick look at some of the types of careers in the industry; most fall into one of three categories—*creative*, *business*, and *legal*. Each career has its own unique set of skills and business questions associated with it.

Creative Industry

What does it really take to create a great song, recording, film, computer game, or live show? Talent (within the industry the term is used to describe actors, technicians, producers, musicians, writers, and the list goes on), money, timing, a modest bit of luck, and you might want to sprinkle on a little magic. Talent, usually working as a team, is very professional at making and selling creative products or your creativity when seeking employment. How does talent look in the industry?

Let's focus on four careers: writers, producers and directors, creative technicians, and performers.

Writers

Customarily, creative and energetic novelists, scriptwriters, and songwriters often appear to be consistently seeking a better way to express a catchy idea or hook. What's hard to do is to write something career creative types will hang their hat on and businesses will want to invest millions in. However, when the royalty streams come together, great writers and everyone else often make serious money. It is possible everything written, signed, filmed, staged, or recorded is worthless when consumers didn't get it or ignored it, or it lacked something that made them want to be repeat customers. But if you do connect, then the money just keeps coming to your mailbox, year after year, if you were smart enough to make a good deal.

Producers and Directors

Producers in the film business make the product happen by providing the money, contacts, and important decisions to bring an idea to life. In the film industry, it's the director who's responsible to guide the lighting, sound, staging, camera, and computer effects techs, and most importantly, the actors to deliver breathtaking performances in a final product viewers will hopefully enjoy by laughing, crying, getting scared or surprised, or experiencing some level of culturally accepted love. It's the same in the theater industry, where the producer is the money and the director is responsible for the performance. In the music business, producers intelligently and creatively combine artists, talented technicians, and business executives to create potential hits. Great recordings require the producer to marry the perfect song to the perfect recording artist, remembering at the same time their image and brand. Hiring the best musicians, vocalists, background singers, and audio engineers who get it and renting the right studio for the feel and sound are still important. Then it's up to the creative team directed by the producer to deliver the "lightning in a bottle" consumers hopefully connect with. In the studio, when you get "goose bumps" you know you've got it, and it's happened to me only once.

What do you have to know to successfully direct a movie or play, or produce a recording session? What do you need to know to stay within your budget? What do you need to know about communication with creative artists, technicians, lawyers, and executives? How can you legally gain the right to use a song or recording of a song in a film? Wow, it's complex, hard work, and you really need to be smart and experienced. But it's also a lot of fun when it becomes an Oscar, Tony, or Grammy award winner. Even better if you get paid! By the way, music is used almost every time there is a visual production, and the production company must acquire a sync license for the use of a song from the music publishing company and a master license from the label to use the recording in the film *before* it is distributed.

Creative Technicians

Technicians are considered creative artists themselves by the producers and directors because they know how to best use lighting, sound, computers, studio acoustics, and special effects to enhance the qualities of the films, recordings, and plays and how to make the singers and actors sound and look their best. It's amazing how different some film stars look in person. Others look almost the same, as they want to dress the part and be recognized. They are often short in stature, thin, and usually naturally beautiful or handsome. It's funny as heck to see Johnny Depp, unshaven, in a T-shirt and sunglasses with an old baseball hat pulled down over his eyes. By the way, leave them alone. They are just like the rest of us, and so never ask for an autograph or picture if you want a career in the industry. Just be yourself and, if you need to, shake their hand, look them in the eye, and compliment their talent and work. Usually, they'll relax at that point and start talking to you. But if you see them eating in a restaurant or they seem to be busy, just wave, smile, don't say anything, and let them have their space.

Audio engineers are responsible for the technical quality of the recording, just as the producers are responsible for the creative quality of the recording. Their job is to make the artists and musicians sound as good as technically possible. Accordingly, they work with the recording artists, musicians, and producers to capture the artists' and musicians' best creative efforts on tape or computer hard disk. Film and stage directors need to achieve the same results for the actors, dancers, and others.

As much as Pro Tools and Logic Pro are changing the recording industry, the creative human element can't be ignored. The creative artists still need to sing the songs and the great musicians still need to perform them. Producers still need to guide the artists and musicians to sing and play their instruments in a manner that reinforces the emotional message in the song. Same thing for film and the theater producers. Computers may replace the equipment in the studio and even the studio itself, yet they cannot provide the expert musicianship, singing, and audio and visual production of real, live human beings. Yeah!

The musicians in a self-made recording are often "a single computer whiz" hitting the keyboard to create sounds similar to "real musicians" in a studio. Often one instrument is generated at a time, and vocalists and then BGVs (background vocalists) last. Knowledge of acoustics (the science of sound), microphone placement, pickup patterns (how sound enters or is blocked from entering the microphone), and live musicians working creatively together is often lost using computer software programs. Most professional producers and engineers use both for master sessions. Live sessions provide many opportunities for emotional creativity. Pro Tools and Logic Pro software replaces the human creative advantages and opportunities with cognitive digital creative opportunities. The results are similar, but there is a difference in the final content and artistic performance of the sound. Consumers do not seem to notice the differences, and the computer sessions often provide labels with significant recording budget savings.

Similarly, computers are often used in the film industry to sync and match location shots and performances so the movie is a continuous story, not something filmed over a period of weeks in various types of weather and other conditions. Which bring us to the question, what do you need to know to be a successful technician in the production process now? The future belongs to the creative audio and video audio engineers who learn how to provide products that use both the traditional methods and computers. The industries are quickly becoming integrated to the point where everywhere there's a visual production, there is also music and vice versa. Learn the production skills with innovation and creativity, and then move toward producing something of quality for less money to really enhance your career opportunities.

Performance Talent

Actors, recording artists, and musicians must breathe life into the products they create by conveying the story's emotional message to potential consumers. Even if it is a three-minute story in a song or a two-hour dramatic stage play, the actors must deliver a believable performance. They must infuse energy into their lines, facial and body expressions, and lyrics, and vitality into the words of the script, or notes printed on the sheet music. BB King (1925–2015) described the emotion in a song in his interview with Ed Vulliamy (2012):

> When he was a boy, BB King used to drive his mule through those fields while his Uncle Jack, up front, sung "the holler", the descendant of the slave chants and responses, wherein the blues began . . . "I remember the holler," says BB. "Holding the reins of a mule pulling a hoe through them cotton fields." The field holler, he

explains, was a lament sung on a minor scale by a single voice. It also functioned as a communication to alert others in the field that the boss was coming, or that water was needed. "Yeah, the holler is where it all started. I think it's in all of us."[5]

Most acts are signed to a label deal as a work-for-hire, which means the label owns their works. There are some exceptions, which will be discussed ahead, but even major recording artists should have a basic understanding of the copyright laws. Some acts have not negotiated a percentage of those streaming incomes with the labels and thus are not owed streaming royalties. The background vocalists (BGV) are the "oohs" and "ahs" of the recording and performance industry whose vocals enhance the main artist's performance. They provide the vocal supporting harmony for major recording artists, along with singing live at theme parks and nightclubs, and on concert tours, cruise ships, and even radio jingles and television commercials.

Musicians are the great interpreters of songs. They reveal their emotions and personalities through their musical instruments and performances. They are the backbone of the recording process, supporting both the lyrics of the song and the performance of the vocalists. BB King also describes his love for his guitar and why he kept on playing "Lucille" in an *Observer* interview (2012):

> *BB King called his guitar Lucille after someone else's fight in a juke joint knocked over a container of kerosene . . . The place caught fire . . . but BB realized he had left his guitar inside and ran to get it . . . "I named my guitar Lucille" . . . "to remind myself not to do something like that again, and I haven't." . . . "I keep wanting to play better, go further. There are so many sounds I still want to make, so many things I haven't yet done. When I was younger I thought maybe I'd reached that peak . . . I know it can never be perfect, it can never be exactly what it should be, so you got to keep going further, getting better."*[6]

Session musicians and vocalists are both considered recording artists. Working in the "arts" is often a part-time gig as there are fewer recording studio positions available than players. However, there are plenty of opportunities for musicians to work in bands, orchestras, churches, theaters, and other various types of venues. They are the best in the world at what they do. They use their emotions/feelings to play the instruments in a manner unique to the style of the song, act's voice, and image. Any musicians can play the same song, but it's the great ones that stand out and make triple recording scales by the way and manner they play the notes. The Wrecking Crew is one of the most famous studio musician groups, who recorded the tracks for the biggest hits of the Beach Boys, Glen Campbell, The Carpenters, Nat King Cole, Frank Sinatra, and many others.

Related Business Models

Once the product has been created the next business questions are to whom and how are you going to sell it? The traditional business side (business system) consists of entrepreneurs and business-minded individuals who publish songs; fund the development of films or computer and Internet programs and games; finance the artists, their recordings, shows, and concerts; and distribute, promote, and market the entertainment products through traditional retail outlets, traditional media, and social and streaming media to various consumer markets. It used to be the purchasing of recordings, films on DVDs, and downloads made the profits; now it's the streaming, ticket sales, merchandise, and rights management royalties that provide the revenues and profits to the industry.

Success as a creative artist, actor, scriptwriter, and so forth is based on how well scripts, songs, and recordings are seen by the industry gatekeepers (e.g., artists and repertoire [A&R], talent agents, booking agencies) as a tool for profit. You can't take rejection personally as decisions are based on what it's going to cost to make, promote, and sell products and what the odds are of making a reasonable profit. It's strictly a business deal. Most sane people when they see the risks and potential financial losses never invest their money or time into the industry unless they are looking for a tax write-off. Here's an example of what I mean. I was flying into New York City a few years ago and sitting next to me was an attorney whose address was 1 Rockefeller Center. Huge attorney. He represented investors who'd put serious money into what became a very successful film. Sorry, I can't give you details. Then he said, "We know the film made millions, but according to their accountants it lost money. Now you know why they say in Hollywood, 'If you don't like it, sue me, baby.'" In business, also create a win-win-win situation by knowing who your customers are, what they need and want, and what price they'll pay. If your customer is another business, then know what their needs for production are and help them solve problems to increase their profits.

Research and Metadata

Metadata is defined as "Data that describes other data, as in describing the origin, structure, or characteristic of computer files, webpages, databases, or other digital resources."[7] Many types of research provide insights into who, what, where, and when consumers buy, use, listen, view, or steal entertainment products.[8] Acts are supplied with tour support money and enough promotion and publicity to give consumers an opportunity to discover them and their recording(s). Labels use the information to direct additional advertisement buys, and promoters use it to research fan bases for hiring acts for staged events, which is often tied to content sales in various regions or zip codes. Providers such as iTunes, Amazon, and Spotify use metadata to determine consumers' behavior and their specific uses, viewing, and listening preferences (buying, selling, and stealing habits) and that metadata may also be used by promoters. The RIAA (Recording Industry Association of America) in the United States and the IFPI (International Federation of the Phonographic Industry), which covers 57 countries worldwide, employ big data to research unit sales for labels they represent. Finally, labels use this and other research and data at their weekly meetings to discuss each act's unit sales and other generated revenue. Surprisingly, it takes only two to four weeks to determine sales trends and to know if the act is a "hit."

Consumers are drawn to entertainment products for totally different reasons. They find the products stimulating, funny, or entertaining (what a concept)! The definition varies, but in all honesty the word *entertainment* has to do with some type of "performance" and "occupation of the mind." And if we like it enough, as an emotional cognitive "feeling," we'll share it with everyone we know, which is what social media is all about. By the way, the word "fan" is derived from the word fanatic, which is a descriptor of the level of intensity of the emotional connection. Entertainment is about perceptions and not necessarily reality, because what we imagine can be a lot more fun and interesting than the certainty of our daily life. So, the business-balancing act is matched with what the label or the company speculates will sell with what it has researched and is sure the consumer wants.

Administration

Who really gets the work accomplished in a business? The executives make the decisions, but who does the work once the decision has been made? The administrators at most labels, film production companies, and other industry-related enterprises are the people who punch the time clock, work eight to five and make things happen—except in the entertainment industry, where the typical office hours are from 10:00 a.m. to 6:00 or 7:00 p.m., plus all the industry functions, parties, and channels of networking you're expected to attend. What the heck—you're young and crazy and it's fun. Well, the first few years it's fun, but after that it is almost a duty or something you need to do to be "seen" and to stay current. The late lunch is almost always the power lunch for the executives and it's catch-up time for what needs to be completed before the boss returns.

3. Who Is My Ideal Customer?

One summer, on a flight to Berlin, I sat next to a tall, attractive woman in her late twenties. I was surprised to discover she was flying to Germany to sing as the featured artist with the Berlin Opera. As we exchanged pleasantries, she was surprised to learn that I teach classes about the business of entertainment and music. It was one of those magical occasions when you meet someone and find out how much you have in common.

As I congratulated her on her success, she explained how her college professors had taught her how to sing, but not how to make a living as an artist in the business. That she had to learn herself. She taught herself about networking and the legal, marketing, promotion, management, and accounting aspects of the business. In the end, she found out that nobody was coming to "discover her," and she realized that to be a financially successful artist, she first had to become an entrepreneur, own her business, and learn how to manage and market herself. It was a hard lesson to learn and it took many years away from her singing career.

Ask yourself who will buy your products or services, or, in our opera singer's case, who hires the artist. She found out that her ideal customers were not the fans; they were second after getting the gig. Once she got the gig, then she had to satisfy the ticket buyers. Without the gig, there were no fans to entertainment and impress. She had to find out who runs the opera world and how to request and receive an audition. Where would she work if she didn't network to find the movers and shakers in that niche industry? They were not coming to look for her, so she had to proactively meet them, built a reputation, and earn creditable reviews. It's the same for film and stage actors: if you don't have powerful agent representation the people looking will not waste their time trying to find you.

Talent Representation

Creating something such as a song, movie script, recording, or even a live show we own is wonderful. However, can you sell it? It takes time and money to self-promote our products and us as songwriters, scriptwriters, performers, or whatever. Most try it and then discover they are not ready. Just think who, what, where, when, and how (see Table 4.2)!

To initiate an opportunity to be noticed, first have your work (product or service) evaluated by an expert in the industry. Is it any good or is it a waste of their time? Who might actually give it a listen, view, or shot? Who can make the decisions at the company? What are they looking for? Where can we find them and when is a good time to connect? Here's a clue: most of the entertainment industry (business executives) are too busy finishing up the year's business by Thanksgiving and are missing in action through the holiday. In addition, they won't answer the phone, return your emails, or book an appointment if they do not know you or have at least heard of you. So, what are you to do? Realize that you have power because you own what you created. Remember our six exclusive rights and because you own what you have created, either you can collect money if someone else wants to use your creative work; or, you can have representation place and monetize your creative works for you. What's it worth? We don't know until the public or major production companies use it in a recording, a film, or a computer game. The best way to make serious money is to find professional representation.

Finding Representation

Once you better understand the administrative structure of the industry and who the important executives are and what they are seeking, connect with the industry-related "agents" who can put a copy of your creative work directly into the right leader's hands (see Figure 4.3). This is a strange industry; so if you're a songwriter, pitch new songs to established music publishers, who may represent you and your material to labels, acts, movie production companies, and many others. If you're a recording artist, music publishers and established personal artist managers may pitch you to A&R

Table 4.2 Talent Representation. It's best to find professional representation (booking agent for live acts), someone who knows someone in A&R at a label, a music publisher if you're pitching a song, a talent agent if you're an actor, director, or production specialist, and a literary agent if you're pitching any type of script or book. They will evaluate, pitch, and place your stuff to the right executives quickly if they perceive your creative work as potentially valuable to the industry. On the other hand, one really great song may create enough consumer buzz even if you give it away on YouTube, and they may start calling you.

Record Label	Song: Mechanical license from music publisher (Harry Fox Agency).	Music Publisher	Job is to find great songs and license them.
	Music Recording: Contract with artist (percentage of revenue after recoupment).		Single song contract.
	Label owns the recordings.		Usually 50% of revenue.
	Live Show (360 Deal): Booking agent (10% of gross revenue) with approval of personal manager (15%–20% of gross revenue).		Give copyright (ownership) to the music publisher. Other types of deals are available once you become successful.
		Booking Agent	Pay is 15%–20% of gross from ticket sales.
			Represents acts to talent buyers and promoters.
			Usually works with approval of personal manager.
			May have to be paid to represent new acts for club appearances.

Figure 4.3 You now have these because you created an original creative work that you own because of the copyright laws.

Table 4.2 Continued

Talent Agent	Best are SAG-AFTRA, Equity, WGA associated.	Literary Agent	Finds best representative for the company in each genre.
	Pay is often limited to 10% of whatever talent is hired or paid.		Pay is 10%–25% of gross authors' advance and sometimes a percentage of royalties paid over several years.
	Alerted by producers & directors for stage, film acting, directing/production auditions.		Don't pay upfront for a literary agent, as it's his or her previous success and industry contacts that provide successful placement. Give only a percentage of what is earned.
	Will know and connect with booking agents, personal managers and literary for representation in other types of media entertainment.		Great agents will know the types of manuscripts, plays, books, and scripts executives are seeking.
			Don't provide associated rights (screen rights as an example) until book has a determined market value.

at major labels. Authors and scriptwriters usually need literary agents to present their creative works to publishing houses, movie directors, and others. If you're a new band trying to get some gigs in your local town or on a nightclub tour, start with a booking agent to represent you with an electronic press kit and sample recordings to club owners.

Connecting the Dots

The companies in the representation of creative works circle, such as the talent agents, music publishers, and others, usually work for a percentage of the sales price, which is sometimes a percentage of royalties, or they may want to "buy" the copyright outright. You can really get nailed if you do not know what you're doing when offered a contract, so learn what you need to know by looking closely at the following chapters. Because the representation companies make their money based on a

Figure 4.3 Connecting the dots.

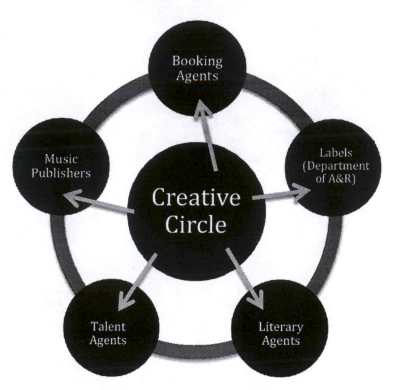

percentage of what your "creative work" may generate, they do not waste their time with creative works pitched to them by creative authors and artists who are not at an appropriate level of professional artistic quality to generate value.

Potential Markets

Here is the place where you apply what you learn through metadata collection. Remember, if you're selling a product, you need to look deeper into lifestyle research that divides the population into segmented markets based on consumer preferences. The basic theory is that "birds of a feather flock together" or people who "think the same" form a unique market to sell to. Think of the money spent on advertising and the time you can save if you can use these resources effectively. In this case, it is all about knowing your fan base, how large it is, and what they like to spend money on, such as merchandise and live shows.

4. What Products or Services Will My Business Provide?

Who is your ideal customer? What's the need or problem that might draw the customer to your product or service? Whatever you are marketing has got to *solve* their problem. Being in business is about solving consumers' problems, with the highest quality at the best price, as quickly as possible. Really bored out of your mind with your current ride? Maybe a really fun, quick, two-seat sports car will solve the problem? Think German cars, such as Porsche. However, if you're tight on money and just want basic transportation from point A to point B, almost any inexpensive car will do. In addition, the more specific you want the product to be, the more it's going to cost. You could easily buy four or five cars for the price of a new Porsche. But remember, we usually get what we pay for and it's important in life to have a little fun. It's the same in the entertainment industry, whether it's the star power, great story, recording artist, promotion, marketing, or something else needed to create and sell an entertainment product. The creative side of the industry is really a package of different types of businesses working together to create products of entertainment. The writers, product creators, producers, technicians, marketers, media, and retail are in the businesses of creating the artistic products. The other hand in the handshake is the industry-related businesses that market, promote, sell, and might pay you (creators) if they sell enough "units" to break even and then profit. Where will you or your company's products or services fit in? Here are some businesses considered vital to create profitable recordings,

TV shows, movies, live shows, computer games, and other money-generating events.

Publishers

Publishers bridge the creative and business systems. On the creative side, they screen treatments and ideas to cultivate writers of print and music. Publishers filter the best materials, often rework them, and then pitch and license them to industry production houses and labels. They find new songs in the music industry from independent songwriters, hire staff writers to write new songs, demo record accepted songs, place them in print (through sheet music), and pitch the songs to artists, record labels, managers and producers, music supervisors, film production companies, and advertisement agencies. They also register copyrights, issue licenses, market songs, and pay writers.

Labels and Music Companies

Labels and now music companies (labels that offer 360 deals) tend to define talent by the money and profits their recorded act's image generates. An act with limited musical or vocal talent is as important to a label as any of its other acts if they can creatively connect with consumers and sell millions of units. A unit is how labels describe one CD, digital download, or vinyl recording sold. Labels desire to make the most profit for the least amount of investment. Thus, the need to use expensive recording studios and sessions that add hundreds of thousands of dollars to the debt of an artist's account is ending. Basic rhythm instruments are recorded with AF of M musicians (American Federation of Musicians) and SAG-AFTRA vocalists (American Federation of Television and Radio Artists). The rest is done with computer programs (discussed previously) to provide quality recordings and lower the cost of the recordings, and may quickly make new artists more profitable. Major labels have quickly moved away from the old business model of developing an act (development deals) in favor of signing only new acts with a proven successful fan base. Only niche labels still cover recording and tour expenses, and then recoup their investments only from record sales. The newer 360 business model gives the labels (now, music companies) better returns on their investments by gaining a share to the artists from concert tours, merchandise, endorsements, corporate sponsorships, and other royalty streams. Recording acts now live in a different world. The burden is on them to create their own success independently before a major label gets serious about a deal. Then the labels analyze the act's financial and fan base psychographic (typologies) to determine the act's potential total investment (cost of doing business with the act) and its potential ROI or return on its investment. It's a business after all.

The Film Industry

Film production companies are also a business; they just create a different product from a simple recording. There are as many deals written as could be both good and bad, but to be successful, movies and film have to compete with the major's products and at the same time contribute to the growth and success of the established industry. So, once again, become part of the system. Get to know the right people, how to hire and use the most talented technicians, find funding, and make a deal for distribution and licensing with Sony, Universal, Time Warner, or an independent. Become part of the industry instead of trying to reinvent the wheel; learn everything possible about how the business works. The financial risks are similar to other business models in the entertainment industry as the companies often have to expend significant capital for the production and marketing before they know if they will make even a marginal, much less, substantial profit.

Think about the business questions behind a movie production. How in the world do movies make hundreds of millions of dollars or in some cases lose millions? Once again there isn't a formula of ingredients as if we were making a cake. However, we know we'll need a script, producer, director, and film technical production experts to shoot, edit, and provide postproduction, unions, distribution, retail, research, marketing, promotion, publicity, and talent agents. Oh, yes, and lots of investors providing serious money.

Authors and Scriptwriters

Just as the foundation of the music business is a wow song, a wow script is the foundation for a wow movie. Remember it's all about the story, even if it's only a three-minute song or a three-hour major film. That's really what the business of entertainment is all about, converting wow stories into different forms of media products. The song is brought to life by the artist's vocal performance, the musicians' hearts, guts, and soul played in every note, the engineer's ability to use the right microphone in the perfect place based on the act's vocal abilities and room acoustics, and the producer, who's got to pull this creative mess into a wow recording. Same thing for the movie production process, except different media, different talent from the actors, totally different knowledge and skills for the techniques, and a director who has to create a wow movie out of the egos, screwups, and final scenes takes. It's also expensive, usually 100 times more than it costs to launch a recording act for a label. When it works, movies profit by hundreds of millions and once in a while by more than a billion dollars. I like the sound of the "b" instead of the "m" on millions! Remember that scriptwriters should join the Writers Guild of America (WGA) so scripts can be registered and copyrights protected by the Library of Congress Copyright Office.

Producers

If your passion is making movies, the first step is to have a script optioned by a producer, who will pay a fee of $1,000 to $5,000 for unknown writers' scripts and $250,000 or more for a hot script from a famous writer. More money is paid if the film is actually made, as the option to make the story into a movie is completed, and the producer then controls exclusive rights to turn the script into a feature-length movie and to control all the corresponding rights for a period of time. In the music business, the producer controls and supervises the recording process. They also pay for the creative team of musicians, vocalists, and engineers. They own the "master recording" and then try to sell it to a major label or form their own label and make a distribution deal with the major labels. Same process (somewhat) in the movie industry, except the producer usually hires a production company to make the movie and pays all the production and actors' expenses. A major film production can be financed one of two ways: the producer can finance and create the film, and pitch to the studios; alternatively, studios that find a great script can then finance the producer and create the film itself. What the film studios prefer is the rich producers spend their money on the product, and then if it doesn't turn out, for whatever reason, they can pass on it without investing in the project. What many people don't know is that foreign rights to show a film are often collected before the first foot of film is shot. That gives the producer or the studio the money to create the film, and then they try to make more money to profit from the film.

Directors

The producer is the money person or venture capitalist who usually hires the director to make the movie. It's the director's job to hire and use the technicians, actors, film crews, studio lots, and locations to deliver a wow movie to the producer. Actually, they just tell their assistants who to hire and what to do. But to be a director it takes an amazing amount of industry-related knowledge, talent, skills, passion, innovation, creativity, and communication skills to make millions to hundreds of millions in profits for the producers and production houses. The financial risk is a little similar as the music industry but at a much higher level of investment, risk, and potential profits.

Movie Companies

Once the movie is made, then the business side of the film company distributes it to cinema outlets, and then later to foreign countries, and simultaneously, to the media networks to cover expenses—all in hope of serious profitmaking. The channels of distribution of DVDs are about the same as CDs, except those retail outlets such as Best Buy and others are dying as streaming continues to increase. Once again, we're witnessing an industry moving through the creative destruction process as much higher-speed broadband is added into cities' fiber-optic networks. A movie on a 5-MB system will take about four times the length of the movie to download or eight hours for a two-hour movie. A song downloads in about five seconds. However, with the new 1-GB broadband, we'll be able to download a two-hour movie in about 36 seconds and a bunch of songs in two seconds. We know what happened to the value of the record labels after the technology developed that allowed hundreds of millions to illegally download their favorite recordings. Labels and creative artists lost billions. Remember the value of a company is often in ownership of the copyright of the master recordings or movies they invested millions in to make. However, if the technology continues to improve, then the film business may suffer the same fight to survive. The real downside is that it costs so much more money to make a film than a recording that it's questionable if the industry will be able to reinvent itself if what they're creating is worthless before or as soon as it's released.

Agents

Agents sell everything from scripts to books, videos to feature-length movies, to the actors, singers, directors, and technicians who create them. The positive side is that they act as a filter system to represent only the best, and therefore the industry executives use them for their casting calls and film shoots. The somewhat negative side is that they take a percentage of the salary of whomever they represent, so prices keep increasing to cover the expenses.

Promotion

Once there is something to sell, then it's time for the business side to let the public know through promotion, publicity, and distribution there is something worth marketing. Promotion is the variety of processes used to alert the public to the label's new recording artists, recordings, and other products commercially available. Providing consumers a free sample through radio station airplay, music videos (YouTube), 30–60-second clips on the Internet (iTunes), and other digital retail websites gives prospective consumers an opportunity to "discover" the act and their recordings. Radio airplay and streaming sites (Spotify, Pandora, Beats, and many others), plus YouTube videos and many websites (where recordings are posted free), provide consumers an opportunity to discover it and then talk/text about it, which may create a buzz. The key to promotion, as you know, is to get consumers talking about the recording.

Younger consumers tend to hype each other about the artists and their recordings, which may increase consumer awareness and potential unit sales or theft. Yes, even units being stolen are an indication consumers are enjoying the recording. Providers such as broadcast radio, streaming services, and others are not in the music business. Instead, they are in the advertisement industry. Any marketing strategy that allows labels to buy airtime, or get views, visits, and hits, as well as advertisers, increases the visibility and the demand for a hip, hot, new product. The larger the audience the more the providers charge for their commercial time, hits, views, and space. Paid advertisements in trade magazines, local newspapers, the popular press, and on the Internet (websites, blogs, news outlets, music reviewers, and chat rooms) provide believability, visibility, interest, and validation that may increase consumer buzz.

Publicity

Promotion is tied tightly to publicity that provides consumers with the act's backstory about their life, connected to their image, such as hero, rebel, lover, good guy, bad guy, cowboy, teen idol, or ideal partner. Once again we are back to talking about emotions tied to products and content sales. Both promotion and publicity are used to exploit the act's "brand," which is then sold as clothing lines, merchandise, endorsements, and so forth. As an example, talk shows book celebrities to gain a larger audience, while at the same time, the label and the acts' branded products are introduced to potential consumers. Even the questions asked and the answers are usually tied to the act's "brand" and image. Welcome to "shameless-self-promotion."

Artist Managers

If an act becomes popular, then another part of the business is cranked up. Managers bring representation, administrative supervision, and surrogate control to a recording artist's complex image and long-term career. A business plan tied to a marketing plan for the business side of the artist's career is regularly developed. Career plans and goals are established based on the perceived commercialization of the artist's image and talents. Managers approve the artist's personal appearances and concert tours through booking agents (who are part of the manager's team). They use the act's and recording's success to emphasize the image of the artist to potential concert promoters, who, in turn, use their money to create and promote tours, shows, and festivals. Once again the "fame and success" of the act are used to create win-win-win situations.

The Management Team

Managers build a team of accountants to balance the books, financial advisors to invest the profits in long-term investments, stocks, and money markets, and attorneys to negotiate, draft, and oversee the process of executing contractual agreements. Event coordinators, security, merchandisers, union stagehands, road managers, roadies, arts managers, and thousands of other niche occupations are available on the business side of the industry to anyone who has the knowledge and the desire to make positive and profitable things happen. The key is what can you (your company or you individually) contribute to the industry to make it better and more profitable? Of course, the manager will represent the artist through a power-of-attorney agreement in their contract. However, all the managers usually do is provide advice, as consultants, to the act on how to build a successful career. It's the manager who recommends to the act who to hire as a booking agent, attorney, business manager (who takes care of the finances), road or tour manager, security, publicity, and other required entrepreneurial business-related decisions. Personal managers usually receive 15% (solo act) to 20% (band) of the gross generated income. Business managers often receive 5% of the act's gross income.

Music Business Agents

Booking and talent agents encourage promoters to produce concerts in various markets. Agents with an AFM booking license generally book union musicians for concerts, night-clubs, and other types of personal appearances and tours. The AFM contract provides protection to the union musicians who are playing the gig. Nonunion musicians and booking agents do not have the full protection of union attorneys if the promoter or club owner fails to fulfill their contractual obligations. Agents work as part of the management team to "pitch" an act's show or performance to event promoters. Prices range from a few thousand dollars per show to millions, depending the act's fame and status. Agents usually receive 10% of the cost of the show to the promoter.

Event and Concert Promoters

Independent promoters and promotion companies, such as AEG Live and Live Nation, provide the money required to fund concerts, festivals, and tours. They are the industry's high rollers, betting big money on the popularity of an act to sell seats in local venues and arenas. Booking agents representing specific acts call the promoters to set up concert tours. A retainer of 50%–100% of the band's payment is customarily required to secure the date for the promoter. The promoter is also required to fulfill the obligations of an artist and provide the technological rider, which is the instructions to the promoter about the act's specific needs for the concert relative to staging, lighting, sound, and load-in and load-out, plus any dressing room requirements and security. Additional riders may detail specific requirements that must be satisfied before the artist will perform, and all riders are part of the contract. Promoters pay everything, including radio advertisements, media promotion, the venue fees, security personnel's and stagehands' salaries, and ticketing costs. Promoters often sign deals with acts and bands based on a straight guarantee, 50% to 95% of the gross income, minus expenses, or some combination of the two. Prices are often based on the fame of the act, size of the venue, availability, timing, culture, types of fans, and lots of other things, including luck.

Tour Support Crews

Acts bring their own support personnel or may hire them locally. Stage crews, lighting, sound, security, road managers, and others are required to put on a successful show. It's not an easy job, but a great experience for young people to break into the business. Most large-venue productions are controlled by the International Alliance of Theatrical Stage Employees (IATSE) union, which represents crews behind most of the entertainment shows, films, and events created and staged in the United States and Canada.

Media

The media consists of streaming networks, radio, broadcast television, cable, the print media, Internet, social media, and others who profit off entertainment industry creative products, such as recordings and films. The entertainment industry profits off of the promotion and publicity of the industry's use of their products as they try to build an audience in order to sell advertisements in the full range of media available. Success for both the media and the entertainment industry is often based on the number of impressions (the number of times a consumer hears, views, or uses a product) tied to the success of and amount of product(s) used or units sold. There are additional royalties generated. As an example, a number-one hit recording on broadcast radio is worth about a half million dollars in blanket license fees split between the music publisher and songwriter(s). There are also fees collected on digital transmissions paid to the label, artists, and recording musicians.

Legal

It's vital to know the copyright law and how to legally acquire a song, recording, book, script, or even an actor to be used in a new entertainment product. Most of the written, filmed, or foundational ingredients owned by others are licensed to production companies, labels, or networks. The business runs on lengthy contracts and almost never on a handshake or spoken word. It's also wise to know the tricks of the trade before you get shafted and have to spend more money to legally get paid what you should have been paid in the first place. It's that simple—understand the business and how it works. Know whom you can trust and have a good lawyer (maybe two) look it over before you sign anything. The more you understand the laws, licensing, personalities, contacts, business practices, and the unique, creative, and innovative structure of this crazy business, the quicker you have a shot at career and financial success.

Here's an example. Say your band is offered a recording deal. Cool! The label's attorney asks, "Who owns the band and the name of the act?" Usually the band members just look at each other in a blank stare. Not smart, as they need an answer *before* they can actually offer a deal. Once again, the band needs to be a legal form of business and the owner of the band (all the members or none) needs to have agreed.

Let's look at the rock band Paramore and three of its members, Hayley Williams, Taylor York, and Jeremy Davis. Now, the band has gone through eight members from time to time.[9] So who owns the band and, in this case, who owns the name? The answer is Hayley Williams. The rest of the band members are hired employees of Hayley Williams and her corporation. So whom did the label sign? Hayley Williams. Who do they pay and deal with whenever they want to change or offer a new deal? Hayley Williams and her legal representation, as they own the name and business of the band. Starting to see how important understanding business and legal issues really is? So, by all means dive in, but first make sure the pool is full of water.

5. Am I Prepared to Spend the Time and Money Needed to Get My Business Started?

Starting your own business as a performer, tech, writer, or producer will be the hardest thing you've ever done for the least amount of money. You'll probably fail several times (as most start-up ventures do) before you score success. Remember that for the creative people in the industry, almost everyone is on their own, in competition with everyone else who wants to do the same thing. You're probably going to simply sell your script, novel, poem, recording, or film to a network or production company or be hired as an independent contractor as a skilled technician if you're a lighting, sound, or stage manager, camera operator, computer effects wizard, or a combination of some or all of these things or something totally different. It takes time, passion, money, and connections within the industry to have a shot at it. You've got to be the best, have a good reputation, be respectful toward coworkers and the bosses, and quickly be able to accomplish the task. Your reputation is everything, your last movie, recording, or book is your resume, and your track record is your calling card that may determine how "hot" you are and how much you can charge for your work.

6. What Differentiates My Business Idea and the Products or Services I Will Provide From Others in the Market?

It is the artist's total package—the image, sound, stage name, backstory, and persona (who and what he or she represents in the minds of the fans)—that drives the industry's profits.

It's called a creative advantage in business terms, and it's usually tied to price, quality, and ego; in the entertainment industry it's tied also to fame, glitz, and glamor. Great singing is more than just hitting the notes correctly. Remember, the music business is really the emotion business. Great songs are the foundation of the business. However, it is the marriage of the great song to the right recording artist that allows consumers to emotionally connect. Actors do the same thing in movies (slight body and face movements, led by eye movement) and stage actors with dramatic body and vocal presentations that would look silly in a film. Superstars often have a recognizable voice, delivering a favorite emotional connection to the fan base. For the business side of the industry, the superstar's image, story, and recognizable face and voice become the marketing tools to generate profits.

7. Where Will My Business Be Located?

Today, you can be in your pjs on the Internet, adding your guitar licks to a master recording, so are we really tied to a location? The answer is yes and no, depending on what part of the industry you are in. However, in most cases you need to be where the deals are made, meaning LA, New York, Nashville, Paris, London, or whatever place is "hot" or happening. As examples, in the past San Francisco, Seattle, Memphis, Detroit, Miami, Tokyo, Chicago, New Orleans, and other cities were the sources of new acts, and rode the wave of creativity over the last 50 years. So, create something so great that the industry moves toward you (almost impossible) or stay there and connect to the established industry (your product or service better be a killer) or move to where the industry is located.

8. How Many Employees Will I Need?

You. Remember if you need an agent, manager, and so forth, they earn a percentage of everything you receive, so word to the wise as you launch your career and business. Some in the trade book publishing industry may say if you're an artist trying to break in, you need a manager, booking agent, attorney, and others. Really? Successful industry personal managers, business managers, and booking agents are paid a percentage of what you make. But you're just getting started and make zip. So why would they want to work for you? In addition, as soon as you hire your first full-time employee, tons of laws and regulations are activated, depending on the country and state you reside in. It costs more than a dollar to pay an employee one dollar. Add another 15 to 20 cents paid to the local, state, and federal government for unemployment compensation, taxes, and lots of fun other stuff. Oh, joy! Don't forget that depending on the product or service your employees are providing to customers, there may be licenses issues and union regulations.

9. What Types of Suppliers Do I Need?

That depends on what your product or service is and how you can best deliver it. Creating products may require a manufacturing process, but that means lots of money invested in something that may or may not sell. Remember, try to solve a problem and you've got a creative advantage toward success. I remember once asking the CEO of MCA Records why they didn't buy a recording studio. Bruce looked at me as if I was crazy and said, "Why in the world would I want to be in the studio business?" Then he explained how most studios are money drains, and besides most of his artists have their favorite studio already, so they wouldn't even use it. Just like almost everything, the answer is usually tied to the budget. Film companies rent their cameras, lighting, sound, and other equipment, and touring acts rent their busses, sound, and lighting or use the venues'. The tax advantages and expenses are much better, and if something breaks, they'll replace it quickly and free.

10. How Much Money Do I Need to Get Started?

Once again it depends on the purpose of your company, but the real answer is usually found when you complete a business plan with financials. Business plans take us step-by-step through the process of starting a business. Most business plan software helps you do the following:

- Create who you are
- Determine what your product or service will be
- Define your competitors
- Know the competitive advantage
- Describe your ideal customers and the size of the possible market
- Analyze trends, industry needs, marketing plans, milestones, and financials

Business plans are really your best guess when you answer the stated questions (as shown in Table 4.1). A good business plan is a guide to the strategic possibility of your proposed business becoming a success or a failure. They are always worth the effort to create and to share with other entrepreneurially minded people when you can. The best templates I have found and use in my classes are at liveplan.com. Good luck.

11. Will I Need to Get a Loan?

You wish you could get a loan! Banks don't want to give a loan unless you back it with collateral. Banks use the business plan financials and your credit application to determine approval. If you own a house, land, or even a valuable copyrights, banks may take them as collateral and provide the loan. Banks are not risk takers; venture capitalists are, but be careful there also. Most start-up ventures are funded with family or your own money. Of course, if your business plan (concept) is unique or a wow, then venture capitalists may jump in, if the profit margins and potential profits are substantial. Venture capitalists or "angels" sometimes lose, but when they back a winner, it's serious money. That's because the details in the deal are often written in favor of the people providing the money. In addition, venture capitalists prefer to take over successful businesses and then sell them for a profit. So the amount of the payments to repay the loan may be difficult to make and if you miss one or two, they might take away your business. Use your own money first to retain 100% ownership of the business, and if you can't do that, try for a bank loan supported by the Small Business Association (through the U.S. government), and lastly pitch your plan to venture capitalists.

The SBA

The U.S. Small Business Administration (SBA) is a great source for helpful information and it may also provide 80% support for loan requests. If the SBA approves your business plan and the added paperwork, you'll need only 20% for collateral to get the entire loan from the bank to start your venture. You can also get a 50% loan quickly, if needed. Starting a business often requires the hardest work and longest hours, for the least amount of money you could ever imagine. However, you're the boss and if it works, the sky is the limit.

12. How Long Will It Take Before My Products or Services Are Available?

It depends on some interesting variables, such as time to set up the business model, develop your product or service, advertise, promote, network to set up a team of advisors or managers, distribute to retail or other businesses (who may actually be your ideal customer), establish a price point, find financing, register with government agencies, and acquire licenses, and then . . . cross your fingers. About 10% of your customers will either be late or not pay at all, which can really kill your cash flow. That's the number-one reason ventures fail, even if they have great products or service: too little money to survive the time it takes from product creation to sales and payment.

13. How Long Do I Have Until I Start Making a Profit?

Usually it takes at least a year, sometimes forever.

14. Who Is My Competition?

Everyone! Honestly, what drives industry insiders crazy is when creative artists who don't have a clue about how the business works or about the corresponding laws, markets, and procedures found in this off-the-wall yet wonderfully fun industry approach them. Here's a page out of my daily life. Often, I am asked to help young artists, songwriters, engineers, bands, and producers with specific concerns they have when they want to sell their "goods" and "services" to the industry. I am not a lawyer (thank God)! So after referring them to an attorney for legal advice, I usually try to understand their issues and offer some basic knowledge. As an example, let's say you are the cowriter of a song, and at a meeting with Warner Music Group, the VP of A&R asks if the song is copyrighted. You should already know the answer *is* yes, as a copyright is valid at the time it is created and fixed into a tangible medium.

However, what they are often assuming is that you do not know the copyright law and what they are really asking is if you have registered your "claim of ownership" with the Copyright Office. You know the answer in both cases; yes, it's copyrighted and yes or no, depending on if you've registered your claim. If you say no, what might they do? Remember the U.S. government supports the first to claim the copyright as the owner. If they try to steal your song, what can you do? The "burden of proof" is then yours to prove in a court of law. However, most of the people in the industry are decent and will not openly deceive you, but they are in *competition* with you, so protect yourself and always try to create win-win-win situations.

Now, let's assume the music publisher offers a single-song contract. Can you make the deal? It depends. If you cowrote the song with another writer, then probably *not* as you do not own the entire song or have the administrative rights. Get it— you have to understand the law in order to make the deal. Making a deal without the legal right to do so (even if you were not aware of it) will not save your tail end from visiting a courtroom. If the other writer has half ownership, Warner Music Group will probably sue you, as you never told them you shared authorship *and* ownership. In other words, it's important for the two writers who own the song they created together to have a written agreement between themselves before the song is pitched. The best solution to competition is to always look at deals from both sides, be honest, protect yourself, and create a win-win-win situation.

15. How Will I Price My Product Compared to My Competition?

Better quality if possible, better customer service, and all at a lower price. Use the business plan to determine profit and losses. That will help determine, as an example, the cost of recording, pressing, distribution, promotion, and creating a website for an album; then simply determine the total cost of the album divided by the number of units pressed, add a profit, and then compare it to the price of the competition. Throw darts at the wall and figure the number of albums you'd sell. Sell them through iTunes and you receive 70% of the listing price, or $7.00 on a $10 sale. There are different price margins for Amazon.com and SoundCloud, and others have their own policies. However, it really boils down to the balance sheet and the profits or losses, depending on how many "units" sold, at what cost to you, and for how much profit.

16. How Will I Set Up the Legal Structure of My Business?

There are three basic forms of business: a sole proprietorship, partnership or limited liability company, and full corporation (see Table 4.3).

A Sole Proprietorship

The type of business owned by most independent contractors working as songwriters, musicians, vocalists, and music producers is a sole proprietorship (you're working for yourself; see Table 4.3). In addition, great musicians usually work as independent contractors often represented by a union, such as the AFM, and artists working in media and recordings are represented by SAG-AFTRA. Yet both the musicians and singers are still in business for themselves as sole proprietors or freelance artists.

Advantages of being a sole proprietor include that you're the boss and you get to collect all the profits. It is the simplest and most cost-effective way to enter the industry as a business, and most legal costs are limited to state and local permits and licenses. You may also write off many business expenses (see an accountant) as income and expenses are considered the same as your personal IRS federal taxes. Of course, you'll have to file for income tax and maybe a couple of the others unless you want to barter in employees trading cows, which is an agricultural form that (for some strange reason) is also included. Setting the humor aside, IRS.gov and SBA.gov are amazing websites that provide most of the answers to your questions when starting a small business.[11] However, you are also personally responsible for all expenses billed to you and your company and you are not protected against lawsuits.[12]

There is not a legal difference between your business income and other income, which means you are personally liable for

Table 4.3 A Sole Proprietorship. Sole proprietors still need to pay taxes and in most states and cities file for state business licenses.

If you are liable for:	Then Use Form:	Separate Instructions:
Income tax	1040, U.S. Individual Income Tax Return (PDF) and Schedule C (Form 1040), Profit or Loss from Business (PDF) or Schedule C-EZ (Form 1040), Net Profit from Business (PDF)	Instructions for 1040, U.S. Individual Income Tax Return (PDF) Instructions for Schedule C (Form 1040) (PDF)
Self-employment tax	Schedule SE (Form 1040), Self-Employment Tax (PDF)	Instructions for Schedule SE (Form 1040) (PDF)
Estimated tax	1040-ES, Estimated Tax for Individuals (PDF)	
Social security and Medicare taxes and income tax withholding	941, Employer's Quarterly Federal Tax Return (PDF) 943, Employer's Annual Federal Tax Return for Agricultural Employees (PDF) 944, Employer's Annual Federal Tax Return (PDF)	Instructions for Form 941 (PDF) Instructions for Form 943 (PDF) Instructions for Form 944 (PDF)
Providing information on Social Security and Medicare taxes and income tax withholding	W-2, Wage and Tax Statement (PDF) (to employee) and W-3, Transmittal of Wage and Tax Statements (PDF) (to the Social Security Administration)	
Federal unemployment (FUTA) tax	940, Employer's Annual Federal Unemployment (FUTA) Tax Return (PDF)	Instructions for Form 940 (PDF)
Filing information returns for payments to nonemployees and transactions with other persons	See Information Returns	
Excise taxes	Refer to the Excise Tax web page	

Source: IRS.gov (2015).[10]

your business and employees' (if you have any) debts. A corporation (form of business) is different as only the assets of the corporation are liable in case of a lawsuit. However, there is double tax as the IRS taxes both the profits of the corporation and your wages or income.

As a sole proprietorship you may also be required to file a DBA (doing business as) form if you use a different name (e.g., a stage name or a business name) other than your given name. It is often difficult to find start-up funding for a sole proprietorship, and although you are your own boss, it may also take lots of work and weekends to succeed. It's best to start a business as your passion, which then makes it a much more enjoyable situation, especially if you write a hit song or sell entertainment products you've also created.

You'll need an EIN number from the IRS to open a bank account in the name of your business (instead of your real name). According to IRS.gov (2011),

> *An Employer Identification Number (EIN) is also known as a Federal Tax Identification Number, and is used to identify a business entity. Generally, businesses need an EIN. You may apply for an EIN in various ways, including apply online. This is a free service offered by the Internal Revenue Service.*[13]

Partnerships and Limited Liability Companies

File for a partnership when more than one person owns the business, such as a family-owned bar, record label, or promotion company. You may also form a LLC (limited liability company), which is a legal hybrid business structure between a partnership and a corporation. It's the best of both worlds, offering the tax advantages of the partnership (no corporate tax) and the legal financial protection of a full corporation. The definition is also somewhat confusing, as one person can also own an LLC, as can two or more people, who are legally called members.[14] However, that's where things change, as an LLC has additional federal and sometimes state requirements. According to the SBA website (2015), while each state has slight variations to forming an LLC, they all adhere to some general principles:

- **Choose a Business Name.** *There are 3 rules that your LLC name needs to follow: (1) it must be different from an existing LLC in your state, (2) it must indicate that it's an LLC (such as "LLC" or Limited Company") and (3) it must not include words restricted by your state (such as "bank" and "insurance"). Your business name is automatically registered with your state when you register your business, so you do not have to go through a separate process.*
- **File the Articles of Organization.** *The "articles of organization" is a simple document that legitimizes your LLC and includes information like your business name, address, and the names of its members. For most states, you file with the Secretary of State. However, other states may require that you file with a different office such as the State Corporation Commission, Department of Commerce and Consumer Affairs, Department of Consumer and Regulatory Affairs, or the Division of Corporations and Commercial Code. Note: there may be an associated filing fee.*
- **Create an Operating Agreement.** *Most states do not require operating agreements. However, an operating agreement is highly recommended for multi-member LLCs because it structures your LLC's finances and organization, and provides rules and regulations for smooth operation. The operating agreement usually includes percentage of interests, allocation of profits and losses, member's rights and responsibilities and other provisions.*
- **Obtain Licenses and Permits.** *Once your business is registered, you must obtain business licenses and permits. Regulations vary by industry, state and locality. Use the Licensing & Permits tool to find a listing of federal, state and local permits, licenses and registrations you'll need to run a business.*
- **Hiring Employees.** *If you are hiring employees, read more about federal and state regulations for employers.*
- **Announce Your Business.** *Some states, including Arizona and New York, require the extra step of publishing a statement in your local newspaper about your LLC formation. Check*

with your state's business filing office for requirements in your area.[15]

Bands are another great example of a partnership if all the members in the band are equal partners and owners of the band's business. According to IRS.gov (2015),

> *A partnership must file an annual information return to report the income, deductions, gains, losses, etc., from its operations, but it does not pay income tax. Instead, it "passes through" any profits or losses to its partners. Each partner includes his or her share of the partnership's income or loss on his or her tax return . . . Partners are not employees and should not be issued a Form W-2. The partnership must furnish copies of Schedule K-1 (Form 1065) to the partners by the date Form 1065 is required to be filed, including extensions.*[16]

Many bands select an LLC instead of a partnership as all the members receive limited protection from personal liability and profits can be shared as the members agree.

Thus, some members may be paid more or less based on contributions to the band as a whole or may own more of the band's debt. From time to time, one member may take advantage of another, so everything needs to be in writing and fully agreed to before being filed. Each member is considered self-employed and must pay his or her own Medicare and Social Security taxes based on the entire company's net income.[17] LLCs are usually dissolved if a member leaves, which in the turbulent world of rock and roll means once you're offered a label deal (if the label is signing the whole band instead of the lead singer) the LLC is converted to a full corporation.

To form an LLC remember to file "articles of organization," which is simply a form requesting your business name, address, and the name of the members, with the Secretary of the State in the state you reside in. The federal government does not view an LLC as a separate tax entity, such as a person or a corporation. Thus, LLCs are not taxed, but each member's income is designated either as personal, partnership, or corporation income by the federal government.[18] You have to decide which! However, some states do tax LLC income, so if your LLC is in one of those states, you're getting taxed twice, first with the federal government and second with the state. Figures! Nonetheless, according to the SBA web page "Choose Your Business Structure" (2015), "Certain LLC income are automatically classified and taxed as a corporation by federal tax law." In this situation you do not get to declare the type of business structure you're being taxed as. Bummer![19] Additionally, different forms are required for various types of LLC declared income. Obviously, I am not mentioning this stuff to blow your mind, just to let you know that to be successful in this industry you do have to play by the rules of the game; thus it is simple when starting out as a creative artist (sole proprietorship), but it can quickly get complicated as you become more successful as an LLC or corporation. It is best to have the advice of an entertainment industry attorney and a business agent when the time arises. Table 4.4 is a simple chart about tax forms; for guidelines about how to classify an LLC for taxes, visit IRS.gov.[20]

Corporations

Corporations are a separate entity owned by stockholders. In other words, the shareholders own the company, not the members, as in the LLC, or a single person who owns his or her own DBA business as a sole proprietorship. Corporations are administratively and financially complex. However, we live in a great country with a very high quality of life usually provided by the investments of stockholders (in hopes of profits) in new and innovative products and services. Guess what types of national governments do not allow corporations, and therefore their citizens often live a much lower quality of life? Think about copyrights! In addition, in most free societies where corporations are allowed, the government requires double tax, one on the profits of

Table 4.4 LLC. Most bands today file as LLCs instead of as partnerships for legal purposes.

If you are an LLC	Use Form	Separate Instructions
To elect a classification for paying federal taxes	Form 8832	Can also be used if you want to change tax classification
Single-member LLC	Form 1040 Schedule C	File as a sole proprietor
Partners in a LLC	Form 1065	
Filing as a corporation	Form 1120	Standard form for corporations

Source: IRS.gov (2015).[21]

the corporation and another on the shareholders' profits and salaries employees earn.[22] Regulations vary from state to state and the company is registered in its home state. Obviously, all the paperwork, licensing, permits, taxes, and name (trademarked) are required. Paperwork and accounting records are available and often filed with the government.[23]

However, there are different types of corporations, which makes things more complex see (Table 4.5). These may seek financial investors (stock or ownership) and according to Greenberg (2011), "S Corporations may transfer profits and/or losses to the individual shareholders' tax return in proportion to stock ownership."[24] Of course, this is more information than most want to know; you just need to be aware that to be successful in the industry, you'll also need to be an entrepreneur who owns your own business of one type or another or you'll be an employee of an industry-related business. Legal action can usually be filed only against the corporation's assets, not the owners' personal property, which is typically record labels, major recording artists, and

Table 4.5 Corporations. Once an act hits professional levels the other forms of business, such as a sole proprietorship, partnership, or LLC, are often turned into corporations for tax and liability purposes.

If you are a corporation C or an S then you may be liable for:	Use Form:	Separate Instructions:
Income tax	1120, U.S. Corporation Income Tax Return (PDF)	Instructions for Form 1120 U.S. Corporation Income Tax Return (PDF)
Estimated tax	1120-W, Estimated Tax for Corporations (PDF)	Instructions for Form 1120-W (PDF)
Employment taxes: • **Social Security and Medicare taxes and income tax withholding** • **Federal unemployment (FUTA) tax**	941, Employer's Quarterly Federal Tax Return (PDF) or 943, Employer's Annual Federal Tax Return for Agricultural Employees (PDF) (for farm employees) 940, Employer's Annual Federal Unemployment (FUTA) Tax return (PDF)	Instructions for Form 941 (PDF) Instructions for Form 943 (PDF) Instructions for Form 940 (PDF)
Excise taxes	Refer to the Excise Tax web page	

Source: IRS.gov (2015).[25]

stockholders in this industry. Thus, the corporation's business structure gives successful recording artists and entertainers a way to protect their personal assets from financial claims against themselves.

17. What Taxes Do I Need to Pay?

All that is required. You've got to pay taxes, which may include city, state, and federal, plus in California there's an entertainment tax. Oh, joy!

The Competition

The competition is everyone who is currently in the established industry, plus everyone who wants to become a part of the business or as a creative indie contractor, owner of a label, club, and so forth, or a promoter who creates his or her own distribution processes, working as an entrepreneur. Your product, be it a recording, book, script, or film, is also in competition with all of the great products that have survived over the last couple of hundred years and all the new stuff the traditional industry and independent creatives are writing, producing, shooting, drawing, and trying to sell or give away to get noticed.

Name of the Company

Once you've decided on your form of business and filed the corresponding paperwork and forms for a partnership or corporation, it's time for the next step. Deciding on a name for the business is important, as the name should say something about what your business is, such as the Rolling Stones as a band or *Rolling Stone* as a magazine. Notice the only difference is the "s" in the band's name. To make sure the name is available, check with the U.S. Patent and Trademark Office to determine if a name is already owned by another company. I once received a phone call from a fellow who owned a business in the name of the domain I had for a totally different business. Let's just say he wasn't very nice. He explained (loudly) that he had the trademark name for his company and that I was infringing on his trademark. I explained that my company was not a bar in Nevada, but a totally different form of business, not related to his in any way. I explained I had no intention of infringing on his business and that a name can't be copyrighted (remember). He had chosen for his business name a commonly used phrase found in many different types of businesses and even in publications. I assured him that if I ever opened a bar, I would not use the name. In addition, I was using the name as a trade name for my DBA business, not as a trademark to protect a "brand name," as he was trying to do for his bar business.[26]

Registration of Business Venture

Don't forget to register your company name with the state you live in if you want to incorporate. If it's a DBA then many states require you to use your own name or file the required form, and of course, you may want to apply for a trademark, which if you do yourself, costs about $300. If you're into certain types of businesses (e.g., broadcasting, agriculture, aviation) file for a federal license or permit.[27] Almost done! Register with your state agencies. The SBA has a great site for all of this information and links to federal and state sites and requirements. You may also have to register your business within the city you live in by obtaining a business license. My guess is that you've already figured out that you'll be taxed on profits and income from your business.

Having a solid business plan is key to your success in entertainment as much as understanding how the entertainment business works. While you really can't change the way businesses do business, you can use your entrepreneurial interests to come up with products and services that people will want and you can protect what you do by making good use of resources in the entertainment industry network: lawyers, the federal government, other artists, business leaders, agents, and the like.

Breaking Into the Industry

So, if you are convinced you have what it takes and you want to find your place and bring your message to people, let's find out what it takes to make a deal. If you're an artist, creative writer, producer, director, or whatever, how can you earn your shot at a career in the industry? Let's start with songwriting and music publishing as an example of the types and complexities of typical industry-related deals.

Notes

1. Osterwalder, Alexander, and Yves Pigneur. "The Business Model Canvas: 'A Shared Language for Describing, Visualizing, Assessing, and Changing Business Models.'" Excerpt from: Alexander Osterwalder and Yves Pigneur. "Business Model Generation." IBook. In *Business Model Generation: A Handbook for Visionaries, Game Changers, and Challengers*. 1st ed., Vol. I Books. Hoboken, NJ: John Wiley & Sons, 2010.

2. Ibid.

3. Osterwalder, Alexander, and Yves Pigneur. *Business Model Generation*. Hoboken: John Wiley & Sons, 2010. Accessed December 22, 2015. www.apple.com.

4. Ibid.

5. Vulliamy, Ed. "BB King at 87: The Last of the Great Bluesmen." *The Observer*. October 6, 2012. Accessed May 30, 2015. http://www.theguardian.com/music/2012/oct/06/bb-king-music-blues-guitar.

6. Ibid.

7. "Metadata." Dictionary. 2015. Accessed May 31, 2015. https://dictionary.search.yahoo.com/search;_ylt=A0LEVvVzhWtV1jwAVqkn nIlQ; _ylu=X3oDMTByMjB0aG5zBGNvbG8DYmYxBHBvcw MxBHZ0aWQDBHNlYwNzYw—? p=metadata&fr=yhs-mozilla-003&hspart=mozilla&hsimp=yhs-003.

8. "US Framework and VALS Types." The Vals Types. 2015. Accessed May 31, 2015. Strategic Business Insights (SBI); www.strategicbusinessinsights.com/vals.

9. "Paramore." Wikipedia.org. Accessed July 2, 2015. https://en.wikipedia.org/wiki/Paramore.

10. "Sole Proprietorships." IRS.gov. Accessed February 20, 2016. https://www.irs.gov/Businesses/Small-Businesses-&-Self-Employed/Sole-Proprietorships.

11. "Choose Your Business Structure." SBA.gov » Starting & Managing » Starting a Business » Choose Your Business Structure » Sole Proprietorship. Accessed July 3, 2015. https://www.sba.gov/content/sole-proprietorship-0.

12. Greenberg, Sue. *Artist as a Bookkeeper*, 2nd ed., St. Louis, MO: St. Louis Volunteer Lawyers and Accountants for the Arts, 2011, 17 and personal interviews with industry insiders (2000–2011).

13. International Revenue Service (IRS), "Employee ID Numbers," IRS.gov. Accessed April 5, 2011. http://www.irs.gov/businesses/small/article/0,,id=98350,00.html,.

14. "Choose Your Business Structure." SBA.gov » Starting & Managing » Starting a Business » Choose Your Business Structure » Limited Liability Company. Accessed July 3, 2015. https://www.sba.gov/content/limited-liability-company-llc.

15. Ibid.

16. USA Government. "Partnerships." IRS. June 29, 2015. Accessed July 2, 2015. http://www.irs.gov/Businesses/Small-Businesses-&-Self-Employed/Partnerships.

17. "Choose Your Business Structure." SBA.gov » Starting & Managing » Starting a Business » Choose Your Business Structure » Limited Liability Company. Accessed July 3, 2015. https://www.sba.gov/content/limited-liability-company-llc.

18. Ibid.

19. Ibid.

20. Ibid.

21. "Filing as a Corporation or Partnership." IRS.gov. June 4, 2015. Accessed February 20, 2016. https://www.citationmachine.net/chicago/cite-a-website/manual.

22. Ibid.

23. Ibid.

24. Greenberg, Sue. *Artist as a Bookkeeper*, 2nd ed., 18.

25. "Filing as a Corporation or Partnership." IRS.gov. June 4, 2015. Accessed February 20, 2016. https://www.citationmachine.net/chicago/cite-a-website/manual.

26. Beesley, Caron. "The Difference Between a Trade Name and a Trademark—And Why You Can't Overlook Either." SBA Small Business Administration, Starting a Business. June 22, 2015. Accessed July 2, 2015. https://www.sba.gov/blogs/difference-between-trade-name-and-trademark-and-why-you-cant-overlook-either.

27. USA.gov. "Obtain Business Licenses and Permits." U.S. SBA Small Business Administration. Accessed July 2, 2015. https://www.sba.gov/content/what-federal-licenses-and-permits-does-your-business-need.

THE SIGNIFICANCE
OF NARRATION   5

There is little doubt the quality of a song's message, the plot points in the movie script, or the twist and turns in a favorite novel connect with consumers. It's the narrative style of writing that magically sparks our elaboration of the essence of a story through a sequence of events over a period of time. Our minds bring it to life as if we are the characters in the story, playing both the hero and, at times, the villain.

We naturally visualize the performance of the vocalists and the musicians, actors, or characters in our minds as we watch, read, or listen. The storyline we create comes together in forms of conflict between love and hate, happiness and anger, joy and sadness, good and evil, or some other emotion tied to our personality. We know that music affects the pulse rate, respiration, and concentration. Other forms of entertainment products also seem to wake us up to "feeling more alive" when we can put ourselves in the story. Do we want to be the lover, hero, villain, or victim? The emotions we're creating as music listeners, readers, and movie fans seem to last only as long as the entertainment product is being heard, words are being read, love scenes being dreamed, or car chases or explosions are being watched. What happens when we replay the sounds of the song, the scenes we want to experience, or the words we wish we had said in our minds? What do songwriters and other artists do to make us want to be a part of the story *they* want to tell?

The Narration Is the Message

Here's where things get important. Throwing a bunch of words against a wall may be interesting, but it probably doesn't make a great story for anyone but the author. Banging on some instruments in an abstract manner is noise, and recording something that sounds good or looks good without thinking about who will hear it is a waste of time. That's where the ingredients (our minds, talents, and devices) are professionally applied to create art and entertainment products that can rock us. Without the wow story, song, film, or game, we're kind of wasting our and the consumers' time. That's how important great songs, scripts, novels, games, and film stories are to the consumers. That's when what you make becomes valuable, when the ways the story is narrated give readers, listeners, viewers, and players an opportunity to have a wow experience; we should get excited, want to talk about it (free promotion), and pay for it!

Our Brains on Music

Researcher and author Daniel Levitin takes it a few steps deeper by claiming that music over the last 10,000 years has helped our brains evolve in six ways, including how we experience knowledge, friendship, religion, joy, comfort, and love.[1] He says,

> at least in part, the evolution of music and brains over tens of thousands of years and across the six inhabited continents. Music, I argue, is not simply a distraction or a pastime, but a core element of our identity as a species, an activity that paved the way for more complex behaviors such as language, large-scale cooperative undertakings, and the passing down of important information from one generation to the next. This book explains how I came to the (some might say) radical notion that that there are basically six kinds of songs that do all of this. They are songs of friendship, joy, comfort, knowledge, religion, and love.[2]

Why do we enjoy this stuff? In other words, how and why do consumers use entertainment psychologically and emotionally? If we could figure that out, maybe we could be more successful in creating the story and profiting from our efforts. Music fans appear to create for themselves the following emotions when listening to various forms of music. How the emotions

Figure 5.1 Susanne Langer wrote one of the most successful books on music and emotions in the 1940s, titled *Philosophy in a New Key: A Study in the Symbolism of Reason, Rite, and Art*. Langer claims listening to music helps us experience the 12 emotions of sadness, relaxation, patriotism, seriousness, amusement, devotion, excitement, sentiment, irritation, happiness, longing, and wanting to dance.

The Emotions of Entertainment

Sadness	Seriousness	Excitement	Happiness
Relaxation	Amusement	Sentiment	Longing
Patriotism	Devotion	Irritation	Wanting to Dance

are turned into meanings specific to an individual is not fully understood. However, Langer suggests some of the emotions felt include those shown in Figure 5.1.[3] I remember visiting East Berlin once and in a small coffee shop got my first experience of German techno pop. Thinking of the bam, bam, bam of the beat of the music I can still remember the drab, empty, colorless room with the young German man bobbing his head to the loud irritating "noise" blasting from the speakers. Most of us can tie our life experiences through some type of recording or listening experience, which ties us to the emotions we were experiencing at a unique time in our life. Now, with all our devices, we often use music and our earbuds to walk through our days in an emotional mood, listening to our favorite hits even if we're stuck in a world of controlled darkness.

The Magic of Entertainment

Electronic nerve signals and brain chemical reactions to music, books, and movies as a cognitive stimulus seem to take place instantly, yet why one person enjoys one type of entertainment and others don't is part of the game. Indeed, what we enjoyed as a kid we probably don't laugh at as we get older. Would you agree that because we are able to enjoy the fantasy of entertainment we live enriched lives? And some of us get really lucky and contribute to it by creating or working in the business!

Constructivism

There are things or events going on around us all the time we do not notice, but our brain may already know what our response "might mean" to us. However, sometimes we are surprised to learn something we were absolutely sure is true turns out to not be true. Here's an example to help explain what I mean. Let's say you are driving your car, approaching a green light. You drive through it because you know the cars approaching the red light will stop. But let's say another car runs the red light and smashes into you. Shocked, nailed, bewildered! Guess what? The next time you approach a green traffic light, you'll probably slow down a little and look both ways before driving on. Why the behavior change? Because you discovered the "reality" that going through a green light may not be as safe as you thought it was! Why did we enjoy Mickey Mouse and Donald Duck when we were very young and now we just don't get it? Why do many college students have a liberal attitude about politics, saving the world, and so forth, and then become politically conservative later in life? There are not any right or wrong answers. No judgment is being prescribed, just a thought about how our minds work and how we can change our minds, and our tastes and actions over time, based on our engagement and experiences in the world.

George Kelly's *Theory of Personality* provides some clues into how and why we change our minds.[4] Knowing this is especially helpful when you are trying to create something that satisfies both what you want to communicate to a consumer and what the consumer will gravitate to as an entertainment product. One of Kelly's theories is we tend to select the types of music, books, films, plays, and events that help us understand, define, or celebrate "who we are" or our subconscious sense of self and how we define reality. Go figure! We use some form of fantasy to figure out reality. Funny, isn't it? Maybe some of the realities of the world are better off not being thought about at all. After many replications called impressions, the selections we make seem to help us shape the pathways in our mind that become our belief system (of what the world is really about) and our personality (how we fit into it).

If we buy into Kelly and construct theory (and many people do), entertainment and music are perceived as "events" used to help us shape, understand, and celebrate our personality, position in life, and other things. Within us, shifts in perception happen at a subconscious level so fast we don't even know it's happening. If we discover or have a need to tie a change in us to an entertainment product (the narrative of the story in the song, film, game, or whatever), we either like it or don't, and we may like it or want it only for a certain period of time, like the cartoons of our childhoods. Why? We seem to be able to know what we need to choose to entertain ourselves as an instant experience of "what if?" Why do we laugh, cry, or become angry when we are part of an entertainment experience, and then get over it? Or not? The emotion we create for ourselves becomes a "self-reflective statement" of "how we really are" as we're constantly attempting to figure out or celebrate "who we are at this moment in time." Since a song or movie can change us, how will you know the value of what you can create and if or how it will reach the people you want to reach?

Psychographics

Let me tell you a story. The executives in the industry are really smart. My friend Beverly and I were having lunch with our friend Cris (yes, it is spelled without the h), from Warner Music Group. I asked Cris why they passed on an act I'd introduced to the label and her response was a wakeup call for me. Cris is a fun, bright, scary-smart young lady who got serious for a moment and said, "Well, we did a psychographic analysis of him and realized his image, talent, and appearance just wasn't what we needed to make him into a superstar. Writes good, performs well, looks great, but he just doesn't fit, there's something missing." I knew what she was talking about, as years ago I'd developed a psychographic research method to analyze artist images for my PhD. But it was the first time I'd heard someone from a major label talking about using psychographics to analyze potential acts' talents and image regarding consumer preferences before they considered signing. We continued the conversation about how the industry has changed due to technology and consumer behavior (demand for free stuff), and I realized what she was saying is that labels now sign acts for different reasons than they used to in the past. Revenue from album sales is in the toilet, so now labels have to offer 360 deals for acts they can *brand* into a public personality and market emotionally through social media, live shows, and merchandise sales. Album sales don't have much to do with the profits in the industry anymore, so the talents needed for artists to blow up have changed.

I think you're starting to see why most artists who simply record a song and throw it up on the Internet fail. If the song, recording, book, or film does not have something to say in way(s) we can use to entertain ourselves (even at the subconscious level), we'll pass. Who needs it? Worse, why spend money on it if it doesn't provide the stimulating narrative in a realistic or believable manner we can use to craft ourselves (at least who we think we are) as part of the story? The ability to determine and predict the reliability of the story (even if it's only a three-minute song) is one of the reasons the professionals in the industry succeed and most of the rest of us have to just get very lucky. Executives hire the people with the talent to write the great songs and story first or they have to acquire the rights to them from the copyright owners, as we've already learned in previous chapters. Here's where the roles of the music publisher, talent agents, book publishing agents, and the department of A&R become valuable to the industry. Film production companies, labels, mass media networks, and computer game companies figure out this type of stuff before they spend millions on creating, marketing, and selling entertainment products. Scott, one of my buddies at Warner Music Group, once told me, "If I run into a person who keeps bringing me hit songs or artists who become huge, we're going to hire them as they seem to have the talent (ears) to successfully predict success." It's a rare talent, but if you've got it, flaunt it!

Instruments of Creativity

Research on how consumers perceive an entertainment product is a serious part of the entertainment industry. No sense in spending millions on creating products nobody digs. But what are even more fun are the instruments we use to create the entertainment event or experience. Think of a world without musical instruments. Boring as heck. Some caveman must have started beating on a stretched animal hide and, all of a sudden, there was an emotional connection to the sound of the beats and rhythm. Once someone discovered sounds, beats, and rhythms, others invented various forms of instruments, including pipes and tubes they could blow through. Nobody seems to know who invented the guitar, but it probably evolved from string instruments first developed about 2,000 years BCE. The piano was invented by Bartolomeo Cristofori (1655–1731), a harpsichord maker, almost 200 years *after* Columbus discovered America.[5] Most of us think Les Paul invented the guitar, but he actually invented multiple track recording. The electric guitar appears to have been invented by musician George Beauchamp and electrical engineer Adolph Rickenbacker in the early 1930s. It was developed out of steel guitars used in Hawaiian music that at the time were simply strings pulled across a hollowed-out piece of wood or an old acoustic guitar, played on your lap with a steel bar.[6] Paper and ink are instruments used by the first book scribes, and also used by some early artists for sketches and eventually painting. I hope you get a chance in your life to see Michelangelo's statue of David. It's fantastic and you'll be shaking your head (as I did), wondering how the heck he scraped and chipped that hard rock to appear to have the texture of human skin. Talent! We've got to use tools and devices to convert our ideas into physical entertainment art and products. Ideas are not owned by anyone, but the song, book, statue, film, or play we make with paper and a pencil, computer, camera, and so forth is ours and we own it, thanks to the copyright law. Yeah!

Professional Writing

Because they are the foundation of entertainment products, the writers of scripts, books, computer software for games, and songs are vital to the success of the entire industry. Without a story, what do we have? Even the artist's image and a major movie star's persona are a part of the story. What do people say? Ninety-five percent of the communication message is nonverbal. So, it's not just the words on the paper— it's also the space between them. A wow story is the order of the words, the colorful yet subtle descriptive manner of a lovely or evil glance in the eyes. They become the supportive details of the story being told that help the consumers discover the hook, cute twist, and the hidden motive, to be satisfactorily entertained. We're starting to understand why it's so hard to write something good enough to be used by the industry to create various entertainment products, such as books, games, recordings, and movies. But it's more than the quality of the narrative; it's also the amount of money (investment) the industry executives will have to spend and the many hours the talented business-minded employees will commit to promote, market, distribute, and sell the final product. Let's take a closer look at songwriting as an example.

Professional Songwriting

The truth is that success for songwriters (as a career) depends on how well consumers connect emotionally to the music products and iconic acts using their songs. It also depends on how well the business system delivers their entertainment products to consumer eyes and ears so they may discover the song's messages for their own psychological and emotional satisfaction. Almost anyone can sit on top of a mountain and write songs. However, if you want to make a living at it, you'll need the industry (producers, artists, labels, promoters, and so forth) to invest its time, money, and careers, to select your song for its next project, event, film, game, or whatever. Just remember that even the greatest songs never make any money until consumers know they exist and start talking about them.

Songwriting Principles

Mark Volman of the superstar group The Turtles suggests the following ten principles for writing great songs: "simplicity, clarity, compression, emphasis, consistency, coherence, specificity, reputation, unity, and genuine feeling."[7] His themes of love include "feeling the need, I think I've just found her or him, the big come on, this is it, I'm in love, the honeymoon is over, cheating, leaving, remember how it used to be." His shape and form of a hit song is "verse, pre-chorus, chorus (the hook), and the bridge."[8]

It customarily takes many years to achieve the competencies and industry contacts required to be successful as a songwriter. I wish I could write songs, but that's way above my pay grade. Of course you should have a natural talent in the ways you express emotions tied to music. The ability to observe social and personal situations and the ability to interpret the social consciousness artistically are important. The ability to convey thoughts and observations emotionally through music and lyrics is paramount. Being able to play an instrument in tune is one thing; being able to create and convey emotional messages is clearly another. Formal education in music, literature, business, psychology, sociology, and world history is enriching. However, great writers often see simple things differently and tie in their own life experiences. Write a song about people walking on city streets, traffic lights, or kids playing and let's throw in some snow. Now put it into a song. Go!

Silver Bells

Here's what Jay Livingston and Ray Evans came up with in 1950 at the request of Paramount Pictures for the Bob Hope film *The Lemon Drop Kid*. The song was originally called "Tinkle Bells" until Livingston's wife asked him the question, "Do you know what the word 'tinkle' means?" (child's way of expressing the need to go to the bathroom).[9]

Jay and I used to talk about his songwriting, and one thing he made clear was that finding the perfect lyric for the right place in a song is "damn hard work." In my mind, I can still see the exhausted strain on his face and hear his trembling voice passionately requesting me to tell students how hard it really *is* to write a great song. There—direct from one of the greatest songwriters who ever lived, straight to any of you who want to write songs. He also expressed how it is a craft and the way to get better is to write, write, write, and cowrite. He said some songs come easily, but most don't, and the hardest thing to do in the world (for him) was to sit down cold turkey at a piano, write from scratch, and write a great hit song.

He was in ill health the last time we were together, but still brilliant as he gave me a grand tour of his home overlooking Los Angeles and Catalina Island. He enjoyed showing me the sunken bathtub where actress Kim Novak took a bath after her house (next door) was wiped out by a landslide. He took all the photos of himself with the most famous stars in the world in stride, just as he downplayed the three Oscars on the living room piano he and Evans won for writing "Buttons and Bows," "Mona Lisa," and "Que Sera, Sera."

There is a lesson for all of us in Jay's story. He was a very kind man who was born in the small town of McDonald, Pennsylvania. Jewish at a time not to be, he changed his name to Livingston to sound English. He met Ray Evans at a school in New York, and the two of them traveled around the world as young men on cruise ships, writing songs, playing piano, and meeting important industry people. They worked at odd jobs and moved to Hollywood on the suggestion of friends that they might be able to find work as songwriters in the film industry. They didn't wait around to get "discovered." They continued to write during the worst of times and moved to where their talents would be marketable. They were passionate about it. They networked and networked even more to form friendships with industry professionals, including songwriter Johnny Mercer and executives in the business side of the industry. After about three years, Paramount's Music Department (film production/music publishing) signed them and the rest is history![10] What can I say—it was an honor to know the gentleman and be part of his family. He didn't sit around—he and Ray made things happen. So can you.

Tie a Yellow Ribbon Round the Ole Oak Tree

To be successful songwriters need to be much more than . . . songwriters. Of course, the ability to write songs that potentially connect with many fans is fundamental. However, getting them sold to a music publisher, industry insiders, and others is an even more important part of the game. Check out the success of songwriter L. (Larry) Russell Brown and we'll learn a lot about things writers often have to accomplish in addition to writing great potential hits. It's a step-by-step process. Things such as attitude, networking, and making things happen are also vital. Larry was born in Newark, New Jersey, to a poor family living in a rough neighborhood project located between a park, train tracks, and industry storage buildings. Larry has a story to tell in each of his songs—mega-hit songs—that is part of his life, and he shapes his experiences into songs with great hooks to connect consumers to his themes.

He was a tough kid whose violent and abusive father often took out his frustration physically on him. As a young kid, Larry formed a friendship with two other guys (who also lived in the projects), Richard the Ox, who played guitar, and Lebo, a future cop who played boxes as if they were drums. Larry played the harmonica. Years later, Ox and Larry become a duet, wrote songs and performed them as a group, called The Duals. With their first song and recording in hand, they pitched it to almost everyone in the established music business in New York. Of course, they were rejected by everyone. But great writers thrive on rejection. Cheesy saying, but true!

They didn't quit. Larry heard about a record label in Harlem called Fury Records, so they grabbed their guitars and found Bobby's Record Shop at the label's address. They played their original song live for Mr. Robinson, who became a famous person in the industry himself. He liked their sound and offered them a recording deal. The guys were thrilled, blown away, excited, and ecstatic about their future success.[11] One of the mistakes many artists make when offered a recording deal is to assume that signing the first deal is going to make them superstars. Later, they often discover that signing a deal is just the beginning of a very long, hard road to—maybe—success.

The following week they went back to sign the deal with Robinson's partner, Fats. He lived in a townhouse near Central Park. They were led to an office downstairs, where they were given a note that told them to sign the stack of papers on a table. When finished a voice directed them upstairs to a bathroom, where they found the executive they had come to see shaving naked in his bathroom![12] What a way to get introduced to an industry executive! They confirmed they had signed the papers and were escorted out. The recording sold two copies. But he didn't quit!

At times, Larry had to steal to eat, but he inhaled the stories published in *Reader's Digest* to escape the anguish he felt about his brutal father and when he could, he listened to radio stations to daydream about a better life. His main interest was girls, yet he also knew the only way out of the neighborhood was a choice between crime, the military, or something really crazy, like songwriting and singing. After trying the first two, he returned from the military determined to give the music business a shot.

Networking

Great songwriters know that to be successful, you have to do much more than write potential hit songs. Livingston and Evans and L. Russell Brown are great examples of how they also network within the industry to connect with the

Table 5.1 Career Breaking Songs. Wow songs are the pathways to fame for an artist. The effect is a chain of events in our minds from our experiences tied to a personal emotion connected to artists because of the emotion we created for ourselves when listening to one of their recordings. If it is a loving feeling, as an example, then we may feel a "family or loving connection" to the act. Thus, we perceive the relationship between the act (singing the song) and us as a personal or friendly, buddy, lover, or family type of connection. The emotions give us an opportunity to support the act's career through buying or stealing their recordings and buying tickets, merchandise, and other things. We often "feel" the act brings us a moment of excitement, love, or any other connected emotions, which seems to make the fan/artist relationship personal. Major artists are aware of the power of wow songs, and they (and their team of people) are always in search of the next big hit by first finding the next great (soon to be a classic) song.[13]

Artist	Song	Year	Songwriter(s)
Marvin Gaye	*I Heard It Through the Grapevine*	1966	Norman Whitfield & Barrett Strong
John Lennon	*Imagine*	1971	John Lennon
Roberta Flack	*The First Time Ever I Saw Your Face*	1969	Ewan Macoll
ABBA	*Dancing Queen*	1976	Benny Andersson, Bjoun Ulvaeus, & Stig Anderson
Led Zeppelin	*Stairway to Heaven*	1970	Jimmy Page & Robert Plant
Vera Lynn	*We'll Meet Again*	1939	Ross Parker & Hughie Charles
The Kinks	*Lola*	1970	Ray Davies
The Beatles	*Hey Jude*	1968	Lennon & McCartney
Frank Sinatra	*I've Got You Under My Skin*	1956	Cole Porter
Amy Winehouse	*Rehab*	2006	Amy Winehouse
Patsy Cline	*Crazy*	1961	Willie Nelson
Judy Garland	*Somewhere Over the Rainbow*	1939	Harold Arlen & E. Y. Yarburg
Don McLean	*American Pie*	1971	Don McLean
Norah Jones	*Cold, Cold Heart*	1951	Hank Williams
The Doors	*Light My Fire*	1966	The Doors
Queen	*Somebody to Love*	1976	Freddie Mercury
Louis Armstrong	*What a Wonderful World*	1968	Bob Theile & George David Weiss

music publishers and others to help them get the right music publishing deals. Livingston and Evans met many famous people, yet it was their move and the friendship they developed in Hollywood that landed them in the right place at the right time. Then their talent as songwriters allowed them to cash in.

L. Russell Brown knew he needed to network to meet people who could introduce him to others who might introduce him to still others in the industry. After his service in the military, Larry and Raymond Bloodworth, whom he had met in the military, returned to writing songs and performing in New York clubs. According to Brown, "they wrote every day and were turned down more often than a cheap motel bed." "This ain't what we're looking for" became their theme song. They sang with others who would eventually become famous artists (Bob Dylan and others) in Greenwich Village clubs. Then a friend of his introduced him to one of his friends, who was dating the sister of a known songwriter. He gave Larry the phone number of the cowriter's friend, Nevel Nader.[14]

Over lunch in 2015, Larry told me, he called Nevel and was invited to a flophouse hotel room in New York. Suavely dressed and sitting on his bed, Nevel asked Larry to play a couple of songs, which he liked. Then he got up and pissed into the sink. No bathroom in that dive! Larry thought, "This guy's going to help me, who's going to help him?" Crazy as it is, Larry liked Nevel and the three of them (Raymond, Nevel, and Larry) starting writing together. Nevel had a friend with a little extra money who invested in a demo recording. The arranger of the music on the session liked the recording so much that he gave a copy of it to his friend, who was Bobby Darin's publicist. She liked it so much she gave a copy to her friend, the famous songwriter/producer Bob Crewe. Crewe was the producer of the Four Seasons, a young group of guys also out of New Jersey, who had just nailed three number-one hits on the U.S. *Billboard* (Amusement Business) charts.[15]

Connecting the Dots

That's still how most of this crazy business works today. So let's count the number of people and contacts Larry had to acquire, get excited, and pass the recording to Bob Crewe and his music publisher/record production company.

1. A friend of his,
2. Introduces him to one of his friends,
3. Who was dating the sister of a known songwriter;
4. He gave Larry the phone number of the cowriter friend, Nevel Nader.
5. Larry visits and meets Nevel.
6. Nevel meets Raymond, Larry's cowriter,
7. Larry, Raymond, and Nevel cowrite together.
8. Nevel's friend invests money in a demo recording.
9. The arranger at the recording session gets a copy of the demo,
10. Gives a copy of the recording to his friend (Bobby Darin's publicist), and
11. She gives her copy to Bob Crewe.

Welcome to the songwriter's side of the music business! The surprise is that Bob Crewe did meet with Larry right away.[16] Often songwriters have to meet many music publishers before they find the one who "gets it." Lots of rejection is the name of the game, and "Thrive on rejection" becomes the daily motto. The advantage of the Internet and digital technology is it gives anyone the power to write songs. The problem is the fact that a writer can use a computer to write, record, and upload a song does not mean that it can be monetized. Millions think that because they have written a song (that sounds good to them) and posted it on a site, it will magically be discovered and become a huge hit. Yeah, right.

The business is about money, not throwing millions down a sink based on ego and assumptions of fame. If you want to make money by creating and recording a song, and lose everything or just about everything you might make on it, post it on the Internet. You could improve your odds by visiting Las Vegas, where the odds are only slightly in favor of the house.

People meeting people who know people who know others who can make things happen has always been used as a "filtering" system to eliminate most of the amateur songwriters and singers who lack the talent to make it. The really bad stuff is never heard by many at the top who have the money and power to mold a songwriter's or recording artist's career. The odds have simply increased, as maybe one in a thousand songs had the potential (if recorded by the right act) in pre-Internet times; now it's probably closer to one in a million. Because the "pool" of new material is so large, it's harder for the industry insiders, labels, and so forth to find great new songs and artists. A potential great hit song that was a needle in the haystack before the Internet has become the needle in huge fields of haystacks, making it almost impossible to find. Sure, digital technology has provided many opportunities to young and developing acts, but most of what is posted can't be monetized. Thus, why would a label invest money in an act's career?

Reputation

A note of caution: songwriters and singers often do not have a clue about how quickly a really bad song or recording can tag them as a failure long before they actually learn how to craft hit songs, recordings, and live performances. Remember what I told you about my lunch with Cris at Warner Records? A&R folks are looking for a complete package, someone with a brand and product that their monetizing will turn into profits. As you can see from our example, Livingston and Evans and L. Russell Brown took years of living, writing, networking, and so forth before they were accepted as great writers. In Larry's case, after the friend of a friend of a friend introduced him to another writer, it took an investment (demo recording) and then a professional in the business (the arranger) to give a copy to his friend, another professional in the business (the publicist), who gave a copy to her friend, Bob Crewe. By the way, it took a long time to get an appointment with Crewe, but once he heard Larry and the guys sing and knew of the songs they were writing, they landed a contract.[17] Getting to the right industry people who "get it" and who can monetize your creative songs and acts is vital for long-term career success. Signing a deal with Crewe instead of the one Larry signed with Fury Records shows the difference between what can happen with an established industry professional and a rather fly-by-night independent label. The independent label was just signing tons of acts, knowing that one or two might break through, while Crewe invested his time and money into the talent and career of his signees. That's a difference for us to consider.

The Rest of the Story: Shameless Self-Promotion

Crewe helped his songwriters and acts develop into better writers and acts. Look for someone who can do that for you; it's not usually an Internet site that wants an annual fee or a percentage of your royalties. Crewe would help with the recordings as a producer and label executive. He set up appearances on key television shows (publicity and promotion), including Dick Clark's *American Bandstand* and then had Larry and Raymond tour as an act with the stage name "The Distant Cousins."[18] Larry and Jay spent many years developing their craft as writers, and they also spent years meeting the right people who could help make their career happen—in their case, Alfred Hitchcock and Bob Crewe.

The Odds of Wow

Here's another lesson for us to explore about the need for exposure through traditional and social media, live shows, and of course YouTube and other video sources that now provide consumers a chance to discover your songs and recordings. Simply posting a recording on the Internet does not drive consumers anywhere but mad. Your recordings posted on the Internet have just become the needle in a very large field full of haystacks! And almost everyone knows that most of the haystacks are not very good, so why even try to find the needle? Consumers *will* search for something great (recording) for enjoyment. If they land on your song/recording and hear something that doesn't work for them, they're not coming back! You've got to do more than just post your stuff on the Internet and expect the planet to explode in excitement. Which is why streaming services such as Netflix, Spotify, Amazon's Prime, and Apple Beats are quickly taking over. Consumers can select what they want to watch and hear instantly for just a small monthly subscription fee or by tolerating a few commercials.

Career Steps

If you want to be a successful songwriter, don't post or give away your songs. Instead, join a professional songwriters organization, meet others, and cowrite like crazy. There are many such organizations across the country where songwriters encourage each other, cowrite, and get advice from the experts. Cowrite with someone who is better than you, who has a different kind of talent and a different kind of story. Writing with others helps you see a lyric and hear music from a different perspective and often improves your songwriting. The goal is to get a song you and others strongly believe has potential and then start connecting with music publishers in the industry (see Table 5.2).

Have a Professional Review Your Creative Works

Most music publishers will not accept unsolicited songs. So how can you get past the armed guards to submit your material? Look for the network, like L. Russell Brown did—someone who knows someone established and respected in the industry whose name you can use to get an appointment if they approve. Do not lie or use someone's name as a referral without his or her permission. Performing rights organizations, such as ASCAP, BMI, and SESAC, usually provide a few minutes to listen to new writers' material. If the writer has great potential they'll want to sign him or her for representation of public performance royalties, such as radio station airplay. They have a staff whose job is to find and help new writers. Let them listen to your work, and if they like what they hear, they will often refer you to established music publishers they think might click with your type and style of songs.

Music Publishers

Select the type of publisher who has a successful track record in your musical style. If you're writing rock music, find a publisher who has been successful in developing (called song casting) rock songs with major labels and acts. Song casting is a term used to describe the process publishers use to figure out which songs they want to pitch to music providers, such as labels, artists, and film production companies. Think of the casting part like casting a fishing line into the sea of musicians and songwriters who want the same thing that you do. Successful publishers know the right people, have the right contacts, and can successfully place songs, where your job is to create them, not sell them. Getting a single-song contract music publishing deal (with the right publisher) is the first step in the filtering process toward becoming a true industry professional. Now you just have to write a good enough song that industry professionals will see it as having the potential for profits.

Publishers put money, time, effort, and their reputation on the line when pitching material to music users and providers. They know only the best will get licensed, and if a publisher pitches too many "dogs," the door to the labels, film production companies, and others is permanently slammed shut. Music publishers and entertainment industry leaders work on trust—they know time is money. Neither expects the other to present material that can't be monetized. However, one person's villain is another's hero, meaning that one publisher

Table 5.2 Creative Works Representation. Different types of representation are usually required to connect your creative works (e.g., songs, scripts, book manuscripts) with industry executives. So the big question is how do you get it to them first?

Type of Narrative/ Creative/ Work	Deal Points	Representation or Agency	License Agency	Type of Licenses	Production Company	Entertainment Product
Song	Ownership of copyright and a percentage of royalties	Music publisher	Harry Fox Agency or directly	Mechanicals	Label	Recording of song
		Music publisher	Directly	Sync	Film, TV production company	Use of a song in film, TV show, or visual media
Recording of song	Negotiated	Direct	Directly	Master	Film, TV production company	Use of the recording of the song in film, TV show, or visual media
Books	10%–20% of advances and royalties	Literary agents	Directly	Literary rights	Book publishing companies	Printed books distributed at retail and digitally
		Literary agents	Directly	Film and media rights	Movie, film, and media production companies	Film, movie distributed

may decide a song is awful and another may throw a contract at you. It's all about guts—feeling and knowing potential acts and their image or branding, customers (fans), and, of course, the business.

Three Agendas

Jay Livingston, Ray Evans, and L. Russell Brown had to develop their songwriting abilities by writing almost every day and watching how industry insiders and fans reacted to both them and their material. The writers wanted validation for their creativity, fans desire to be emotionally connected, and the industry wants to make money. Three different agendas all still found in different creative and business elements of the digital entertainment industry. As an example, industry music publishers and labels can stroke egos all day long, but what gets them to the point of offering a contract is the possibility of monetizing what you've created. It's going to cost them money (as an investment in your creativity) to license your songs to acts, labels, advertisers, computer game production companies, and film and TV production companies. Others who have worked their entire life to "make it" are often frustrated by the failure to convince labels, promoters, and other industry insiders to sign them to a related music publishing deal. However, we are talking about apples and oranges, as the creative people want validation and the business suits want profits. Rejection does not mean your songs and recordings are not good, or even great. It just means that the industry labels and so forth have run the accounting numbers and do not see a way for their financial investment to make them and you (if you created it) a profit.

A Stable of Talents

Of course, the special talent of narrative writing is tied to many types of creative writers. Great works of fiction often use real characters to bring the reality of a story to life. According to an article in the *New York Times* (1999), Mario Puzo was born in Hell's Kitchen in New York, a place for Italian immigrants.

His parents left Naples, and life in America was made even more difficult when his father deserted the family, leaving his mother, Maria, to raise seven kids. As a fatherless kid, Puzo had the choice between good or evil, which was all around him in the gangs in his neighborhood. However, his mother was (as he described) "a wonderful, handsome woman, but a fairly ruthless person."[19] She seemed to keep him in line as he instead turned to writing stories about family, people, and crime, which kept him busy and out of the street gangs. Her essence was used by Puzo to model Don Corleone, the godfather in the first movie, played brilliantly by Marlon Brando and equally as well by Robert De Niro (as a young man) in the second. He served in World War II. He used his war experiences and those of other soldiers to write a book and short story about returning veterans as he worked as an assistant editor. Puzo's love of writing narratives about human behavior paid off handsomely when the paperback rights to his first really successful book, *The Godfather*, sold for $410,000. Puzo considered himself "an Italian peasant" who was a "romantic writer with a sympathy for evil." He claimed to "hate violence" and claimed the success of the *Godfather* books and movies displayed consumers' "disenchantment with the American justice system" and their "craving for close family ties."[20] Brilliant!

Puzo knew Marlon Brando could play the role of a wonderful, handsome, fairly ruthless person and actually sent him a personal letter and the book, requesting he consider the role.[21] This does not happen very often, but when it does it's usually one famous successful creative person contacting another through a mutual friend established in the networking process. The mutual friend probably contacted Brando first for his approval before providing any forwarding information to Puzo. Puzo's timing and request were perfect as the book was already a huge success and the movie rights were already sold. Brando knew that it would probably be a long shot, that the author wanted him, and that he had some power before the deal with him to play the role was even considered. If you don't have the successful track record, don't send your song, material, or whatever to famous artists, labels, and so forth, as they will probably not even look at it. Another reason is that they don't want to be sued if they put out something that is even close the story line, song, and so forth you submitted.

Professional Organizations

The Songwriters Guild of America (SGA) was formed in 1931 by songwriters Billy Rose, George M. Meyer, and Edgar Leslie.[22] As a voluntary songwriters association run by and for its members, it provides services and activities songwriters may need to be successful in the business. The SGA offers workshops, critique sessions, pitch opportunities, access to catalogs, award events, publishing company audits, catalog administration, financial evaluations, medical and life insurance, and legislative and legal support. The guild's support provided several important legal provisions in the 1976 Copyright Act.[23]

The Nashville Songwriters Association International (NSAI) is a good example of a nonprofit organization for members who are interested in meeting other writers. The NSAI sponsors regional workshops in various locations in the United States and Europe. The workshops provide opportunities for writers to improve their knowledge and craft. The NSAI also offers a song evaluation service to its members. Songs submitted are evaluated for their commercial appeal based on the theme, lyrics, melody, and overall impact. Once a year, the NSAI also sponsors Tin Pan Alley South, which is a series of shows, workshops, and opportunities for songwriting members to network and write with professional and successful writers. There are other types of organizations in many cities, so find one and get involved.

The Purpose of Membership

How important are these organizations in supporting songwriters and others in the entertainment industry? Let's take a quick look at the Writers Guild of America (www.wga.org), which has two main offices, one in New York (east) and one in Los Angeles (west). The WGA represents authors of theatrical, film, and television shows and provides a memorandum of agreement (called a memorandum basic agreement, or MBA) of which there are two versions, based on long-form television formats. Let's write a script! The MBA is a 610 page (with side letters) three-year agreement between the WGA and the members' writers "in good standing" and the production companies, theater, and networks producing and showing, airing, screening, broadcasting, transmitting, and streaming plays, movies, TV shows, radio programs, literary material, syndication, and computer game productions. Payments are based on the length of the final master and the number of times it is performed, aired, broadcasted, transmitted, played, or streamed.

To be defined as a "professional" writer the basic qualifications include selling, licensing, or optioning literary material written and owned by the writer(s) to various types of production companies. The time or length of the employment and the number of corresponding credits on movies, television shows, stage plays, and so forth determine the qualifications for membership. Minimum scales for payments are about $6,500 for 15 minutes or less of actual running time and about $40,000 for 90 minutes to two hours on schedule A and about a thousand more on schedule B. Writers are also paid a small percentage of gross revenues for all products aired in foreign distribution. Territories are defined as *domestic*, which is the United States and Canada, *foreign*, which is most of the rest of the world, *and other* for undeveloped societies. Media is defined as theatrical, home video, pay TV, domestic free TV, foreign free TV, cable, new media, or other. Writers are defined as freelance, head writers, associate writers, and associate rewriters. Had enough? It's really a business and even if you're a wow writer, understand and join the industry-rated businesses and organizations if you want to actually work in the industry, have the biz use your creative works, or (most importantly) make any money. Just a thought!

The Value of Wow

Record producers and smart recording artists search for great songs that will establish or enhance their careers. Same thing for movie producers and TV, game show, and other production companies. When presented with the pitch of a script, book treatment, or song, their job is to consider the quality of the narrative and how successful it might be told as a recording by a major recording artist, played as a movie, or staged in theater presentation.

They are the first filtering system of the business side of the industry, just as the agents and music publishers filter through all the wannabe actors, songwriters, recording artist authors, and so forth. If (and it's a big if) it's passed along, then there's usually a team of business executives who together approve or pass. It's different people in different types of companies doing the same basic thing—finding a wow song, story, or film that might be converted into a profitable entertainment product. Then the company's executives and accountants determine the amount of money to invest in the product and compare that figure to the potential gross revenues the final master recording, movie, TV show, play, or computer game might generate.

The odds and risks of winning the game we are playing are better in Vegas than in this industry, if you don't have a clue about how it works. Indeed, all artists are on a constant search for new songs that will sustain, enhance, and grow their fame and careers. It's usually a song that becomes so popular that the artist is always associated with it, such as Bing Crosby and "White Christmas," Carl Perkins and "Blue Suede Shoes," Tony Orlando and Dawn and "Tie a Yellow Ribbon Round the Ole Oak Tree," or "Take This Job and Shove It," sung by Johnny Paycheck, notes Neil McCormick.[24] It's the same for the computer game business, film and TV production companies, and concert promoters—and the list goes on.

Use the System

If you want to write books, scripts, plays, games, songs, or musical compositions, you should understand a couple of industry requirements. The first is the fundamental ingredient of the story in the form of a pitch for a book, movie, or song, and the second is the evaluation process by the financial side of the business to determine the odds of making a successful product. Remember there are many moving parts, as shown in Table 5.2, that make the ingredients in the industry work. Look at the parts of the industry that are connected to songs, recordings, and books. Literary agents represent book authors; music publishers represent songwriters; and Harry Fox, ASCAP, BMI, and SESAC represent the music publishers, who also represent the songwriters. Think of the game of dominoes and how the game won't work unless you have the parts connecting the other parts—it is the same in the industry.

The Pitch

If you're a songwriter, find a music publisher who is well connected to the most successful people in the industry and who can place your songs with those individuals. If you're an author find an agent. The people in your network should be a progressive agent between the songwriter (you) and the record labels, producers, artist managers, and recording artists who are looking for songs from your genre. Find a publisher who will help you become a better songwriter by offering you constructive criticism and positive, creative, and financial support. Make sure they are honest business people who have a good reputation and passion for your success, besides their own.

In general, music publishers find songs—wow songs—rework them, register them with the Library of Congress, song cast them (analyze who can use them), pitch the songs to music users (labels, acts, film production music supervisors, advertisement agencies, and others), issue licenses or have their corresponding agency, such as Harry Fox, ASCAP, BMI, or SESAC, issue the license and collect the royalties, and then . . . pay the bills and count the profits. Here's a great example where things get downright weird. Book publishers are really the book publishing houses, such as Random House, that pay for the printing of books and then make their profits marketing and selling them to consumers. In other words, book publishers do the same thing as record labels for recordings, such as the labels of Warner Music Group, and movie companies, such as Lions Gate. But music publishers analyze songs, rework them, and then pitch them to music users (labels), which is basically the same thing literary agents do in the book industry for writers and the same thing talent agents do for actors. Getting confused yet? It's just the way it is: same titles, same language, totally different meaning and processes. Figures. So new authors usually need an agent to represent them before their book manuscript can be pitched to a respected book publishing house. Of course, there are exceptions to everything and smaller book publishers may at times review manuscripts directly. Some will even print the book if you pay them, which is usually a rip. Everything sold, acquired, and produced will require legal contracts, such as between the songwriter and music publisher, author and agent, agent and book publishers, actors and agents, and agents, actors, and film production companies.

Striking Gold

What happens when you find an agent or music publisher who wants to provide representation? As an example, let's take a look at the different types of deals offered in the music industry. How much you know and how well you negotiate your part of the deal will determine how much money you'll be paid. For a couple of minutes, put on your legal hat. Then read the deal twice, and have one or two attorneys look at it and explain what each term and sentence mean from a legal perspective. Then decide whether you want to sign it or seek a better offer.

Notes

1. Levitin, Daniel J. "Chapter 1: Taking from the Top." In *The World in Six Songs: How the Musical Brain Created Human Nature*, 3–4. 1st ed. New York, NY: Dutton/A Penguin Group, 2008.
2. Ibid.
3. Langer, Susanne K. "The New Key." In *Philosophy in a New Key: A Study in the Symbolism of Reason, Rite, and Art*, 1–19. 3rd ed., Cambridge, MA: Harvard University Press, 1957.
4. Kelly, George A. *A Theory of Personality: The Psychology of Personal Constructs*. New York: W.W. Norton, 1963.
5. Cazaubon, Mantius. "Who Invented the Piano and When? . . . And Why?" Piano-Keyboard-Guide.com. Accessed September 11, 2015. Piano-Keyboard-Guide.com.
6. Theodoros II. "How the Electric Guitar Was Invented." Gizmodo. October 29, 2013. Accessed September 11, 2015. http://gizmodo.com/how-the-electric-guitar-was-invented-1453939129.
7. Volman, Mark. "Ten Principles for Writing a Great Song." Lecture, EIS 1220. The Entertainment Industry from Belmont University, Nashville, January 1, 2012.
8. Ibid.
9. Livingston, Jay. Personal Interview by author. January 1, 1998.
10. Livingston, Jay. Personal Interview by author. January 1, 1997.
11. Brown, L. Russell, Sandy Linzer, and Larry Wacholtz. "Non-formatted." In *I'M L. Russell Brown and I Wrote. . . . "TIE A YELLOW RIBBON ROUND THE OLE OAK TREE" . . . and This Is My Story . . . & Smash*. Unpublished ed., Vol. 1. Nashville, TN: Non Published, 2015.
12. Ibid.
13. Bronson, Fred. "Hot 100 55th Anniversary: The All-Time Top 100 Songs Billboard." *Billboard*. August 8, 2013. Accessed February 20, 2016. http://www.billboard.com/articles/list/2155531/the-hot-100-all-time-top-songs.
14. Brown, L. Russell. "Lunch with L." Interview by author. February 12, 2015.
15. Ibid.; Brown, L. Russell, Sandy Linzer, and Larry Wacholtz. "Non-formatted." 2015.
16. Ibid.
17. Brown, L. Russell. "Lunch with L." Interview by author. February 12, 2015.
18. Ibid.

19. Victor, Thomas, and Mel Gussow. "Mario Puzo, Author Who Made 'The Godfather' a World Addiction, Is Dead at 78." *The New York Times*, February 7, 1999, Movies sec. Accessed September 14, 2015. http://www.nytimes.com/1999/07/03/movies/mario-puzo-author-who-made-the-godfather-a-world-addiction-is-dead-at-78.html?pagewanted=2.

20. Ibid.

21. Reed, Ryan. "'Godfather' Author Mario Puzo's Archives Headed to Auction." *Rolling Stone*. 2016. Accessed January 30, 2016. http://www.rollingstone.com/movies/news/godfather-author-mario-puzos-archives-headed-to-auction-20160129.

22. Songwriters Guild of America, "SGA Provides 75 Years of Advocacy, Education to Songwriters." Accessed July 15, 2010. http://www.songwritersguild.com/history.htm.

23. Ibid.

24. McCormick, Neil. "100 Greatest Songs of All Time." *Telegraph.co.uk*. June 3, 2015. Accessed July 13, 2015. http://www.telegraph.co.uk/culture/music/11621427/best-songs-of-all-time.html.

MUSIC PUBLISHING—
WHAT'S THE DEAL?   6

The Wannabe Test Market

There's an Internet market for entertainment products for the wannabe amateurs. Nothing wrong with it except for two little, tiny things: there is often no money in it and unless you create a great buzz, the established industry isn't interested in your products or career. In other words, it's a great place to take your shot to learn what works and fails in your computer-based recordings and videos posted on Sound-Cloud and YouTube. However, consumers are pretty darn smart as they've spent hours and hours looking for something great, and guess what—yours usually isn't it. But it's a great place to learn to thrive on rejection. Which is what usually happens when you start pitching "material" to industry executives anyway. And now that the technology (Pro Tools) and distribution (Internet) exist where anyone can create and deliver creative works to the planet, the rules of the game have forever changed.

The Free Internet Marketplace

Now the music labels won't even consider listening to your material, agents won't read your manuscripts, or even watch your self-made movie or video because it's already been released and failed to get millions of hits. We've already talked about our egos and trying to leapfrog the established industry by doing it ourselves. Nobody knows where the next big thing, whatever that might be, is coming from a lot of the time. However, even a blind squirrel finds a nut sometimes, so go for it. But also learn from the experience and know that you'll have to create a higher level of creative work before you pitch it to executives. One last thought: the industry has lots of interns and lower-level employees scanning the Internet all the time, seeking that new act, song, recording, or whatever they could develop into the next money machine. So, if they don't see a great buzz or see a burst of fans digging it, they're going to pass.

The Traditional "Filter" System

Other great filtering systems include film festivals, agents, and even nightclub performances or buy-on shows or tours. Most of us have recently gone to a major act's concert and noticed the first or opening act is often an unheard-of band that isn't even listed on the bill or advertisement. That's usually a buy-on who's trying to get exposure by paying to play or perform. Most fail, and the $100–$5,000 or more it costs to be an opening act with a known artist turns out to be an expensive ego trip. If you've got to pay to get a shot, you better be great because the second time they won't even take your money. By all means, exposure for whatever you are creating and doing is the name of the game, so give it your best shot, but do it wisely and realize if you want a real career in the industry, you'll probably need to plug into the real thing. However, having a passion to perform, write, run the audio, direct, and so forth is wonderful, and so gain the skills and talents you need to contribute to the industry. At the same time, realize we all need to start somewhere, so don't be afraid to make the mistakes and accept failure as a learning process of what to do better. Legendary songwriter L. Russell Brown, when performing once for some students who want to be in the industry and he said, "Most of you in this room will face rejection, but I can tell you the person who is rejected the most will also be the most successful."

Figure 6.1 There is a two-step process in the established industry where the copyright owner who first created the creative work, such as a song, book manuscript, or play, pitches to agents and music publishers who are known by industry executives. If the intermediaries find promise in the creative work, then there is usually a deal made between the creator/copyright owner and the agent/representative, who will then pitch the creative work to industry connections.

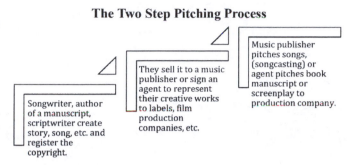

The Two Step Pitching Process

Songwriter, author of a manuscript, scriptwriter create story, song, etc. and register the copyright.

They sell it to a music publisher or sign an agent to represent their creative works to labels, film production companies, etc.

Music publisher pitches songs, (songcasting) or agent pitches book manuscript or screenplay to production company.

A Shot in the Dark

If you possess the ability to write a wow song, story, script, or play, your best shot, no matter how long it may seem, is to pitch it to the right people in the business. Remember, you own it because of the copyright law, right? Why? If you missed it, go back to the legal stuff in the creative process chapter and review what you missed. Usually we love what we've created and think it wonderful, and so do our parents and friends. So what! Also remember if you post it on the Internet, you're basically giving it away, so when you get good enough, pitch it to and through the industry connections. The only opinion that really counts *now* is the industry as their long-term careers depend on their ability to find and pick the wow creative work. It's the vital ingredient they need to acquire before the labels can convert it into a major recording, publishing houses can turn it into a great book, film production companies can turn it into a script to shoot the movie, or the director can hire the stage actors to play their parts. The industry is built on top of wow creative works; the question is, can you create one, and then can you get it to the right people?

What do industry executives use to judge a creative writer's material as potentially successful? It might be helpful if you have a basic understanding of what they're looking for in order to provide it.

I wish the answer was simple, but it's not. The answer is a wow, of course; however, nobody really knows what the heck that is until they've read it, heard it, seen it, and of course, emotionally experienced it. One of the magically fun things about working in the industry is that nobody really knows for sure which of the stories, songs, and so forth consumers will get excited about. They usually make their decisions in the movie business based on their past experiences of success, contacts, knowledge of what's worked in the past, gut feelings, and what consumers might dig in the future. It's really a mystery, except if it's been a successful book or major act, then the odds are better, but there's *still* no guarantee. The bottom line is that most of the "stuff" pitched is good to great, but it needs to be a wow and they are rare. Why does it need to be a wow? Because the biz will probably spend between $1 million and $100 million to acquire the rights (copyright) and turn it into a major film, TV, computer game, stage play, or recording. Sometimes it comes down to a gut reaction between two or three potential wows that determines which creative work is finally selected for production, but even at that point the final decision is often based on which one might be more profitable. Maybe only two people out of three think it's a wow, as we all think differently, and most major labels now need a unanimous decision by all board members before they sign a new act. To get a better understanding of what the creative writer needs to know from a business and legal perspective, let's take a look at the types of deals music publishers usually offer songwriters (see Table 6.1).

Types of Music Publishing and Songwriter Deals

There are several types of songwriter/music publishing deals and hundreds of variations that can be negotiated. The main negotiation points are authorship, ownership, and splits and shares of the royalties generated. The important thing is to see *how* the different types of deals affect your overall income.

Table 6.1 Projected Values. The representation (or agents) or company determines the projected value of a new creative work to determine the likelihood of placing the work with an entertainment industry executive, who may then want to use it to make an entertainment product, such as a recording, film, book, or computer game. How much should you pay the representation company or agent to get your creative work to an industry executive? Professional talent agents usually receive a percentage of the price (outright sale) or a percentage of the projected revenues the copyrighted work may generate. The film industry has minimum payments based on the amount of time used in the final product (e.g., 30 seconds) or union scale work. Music publishers offer single-song contracts for the ownership of the copyright in exchange for 50% of generated revenue, which is then called the "publishing royalties." The payment to representation companies and agents may be called royalties, wages, or commissions. Beware of the agents and publishers who want up-front money, as that is often a red flag of a shark deal, where the snakes representing you are doing it for the money you paid them instead of placing your products with industry insiders.

Projected Value of Creative Work	Song Writers	Music Publishers	Artists	Recording/ Filming Production Teams	Labels	Artist Performance Agencies	Promoter/ Touring
Product Services Value The "Perceived Value" of Your Products and Services to the Industry	**Royalties** (from sales and licensing of songs)	**Royalties** (from sales and licensing of songs and sale of company due to value of owning copyright's future potential royalties)	**Royalties** (percentage of gross from sales of recordings, merchandise, corporate sponsorships, live shows, endorsement, and other sources)	**Wages** (from independent contractors, scale wages from unions, rental or lease fees for equipment and studio rentals)	**360 deal** (profits and rights management fees and royalties from digital downloads, streaming, merchandise, artist live performances, endorsements, and other sources)	**Commissions** (percentage of artist income, such as 10% of gross for booking agents, 15%–20% of gross for personal management, 5% of gross for business management and other sources)	**Profits** (depending on the type of deal and break-even points after all expenses are paid)

Remember, three issues are being negotiated in every deal: (a) the songwriter's royalties, which are a percentage of the total income; (b) the music publisher's royalties, which are a percentage of the overall income; and (c) the ownership of the copyright, which may be a valuable and a sellable piece of personal property with its value based on how many companies want to use it and how many consumers want to buy or listen to it.

Splits and Shares

The industry tends to make the terminology of songwriter deals confusing (see Figure 6.2). Listen carefully, as I ask you this question. Let's say you are selling your song to my music publishers and I offer you a 50% deal. Fifty/fifty sounds fair—you wrote the song and you're expecting my

Figure 6.2 The typical songwriter–music publisher deal gives the music publisher the ownership of the copyright (song) in exchange for 50% (royalties) of all the money generated from licensing the song. The downside of the deal is the music publisher will own the song until it falls into public domain, which is your life plus 70 years. The first time you may request the copyright be returned is between 35 and 40 years of ownership. In Figures 6.1–6.6, you will note the industry uses the terms Songwriters' Share and Music Publishers' Share to denote the percentage of income for the song, whether there is one or more songwriters or music publishers involved in the production of the work.

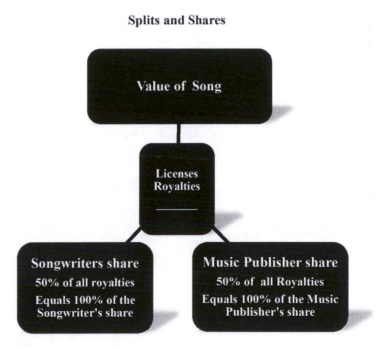

Splits and Shares

Value of Song

Licenses
Royalties

Songwriters share
50% of all royalties
Equals 100% of the Songwriter's share

Music Publisher share
50% of all Royalties
Equals 100% of the Music Publisher's share

music publishing company to license it whenever and wherever possible. Shouldn't you receive at least half of the money (royalties) generated? Should you go with the deal or ask other questions? Here's the deal: if you answered yes to my question, you got zapped, as you'll get only 25% of the royalties. But you thought it was a 50/50 deal! Right?

The Fine Print

Consider another question, such as "50% of what?" Remember you are negotiating for three things: how much of the *songwriting royalties* and how much of the *music publishing royalties* you will receive and the percentage of authorship (ownership) of the song (copyright) you created. The value of a song is based on how much money it generates over the duration of the copyright. However, the royalties are usually divided into two: the songwriter's share and the music publisher's share. So, when I was asking you if you'd accept a 50/50 deal, I was talking only about the 50% of the 100% of the songwriter's share, which is only half of the total royalties collected. After all, you created the song and you're now selling it to my music publishing company because I have the contacts and administrative licenses required to get it into the hands of the major artists, labels, film production companies, blah, blah. That's worth something and to be honest, half of the royalties is fair and accepted as a standard in the music publishing business. But if I get 75% of the total and you get only 25% that may not be fair or ethical. However, it's amazing how people will take money if you throw it at them, so be careful and have an attorney explain the details.

Just remember when negotiating a songwriter/music publishing deal to consider all three points. You think the song is priceless, and the industry doesn't, so they have to estimate how much money licensing it might make based on their contacts, industry needs, and gut feelings. That's the value of your song at this place and time. From zero to who knows? After listening to a few hundred songs a day over several years and by taking the time to meet and really understand the industry insiders' wants and needs for material, publishers know better than most of us who to pitch it to and how much money it might generate. But nobody knows for sure.

Crazy, but here's how the industry often views it. A song is really worth 200%, which of course is nuts, but they usually state in songwriting contracts a percentage of the songwriter's share (based on 100%, equal to 50% of the total money generated) and a music publisher's share (as 100% based on the remaining 50% of the total revenue generated). Confused yet? In other words, any time you see or hear the word "share" you're probably talking about a percentage of only half the total income. That's why you ask the second question, "50% of what?"

Copyright Ownership

As a new writer, your copyright is usually transferred to the publisher offering the deal. What you're really doing, in most cases (except a work-made-for-hire or shark deal), is trading your ownership of the song (and usually all the exclusive rights) to the music publisher in exchange for a percentage of the songwriting share of the royalties (called mailbox money) the song generates, minus certain music publisher expenses. At first you're a rookie, and thus the publishers consider you and your song (their financial investment) as risky. Once you've written a few hits then you're less of a risk, and if you become really successful, then you'll probably form your own publishing company and attempt to keep a part or all of the songs' copyrights. Why would you want to do that? Because ownership of the song may become very valuable in the future, depending on how big of a smash hit it becomes. Remember the two ways a song makes money: royalties for its current use, which are divided into a songwriter's share and a music publisher's, *and* (and this is important) future royalties the song may generate with continued use over the life of the copyright's duration. We know a song is a piece of property and just as a house or some land increases in value over time, so does a successful copyright. Larger music publishers and other companies may want to buy the copyright outright to expand its uses and markets as a way to increase their profits. That will increase the value of the copyright, so they may sell it again and make even more money.

Work-Made-for-Hire Deal

A work-made-for-hire deal returns us to the days of Tin Pan Alley, when songwriters were paid a lump of money for their songs once accepted by the publisher. Traditionally, writers and music publishers split all royalties generated from licensing the song 50/50. But this is different. The music publisher is gambling you'll have the talent to write a song that will satisfy its needs. Examples of a work-made-for-hire include specific "one-time jobs," such as writing jingles or music beds for commercials for radio or TV. The employer (person who hired you to write the song) is usually considered the *author* and *copyright owner* of the song you just finished. All the money you (the songwriter makes) is paid up front once the song is accepted. However, what if the business/owner decides to use the song in other ways, such as having a major recording artist record and release it?

In the music publishing business, a work-for-hire songwriter is usually given a one-time payment and the employer receives the copyright (ownership of the song) and all revenues (both the songwriter's and music publisher's shares). Just remember songwriters can maintain their copyright if they satisfy the "type of employment" details in writing. What if it sells a million copies? You make zip unless the work is specially commissioned in writing stating you retain ownership of the copyright. When determining what type of music publisher/ songwriter deal to make, knowledge of copyright law is essential when negotiating a first deal. Writers are often more excited about the validation of their creativity than thinking about if it's a fair and decent deal or a rip-off. If you sign it, you've got to live with it, so be careful and review all contract offers with an entertainment attorney. Here's the bottom line: music publishers pay you; you should never pay them.

Figure 6.3 If you work as an independent contractor and someone hires you to write a song or book, or paint a figure on the wall, who owns the copyrighted creative work? It is usually the person or company that hired you and all you receive is the money paid up front to accomplish the work. The hiring company or person owns the copyright and will be paid all royalties for the licensing of the work. The exception is a work-for-hire when the artist (being hired) has a clause written into the contract that he or she will retain the copyright.

Work for Hire Deal

Value of Song
Work-Made-For-Hire

License Royalties
——————
Copyright Ownership

Songwriters' Share
0% of all royalties
0% of Copyright Ownership

Music Publishers' Share
100% of Songwriters' Royalties
100% of Music Publishers' Royalties
100% of Copyright Ownership

That's a shark deal—let's not go there. However, the first deal you may be offered may be a work-made-for-hire. What does that mean to your mailbox money? The statutory definition of a work-made-for-hire is

> *A work prepared by an employee within the scope of his or her employment, or a work specially ordered or commissioned for use as a contribution to a collective work, as a part of a motion picture or other audiovisual work, as a translation, as a supplementary work, as a compilation, as an instructional text, as a test, as answer material for a test, or as an atlas, if the parties expressly agree in a written instrument signed by them that the work shall be considered a work made for hire . . . A supplementary work is defined as a work prepared for publication as a secondary adjunct to a work by another author for the purpose of introducing, concluding, illustrating, explaining, revising, commenting upon, or assisting in the use of the other work, such as forewords, afterword, pictorial illustrations, maps, charts, tables, editorial notes, musical arrangements, answer material for tests, bibliographies, appendices and indexes.*[1]

If the "creative work" is part of your job, then your employer owns it. However, if you are commissioned to create a copyrightable work as an independent contractor and have the person or business providing the work sign a written agreement (stating that you retain the copyright and authorship) they get to use it, but do not own it. *Circular 9, Works Made for Hire*, printed by the Copyright Office, states (2012),

> *Copyright law protects a work from the time it is created in a fixed form. From the moment it is set in a print or electronic manuscript, a sound recording, a computer software program, or other such concrete medium, the copyright becomes the property of the author who created it. Only the author or those deriving rights from the author can rightfully claim copyright. There is, however, an exception to this principle: "works made for hire." If a work is made for hire, an employer is considered the author even if an employee actually created the work. The employer can be a firm, an organization, or an individual . . . The Supreme Court's decision in Community for Creative Non-Violence v. Reed . . . held that one must first ascertain whether a work was prepared by (a) an employee or (b) an independent contractor. If an employee created the work, part 1 of the definition above applies, and the work will generally be considered a work made for hire. . . . If an independent contractor created the work, and the work was "specially ordered or commissioned," part 2 of the definition above applies. An "independent contractor" is someone who is not an employee under the general common law of agency.[2]*

Additionally, the definition of work-made-for-hire under the second section or part 2 is limited to the following:

> *Section 101 of the Copyright Act (title 17 of the U.S. Code) defines a "work made for hire" in two parts: (part 1) . . . a work prepared by an employee within the scope of his or her employment or (part 2) . . . a work specially ordered or commissioned for use*
>
> 1. *as a contribution to a collective work,*
> 2. *as a part of a motion picture or other audiovisual work,*
> 3. *as a translation,*
> 4. *as a supplementary work,*

5. *as a compilation,*
6. *as an instructional text,*
7. *as a test,*
8. *as answer material for a test, or*
9. *as an atlas*

> *. . . if the parties expressly agree in a written instrument signed by them that the work shall be considered a work for hire.*[3]

You have written a song by yourself that generates $100,000 in revenues or royalties. In the *work-for-hire deal* you are paid $2,000 to write the song and the company that hires you owns the copyright. Years later the music publisher sells the copyright of the song for an additional $500,000. How much total money did you make from the work-for-hire deal?

Indie Deal

Independent writer deals are usually a *single-song contract* offered to new writers who have written a great song a publisher wants to own (see Figure 6.4). If the writer shows talent, then they may be offered a staff writer deal, where an advance is provided for a number of acceptable songs and yet the single-song contract with the 50/50 split is still used. This

Figure 6.4 An independent deal is when the songwriter or creative artist is offered a single song contract (if we're talking about songwriters) in exchange for 50% of the royalties which are called (crazy as this is) 100% of the songwriting share.

Independent Deal

Value of Song
Independent Deal

License Royalties
50/50 Split
Copyright Ownership

Songwriters' Share
100% of Songwriters' Royalties
0% of Copyright Ownership

Music Publishers' Share
100% of Publishers' Royalties
100% of Copyright Ownership

time you've written a song by yourself that generates $100,000 in revenues or royalties. In the *indie deal* you'd receive no advance but are scheduled to earn *100% of the songwriter share that is equal to 50% of the total royalties generated.* What you're doing once again is giving up 100% ownership in the copyright of the song in exchange for the 50% of the total royalties. Later, the music publisher sells the copyright of the song for an additional $500,000. How much money did you make in their deal?

Staff Deals

Staff writer deals are desirable positions at music publishing companies because they provide songwriters with an advance on their royalties plus the opportunity to cowrite with other major writers, improve songwriting skills, make industry contacts, network with others, and make a reasonable living all at the same time. *Staff writers* are paid *a draw* in exchange for a quota of acceptable songs (usually 8–12 annually) and their royalty share *after* and *if* there are the advances is recoupment. They are usually considered *exclusive writers* for the music publisher. Novice staff writers are often surprised to learn that the songs they write are owned by the music publisher that is paying them. However, from the music publisher's perspective, it is in effect betting the advances (a draw of the money to be paid) that the writers will be able to create songs that will recoup the writers' advances, plus a profit.

The *advance* or *draw* is often paid semiquarterly, which the music publisher hopes to recoup from royalties, if indeed there are any. Once the song's "royalties" repay the writer's advances, the remaining royalties are split into a songwriter's share and a publisher's share. The more successful the writer, the more the payment schedule favors the writer. A powerful publisher may have a contract that is skewed favorably toward the company. Beginning salaries often range from $18,000 to $30,000, while very successful writers can earn hundreds of thousands of dollars per year in advances, salaries,

Figure 6.5 Staff deals are an indication the writer has arrived to a professional industry level as the music publisher will pay him or her a small annual stipend of money ($18,000–$24,000 a year paid quarterly) for the writer to write. How cool is that! If the songs make money, the publishers recoup their advance payments to the writer out of the songwriter's share of the royalties generated. However, if none of the songs make any money, the writer owes the music publisher zip.

and royalties. In addition, if more than one writer or publisher is involved in the deal, the share is divided accordingly. Thus, 100% of the songwriters' share (which equals 50% of the total revenue income) is split 50/50, so in reality, each writer (or publisher) is really receiving only 25% of the total revenues generated. In other words, a 50% share of the songwriters' or publishers' share is equal to only 25% of the total revenues generated by the song. Crazy, yet it is the language often used in the industry, so be careful to comprehend if you are talking about a "share" or "total income" when negotiating a deal. Now, how much money did you make? This time you've written the song by yourself and it generates $100,000 in revenues or royalties. In the *staff deal* you were also paid a $20,000 advance, and earned *100% of the songwriter royalties or half of the total royalties generated*. You gave up 100% ownership of the copyright in the song to make the deal. Later the music publisher sells the copyright of the song for an additional $500,000. How much money did you make?

Co-publishing Deal

Co-publishing agreements help songwriters capture more of their song's equity by establishing co-ownership of their song's copyright at the time of publishing. Copublishing ordinarily occurs during the acquisition of a song when more than one publisher is involved. Songwriter, music producer, and recording artist–owned publishing companies make copublishing deals with the major publishers' world distribution. In addition, producers, recording artists (with their own publishing companies), and smaller independent music publishers generally make copublishing deals with larger, better-connected publishers to administer paperwork, such as licenses and royalty collections, in exchange for a percentage of the publishing royalties and 50% of the copyright ownership. In a co-pub deal, the songwriter, producer, label, or recording artist (whoever owns the song) gives up half of their copyright of the song (ownership). In addition, the total royalties are split 75/25, with 75% of the total generated revenues going to the original songwriter/publisher (100% of the songwriter's share and 50% of the publisher's share) and the remaining half of the music publisher's share (25% of the total) going to the copublisher.

This, of course, means that anyone can be a one-person music publishing company. All you need to be successful in the

Figure 6.6 Once the songwriter has written a few hits and maybe a wow or two, then he or she has more power to negotiate a better deal. Most famous writers declare themselves a music publisher as well as a songwriter. They own two businesses that give them the power to seek additional royalties from the music publisher who wants to make a deal. The co-pub deal is when the writer/publisher gains half of the music publisher's royalties and 50% of the copyright ownership. Thus, in reality the language may be a little confusing, but the split is now 75/25 in favor of the songwriter/publisher. In addition, they will now own 50% of the song's copyright.

Co-publishing Deal

business is a great song and a copublishing deal with a worldwide publishing company that can take care of all the paperwork in exchange for 50% ownership of the song and 50% of the publishing share of the revenue stream. Industry terminology and the practice of converting percentages to shares tend to make copublishing agreements complex. Here is an example of what is really happening. The typical single-song contract has a 50/50 split of all royalties (except printed sheet music and folios, where the writer(s) may be paid as much as 10% of the suggested retail price). The maximum percentage of any whole is 100%. Yet, the music business splits the 100% of a song's potential collected revenue into 100% songwriter and 100% publisher shares, which of course is impossible. So, here's a clue: if ever another music publisher you're working with says they will give you 100%, always ask 100% of what? If you hear the word "share" that means they are probably talking about only half of the royalties. Be careful! Sometimes, they'll say "we'll split the royalties 50/50." They may be thinking 50/50 of the songwriter's share and not even be talking about the publishing share that they plan on retaining at 100%. Always make sure you are clear—total royalty income or shares?

Copublishing works to the advantage of both the songwriter (who is also a publisher) and established industry music publishers only when you've got a wow song. The music publisher receives a share of the equity (copyright ownership), quickly increasing the value of their catalog, and the songwriter/publisher receives the major connections, distribution, and industry power of the major music publisher, plus more money as he or she now also receives half of the publisher's share and half of the selling price of the copyright, if sold. By the way, if the copublisher is located in a foreign country, then this type of deal may be called a *subpublishing deal* as it recognizes that the overseas publisher is getting part of the royalties in exchange for collecting them in their territory of the world.

233

Here we go again. You've written the song by yourself and it generates $100,000 in revenues or royalties. In the *co-pub deal* you were not paid an advance, but you earned *100% of the songwriter royalties and 50% of the publishing royalties, as you're also a writer/publisher*. You're wearing two hats at the same time—a songwriter hat and a publisher hat. Now you're really starting to understand how important it is to know something about how the industry works. It'll put extra money in your pocket! You also gave up 50% ownership of the copyright in the song to make the deal. Later you (who kept the administrative rights to the song) sell it for an additional $500,000. How much money did you make because you made this deal a co-pub deal?

Administrative Deal

An administrative publishing deal is the same as the co-pub except the songwriter (who is also publisher) retains 100% ownership of the copyright. It allows songwriters who own their own publishing companies to keep absolute ownership of their song's copyright. The royalty splits are the same as in the co-pub deal. An administrative publisher or exclusive agent simply issues mechanical licenses, registers songs with the corresponding performance rights organizations, and commonly collects royalties in exchange for 50% of the publisher's share of the revenues collected. Administrative agreements are advantageous to songwriters who demo record their own songs and have the networking contacts to place them with major recording artists, labels, and film production companies themselves. Administrative deals are rare, but when they occur it's usually with a very powerful, successful writer/publisher who may license an administrator to exploit the song's potential equity worldwide. Publishers seek administrative publisher deals to handle various aspects of the business, including the paper business (printing of sheet music and sales), legal, documentation, and demo recording.

Now look at how much money you'd have made if the deal you'd made was our administrative deal? You wrote the song

Figure 6.7 The administrative deal is the best songwriters who are also acting as their own music publishers can hope to make. It's really how many famous songwriters retire, as they own 100% of the copyrights of the songs they have written. The advantage is to sell their entire song catalog (since they own the copyrights) for 5%–15% of the five-year average income, which may add up to millions. On top of that, they still make 100% share of the songwriting royalties on all future royalties generated. Rockin'.

by yourself and it generates $100,000 in revenues or royalties. In the *administrative deal* you are not paid an advance, but you earned *100% of the songwriter royalties and 50% of the publishing royalties, as you're now a writer/publisher*. However, you've kept 100% ownership of the copyright of the song. Later you (who kept the administrative rights to the song) sell the song for an additional $500,000. How much money did you make this time?

Shark Deal

Whenever a publisher compliments your song and then asks you to put up some money to share the expenses, you are probably being cheated. Just remember: you don't pay them; they pay you! (see Figure 6.8).

It is your song—*your* property, not theirs. Be careful! There have been many examples of songs sold to music publishers (mostly in the 1950s and early 1960s) for a set fee of a few hundred dollars or less. Many of the songs became hits, but the writers did not receive any additional revenues as they had already sold all of their royalty and ownership rights. Unethical music publishers who offer shark deals are considered rip-off artists, as they are in the business of making

Figure 6.8 This is easy: if you pay them money, you're getting the shaft. Publishers pay you—you do not pay them. You can run around telling friends you've got a publishing deal, but notice even if the song becomes a hit (and it won't) you get zip.

money off of you, the songwriter. The only money shark publishers usually make is what you paid them to publish your song. In most cases, they do not have the industry contacts to place your song with a major artist, producer, or label, and correspondingly, you will almost never receive any royalties.

In the shark deal they'll sometimes say, "We'll put up $1,000 if you match it." The song made zero and the last time I checked, 0% of 100% is still zip. What if they said they'd give you 100% of the songwriter's share but still never "place" the song with any song users? How much money did you make with this type of deal?

Licenses

Now that you've made a deal, let's put the music publishers to work and make some money! The only reason music publishers should offer a deal is because they believe they can make a profit with the placement of your song with industry professionals. It's also why you, the songwriter, made the deal with the *right* music publisher. Both of you will now lose or profit based on the music publisher's ability to exploit your song/material with top industry acts, labels, music supervisors (for placement in films), and others. The word "exploitation" has a negative rap with many, yet when the checks start to hit your mailbox, you may learn to love it. According to the National Music Publishers Association of America (NMPA 2015),

> *Music publishers are essentially songwriter promoters. They work to pair talented songwriters with artists, which results in exposure to the public and income for the songwriter. Music publishers "pitch" songs to record labels, movie and television producers and others who use music, then license the right to use the song and collect fees for the usage. Those fees are then split with the songwriter.*

Types of Licenses Issued by Music Publishers

1. *Reproduction (Mechanical) Licenses*: Music distributed in physical and digital form. The royalties are generally collected and paid by the Harry Fox Agency.
2. *Public Performance Licenses*: Music broadcast on radio (terrestrial and satellite), in live venues, and other public places. The royalties are collected and paid by public performance societies (ASCAP, BMI, and SESAC). Each broadcaster receives a blanket license from each performing rights society, in exchange for a royalty fee.
3. *Synchronization Licenses*: Music used in film, television, commercials, music videos, etc. Publishers enter into direct licenses with users.
4. *Folio Licenses*: Music published in written form as lyrics and music notation either as bound music folios or online lyric and tablature websites. Publishers enter into direct licenses with users.[4]

Permission to use a copyright (e.g., a song for a film, or a recording) is usually granted through the issuing of a certain type of license. You need two licenses to drive a car, one for the vehicle and the other for the actual driver. So, this is not a new concept.

The Licensing Process

You've learned how to start your venture company, as a sole proprietorship, partnership, LLC, or corporation; now let's make some money. As a music publisher you've got to acquire wow songs the industry will pay you to use in exchange for whatever deal you've made with the songwriters, recording artists, musicians, producer, and singers. How can you do that? We'll talk soon about who you hire to create the product, but first let's examine how you can license your rights to labels, film production companies, media, and businesses using your products as part of their environment or programing.

If you want to use someone's song in your business, better get the right license. However, four basic licenses—mechanical, public performance, synchronization, and print—also cover different types of media, and thus some of them are for broadcast, others digital transmissions, some are combined under a different name, such as transcription, and still others are used for different forms of entertainment products. Some are called the same in other countries, and others are not even used. An original song used in a stage play or musical has restricted rights unless it is also released as an audio recording. If the song is performed only on stage, its "first-use rights" have not been used and cannot be recorded by others. The exception, of course, is the compulsory mechanical license that must be granted by the publishing company after it's first released for sale. Then anybody can record it, but they still need to acquire a mechanical license, usually online through the Harry Fox Agency.

Want to hire a band that will be performing some monster hits from rap acts and songs written by Taylor Swift in your nightclub? Better get a *blanket license* from ASCAP, BMI, and SESAC. Compensation for the use of the song is paid as a licensing fee directly to the publisher or through agents, such as the Harry Fox Agency for mechanicals, and ASCAP, BMI, and SESAC plus many others in foreign countries for radio station airplay, clubs, or live performances. Mechanicals are issued for hard platforms and digital downloads. Public performance (blanket) licenses are issued for broadcasting, public performances, and digital transmissions, depending on the interactivity. Print or folio licenses may be used in individual sheets or folio or book combination. Transcription licenses are used for the wonderfully boring music often found when we step into an elevator with strangers (which is the reason for the music). Dramatic or grand licenses are issued for the songs in stage or theater plays and presentations.

Think again about the copyright owner's six exclusive rights. Remember, music publishers and songwriters gain their ownership rights from the Copyright Law and control their corresponding six exclusive rights (including reproduction, distribution, public performance, public display, derivative, and digital transmission), unless sold to a music publishing company in a deal, and then it controls them. Thus, licenses offered by the copyright owner (music publisher, depending on the type of deal) in exchange for a fee (money) provide the legal clearance for the label or production company to use the song in its businesses.

Just remember the most notable are: (a) mechanicals for recording labels; (b) synchronization or "sync" used in the production of a visual presentation, such as a film or movie; (c) master licenses issued by labels not for the song but for the actual recording of the song; and (d) blanket or public performance licenses issued by ASCAP, BMI, and SESAC (in the United States) and many similar organizations in the rest of the world. They collect royalties for a "public performance" of a *song* performed live, and in a recording, film, and so forth, airplay on a radio station, live in a concert, and embedded in a movie being watched on cable.

Additional licenses include: (e) digital mechanicals, paid by the label to the music publishers, required for recordings *sold* on iTunes and other digital sites; and (f) digital public performances fees, required for *transmissions* of songs in digital transmissions, webcasting, and terrestrial broadcast of radio programs being simulcast over the Internet. The Harry Fox Agency collects mechanicals and digital mechanicals, and ASCAP, BMI, and SESAC (in the United States) collect the public and digital performance license fees for streaming.

Figure 6.9 The current licensing system for songwriters and music publishers, plus labels and sound recordings, is confusing to most of us. The good news is the established system collects royalties for the six exclusive rights of the copyright holder and payments are made to both the songwriters and copyright owners. This is the system skipped and the royalties not collected or paid on the approximately 2 billion recordings illegally downloaded in the last few years. The losses also include music and recordings used in films and other entertainment software and devices that have been acquired without paying the copyright holders and creators.
Source: www.loc.gov (2015).[5]

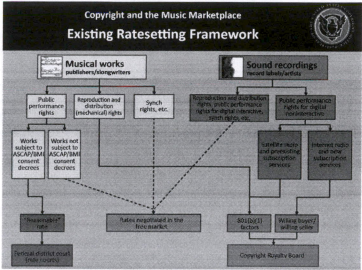

There are also print licenses required for selling sheet music, dramatic or grand licenses for songs in theater productions, and transcription licenses used for radio commercials and prerecorded elevator music.

Organizations such as the National Music Publishers Association (NMPA) represent music publishers who lobby Congress in the U.S.—for instance, for protection of their rights and businesses.

The National Music Publishers Association (NMPA)

The NMPA is a trade organization that represents songwriters and music publishers in the United States. Similar organizations represent them in other territories of the world. According to the NMPA (2014),

> *Founded in 1917, the National Music Publishers' Association (NMPA) is the trade association representing all American music publishers and their songwriting partners. Its mission is to protect,*

promote, and advance the interests of music's creators. The NMPA is the voice of both small and large music publishers, and is the leading advocate for publishers and their songwriter partners in the nation's capital and in every area where publishers do business . . . The goal of NMPA is to protect its members' property rights on the legislative, litigation, and regulatory fronts. In this vein, the NMPA continues to represent its members in negotiations to shape the future of the music industry by fostering a business environment that furthers both creative and financial success.[6]

The Harry Fox Agency (SESAC)

Most American music publishers use the Harry Fox Agency (HFA) to issue mechanical licenses and collect (mostly from record labels) royalties for the use of their songs (the publishers own the copyright) on labels' recordings. HFA originated in 1927 and used to represent about 48,000 music publishers in the United States.[7] However, in 2014 its revenue was declining significantly as consumers switched from buying albums and singles to streaming. According to *Billboard*, the performance rights organization SESAC purchased it for about $20 million and plans to use it for future development in digital licensing:

"Licensing is fragmented across both multiple types of rights, as well as multiple territories for the streaming services that represent the future growth opportunity of the music industry," SESAC chairman and CEO John Josephson said in a statement. "The result is a complex, opaque and currently inefficient licensing regime that fails to deliver the best outcomes for creators and publishers, as well as end users. What excites us about this transaction is the ability it provides to make the licensing process both simpler and more efficient, and in so doing create additional value for music creators and publishers, as well as the digital music platforms."[8]

Mechanical Statutory Rates

Once an artist (in cooperation with his or her producer) decides to record a song, the labels usually acquire the legal right to record and distribute it by acquiring and paying for a mechanical license. The history of the mechanical license starts with the advent of the player piano that needed a roll of paper with holes in it (the piano would sense the holes and play the corresponding piano key automatically). This actually caused a problem as the holes in the paper rolls allowed the music to be reproduced without payment to the copyright holders. Here's where we have the same basic problem as the Google book lawsuit, a new technology (at that time) providing an opportunity for entrepreneurs to profit off of the songwriters'/music publishers' property. This sounds familiar, as it was determined to be a clear violation of the owner's rights. According to Kowalke, in her paper in the *Cybaris Intellectual Property Law Review*,

> *Copyright protection has been a matter of relentless controversy throughout modern history. Technological advancements have also greatly affected copyright protection and fair compensation for artists. Debates concerning copyright protection in an evolving musical industry first arose in 1908, when composers began to lobby Congress for a legislative change granting them exclusive rights to authorize the mechanical reproductions of their works in response to White-Smith Music Publishing Co. v. Apollo Co . . . At this time, the player piano was popular in the United States, and copyright owners had concerns about their right to control the reproduction of their works on piano rolls. The technology used to broadcast audio transmissions was also not widely available at the time the 1909 Copyright Act was debated. Congress did not extend copyright protection to sound recordings on records even though this medium was becoming more widespread because the average consumer did not have the ability to copy records.[9]*

The *statutory rate* is recommended to Congress by three federal judges for approval. The fee is currently 9.1¢ per song ($.091) (for a song under five minutes in length) multiplied by the number of units pressed, shipped, downloaded, or sold. Thus, if a label presses or downloads a combination of 1,000,000 units, the cost for the license is 9.1¢ times 1,000,000 or $91,000 paid by the label to the music publisher(s). A ten-song album (with all songs under five minutes in length) costs the label $910,000. If the song's recorded time is *over five minutes*, then the *statutory rate* shifts to a *per-minute rate* of 01.75¢ per minute ($.0175) times the total minutes. As an example, if the recorded length of the song is 5:01 (one second over the limit), then the entire rate shifts to a per-minute rate of 01.75¢ which in this case is six minutes, as we always round up to the next minute. As an example, a song with the recorded time of 8:01 is considered a nine-minute song, so the mechanical is figured as 01.75¢ times 9 for a total of $0.1575 cents per unit pressed, ship, downloaded, or sold.

Just when we think we know what it costs a label to pay a music publisher for each song recorded and sold, we discover an exception to the statutory rate. There are at least two copyrights in any recording, one for the song, and the other for the recording of the song. Music publishers own the song and the label owns the recording of the song. Labels often place a statement into their recording deals with acts that all songs to be recorded will be paid at the controlled composition clause rate instead of the statutory rate, so why does that reduce the amount of money paid to the music publisher? Because in 1976 the labels persuaded Congress for the exception as long as the main recording act met one of the four qualifiers.

The Controlled Composition Clause

I know, how weird is this! Congress at work again! The qualifiers include any of the following. If the recording artist (a) has a publishing company (owns all or part of the publishing share of the song), (b) is the songwriter, (c) is one of the co-writers of the song (receiving part of the songwriter share), or (d) if the act is the first to record and release it or has any corresponding interest on any of the songs on the corresponding album (controls first-use rights), *then* the label can declare the song a *controlled composition*. Now labels have to pay only 75% of the statutory rate for mechanical licenses.

At the current full statutory rate of $.091 per song, per unit sold, the music publisher and the songwriter each receive $.045, or a little less than a nickel per song sold or

Figure 6.10 The controlled composition clause is an exception to the statutory rate of the mechanical license. Labels were upset about the rapid increase in the statutory rate and they wanted an exception to the rate. If the main recording artist is the songwriter or cowriter, or owns (meaning some of the publishing) or controls the first-use rights, then the labels pay only 75% of the statutory rate for the mechanical licenses. Labels love it and publishers hate it. Instead of making 4.5 cents per unit sold, the writer's share is reduced to 3.4 cents. If the artist is declared a cowriter, then the writer's royalties are reduced to only 1.73 cents per unit. That is a 62% reduction in royalties paid.

The Controlled Composition Clause

> **Value of Song**
> **Controlled Composition Clause**
> **Mechanical License on Indie or Staff Deal**

> **Licenses Royalties**
> **Statutory Rate =$.091 Divided 50/50**
> **CCC Rate = $.06925 Divided 50/50**

> **Songwriters' Share**
> **= $.045 @ SR**
> **= $.03462 CCC**
> **= $.01731 First Writer**
> **= $.01731 Artist/Writer**

> **Music Publishers' Share**
> **= $.045 @ SR**
> **= $.03462 CCC**

downloaded. Under the controlled composition clause (CCC) the rate is reduced to 75% of the $0.091 or $0.06825, a little less than seven cents total, and $0.034125 or about three and a half cents per song sold. However, the recording artist is *now* also considered a songwriter, so guess what happens? The exclusive rights of ownership (remember there are six) are now usually owned and controlled by the music publisher, as you sold your song to them. That means they get to decide "who" will split the reduced royalty rate with the artists, who is now usually declared either a cowriter or co-music publisher. It's the songwriter who usually takes the financial hit and has to split his or her share of $.034125 to $.0170625 or a little fewer than two cents per unit sold. Ouch!

Look at it another way. On a million-selling song the writer is paid half of the statutory rate or $45,000 ($.045 times 1,000,000 units sold). However, if the CCC is used, then the shares are reduced to 75% of that or $.034125 per unit sold. But the CCC also requires the recording artist to be declared a cowriter, which means you (the original songwriter) have to split your share of the songwriting royalties, which brings your wonderful check to a whopping $0.170625 per unit sold. In this case the labels got a great deal and the original songwriter got the shaft. Be careful: try to maintain a close relationship with your music publisher so they can try to share the CCC or protect you from it.

*Statutory rate per song, 5 minutes and under equals $.091 statutory rate × 10 songs **or 91¢ per album.***

*Controlled composition rate per song, 5 minutes and under equals $.0685 statutory rate × 10 songs **or 68.25¢ per album.***

How much money does the label save on the music publishing cost for one album with ten songs that goes platinum?

> **91¢ per album** × 1,000.000 units = **$910,000.00**
> ***68.25¢ per album** × 1,000.000 units = **$682,500.00***
> **Saves the label paying the license = $227,000.00**

If the music publisher agrees to the CCC license, their royalites are reduced on the million-selling album by $113,750.00, the same reduction for the songwriters. Bummer. So why in the world would a music publisher agree to this? Because of the power of the recording act, and here's an example. If it's Taylor Swift, no problem—another act not as famous, maybe not. However, this problem is declining as album sales are quickly being flushed down the toilet.

Cover Songs/The Compulsory License

In 1909 and because of the popularity of player pianos among consumers, Congress passed its first mechanical reproduction licenses for musical composition allowing piano roll companies to make and market their products *without consent* by the copyright owner. According to a statement by Marybeth Peters, former registrar of copyrights before the Subcommittee on Courts, Intellectual Property, and the Internet of the House Committee on the Judiciary,

> *The impetus for this decision was the emergence of the player piano and the ambiguity surrounding the extent of the copyright owner's right to control the making of a copy of its work on a piano roll. The latter question was settled in part in 1908 when the Supreme Court held in White-Smith Publishing Co. v. Apollo Co. (2) that perforated piano rolls were not "copies" under the copyright statute in force at that time, but rather parts of devices which performed the work. During this period (1905–1909), copyright owners were seeking legislative changes, which would grant them the exclusive right to authorize the mechanical reproduction of their works—a wish which Congress granted shortly thereafter. Although the focus at the time was on piano rolls, the mechanical reproduction right also applied to the nascent medium of phonograph records as well.*[10]

It's the same today: if you want to record a cover version of a song previously released for sale, let's say a hit by Lady Gaga, then you can either negotiate directly with the copyright owner or just pay for the license. You do not need to request permission to record it from the music publisher. However, you still must pay for the mechanical license, which can be obtained from the Harry Fox Agency online, if you are selling 2,500 copies or less to consumers for private use. The payment of the license is only for the song and not the original recording by Lady Gaga, as her recording is another copyright owned by the label and it is not legal to make copies of the recording and distribute them.

Notice of Intent

In addition, remember to file a Notice of Intent to Obtain a Compulsory License with the music publisher of a cover song you'd like to record yourself and sell. According to section 201.18 of the copyright law, a clear statement notice of your intention is required:

> (1) A "Notice of Intention" is a Notice identified in section 115(b) of title 17 of the United States Code, and required by that section to be served on a copyright owner or, in certain cases, to be filed in the Copyright Office, before or within thirty days after making, and before distributing any phonorecords of the work, in order to obtain a compulsory license . . . (2) A person is entitled to serve or file a Notice of Intention and thereby obtain a compulsory license . . . (3) . . . a "digital phonorecord delivery" is each individual delivery of a phonorecord by digital transmission of a sound recording which results in a specifically identifiable reproduction by or for any transmission . . . regardless of whether the digital transmission is also a public performance of the sound recording or any nondramatic musical work embodied therein . . . (4) A Notice of Intention shall be served or filed for nondramatic musical works embodied, or intended to be embodied, in phonorecords made under the compulsory license.[11]

Foreign Mechanicals—Subpublishing

Unlike the American practice of individual song mechanical licenses, most foreign countries require one mechanical license for all the songs on an album and rates vary in each country or territory. It's confusing, but thanks to the way Mazziotti describes it in *Music Business Journal*, we can get a little clarity:

> In Europe the situation is more complex than the U.S. since the European Union (E.U.) does not have a single copyright system and the enforcement and management of exclusive rights in musical compositions, music performances and sound recordings still takes place mostly on a country-by-country basis—despite the international reach of the Internet. The launch and full development of innovative services and an E.U.-wide online music business that could create a credible alternative to digital piracy is

being hindered, at a time when there is a shift of content from end-user devices to the "cloud."[12]

Public Performance or "Blanket" Licenses

Royalties collected for mechanicals have decreased dramatically with the steep decline in record sales. However, the royalties collected for public performances have now become the "serious" money for songwriters, music publishers, and composers. ASCAP, BMI, and SESAC in the United States, SOCAN in Canada, and many others in the rest of the world collect the royalties as license fees and then pay the administration bills, and split the remaining in half and send one check directly to the writers and the other to the publishers. Congress authorized performance rights in 1897, yet royalties were not significant until they could be collected. Royalties are collected through the issuing of "blanket" licenses to businesses such as radio stations, TV and cable networks, websites, concert halls and venues, sports arenas, clubs, bars, malls, skating rinks, and others. The blanket license covers all the songs covered by each organization, such as ASCAP, BMI, and SESAC in the United States. Songwriters may affiliate or be a member of only one of the three agencies. Publishers use three unique company names usually under an umbrella company name, such as Famous Music. According to Kowalke (2015),

> Songwriters have the exclusive right to publicly perform their own compositions. "Traditionally, this right was primarily implicated by broadcast analog radio and television. . . ." However, in today's ever-evolving tech-savvy society, digital broadcasters, webcasting, satellite radio, and some online music services frequently implicate this right . . . In order to better administer these rights on behalf of songwriters, entities known as Performance Rights Organizations (PROs) were established. Currently there are three operating PROs: the American Society of Composers, Authors, and Publishers (ASCAP), Broadcast Music, Inc. (BMI), and the Society of European Stage Authors and Composers (SESAC). Combined, ASCAP and BMI

represent roughly 97% of all American compositions. All three PROs offer radio stations the use of a blanket license, which is a set fee that allows each station the use of any composition represented by the PRO.[13]

All three PROs use their own analysis method to determine the value of a performance (which in this case means the use of a song in the recording), not the actual recording. Fees for blanket licenses are determined by weighing various factors, including the type of media (radio, TV, etc.), the size of the prospective broadcast audience (national, local, etc.), and the type of performance (music in movies, TV shows, theme music, etc.). Night clubs and other types of stores and business owners who use music in their establishments are also obligated to purchase a blanket license to pay the songwriters who created the songs they are using and the music publishers who own the songs in the daily operation of their businesses. Many become frustrated when confronted with a bill for the songs they thought were free. However, once the law is explained, most understand why the writers and copyright owner should be paid. Paying a live band to perform or entertain an audience does *not* pay the writers and publisher of the songs being performed. License fees are estimated and issued based on the size of the facility's potential occupancy rate, number of speakers, and how the songs are being used, such as CDs being played, downloaded, or live bands.

Analysis

Things are different today as all three PROs also use digital monitoring of actual broadcasts to determine fee structures and royalty payments. MusicMark and Broadcast Data Services (BDS) and other similar technology-based companies monitor actual airplay, digitize it, and convert the data used to pay corresponding blanket royalties.

Music publishers use the information to determine the where, when, and length of all broadcasts. Music pattern recognition technology is also used to collect short-duration performances of 60 seconds or less, often used in commercials; however, *no* royalties are collected for background music, partial performances, station IDs, and promotional announcement.[14]

Artist managers, booking agents, and concert promoters review the information to schedule tours. Heavy airplay, plus streaming and albums sales, indicates profitable target markets for tours by cross listing the collected information to the artist who recorded the material. All three use different types of analysis, yet they pay similar royalty rates to writers and publishers based on use, bonuses, printed information from TV (cue sheets), and live performances in clubs and tours. Here's an example of fees collected by ASCAP in 2014 in the United States from different types of media and sources.

Nightclubs and bars purchase licenses for songs used in their establishments for live, recorded, and transmitted music, with the exception of music broadcast on TV screens as defined in the Fairness in Music Licensing Act. Additionally, there are many other niche sources, everything from greeting cards to sporting events. Remember recording artists are not paid blanket license royalties unless they are also the songwriters, composers, cowriters, or music publishers of the songs. Only songwriters, composers, and music publishers receive public performance blanket royalties. How much they get paid in royalties depends on the media or type of performance and the amount of usage, divided by the total license fees generated and collected minus administrative fees. All the variables change every year; however, all three PROs pay about the same rate to songwriters and music publishers for chart placement and usage.

Payment Rates and Schedules

Over the last few years, ASCAP and BMI have each collected about $900 million to $1 billion, each issuing blanket licenses representing songwriters, composers, and music publishers.

Both are nonprofit organizations, with ASCAP having about half a million members representing approximately 9 million songs.[15] SESAC is a much smaller organization and Global Music Rights (GMR) is a start-up without much of a history, although it appears to be positioning itself as a world organization representing creative rights management. ASCAP distributed $851 million to members in 2013 (the latest annual report released in 2014),[16] and BMI collected $977 million for fiscal year 2014 (ending June 2014), with $840 million paid in royalties to its affiliates.[17]

ASCAP collects public performance money from songwriters, and then splits the money between songwriters and music publishers. Table 6.2 gives you the latest information from ASCAP's corporate reports on the amounts they collected from songwriters and music publishers they represent (worldwide), and the costs of their administration and operations. The final amount of money that was sent to their representative songwriters and music publishers is also shown. Is that the split you were expecting?

Each songwriter and publisher are paid based on a mathematical formula determined by ASCAP that is then multiplied by the number of times the song has been used in a public performance.

Songwriters, composers, and music publishers are called "members" when they join ASCAP, as they own part of the nonprofit organization. However, songwriters, composers, and music publishers who join BMI and SESAC are called "affiliates" as BMI is still owned by the original radio stations that created it and SESAC is a for-profit organization owned by venture capitalists and a bank. Notice the total receipts for ASCAP were about $945 million and the operating expenses totaled more than $114 million. The remaining $830 million plus is then distributed (quarterly) to the domestic songwriters, composers, music publishers, and foreign distributors. It is expensive to run the organizations due to the accounting, printed information, formal celebrations, and buildings, but they really provide a wonderful service to the creative

Table 6.2 ASCAP Blanket Licenses Collections, Expenses and Paid Royalties. As an example, ASCAP collected almost $1 billion in 2013, reported for fiscal year 2014, and paid most of that minus expenses to songwriters and music publishers who are their members.

Gross License Fees Collected			Expenses and Royalties Paid to Members	
ASCAP 2013 Receipts (2014)			Total receipts for 2013 (2014)	$945,285,000
Domestic license fees	**Revenue**		Total operating expenses for 2013 (2014)	$114,793,000
Television	$101,634,000		Excess of receipts over expenses	$830,592,000
Cable	$196,922,000		Total receipts for 2013 (2014)	$945,285,000
Radio	$166,291,000		Total operating expenses for 2013 (2014)	$114,793,000
General and background music	$109,664,000		**Excess of receipts over expenses**	$830,592,000
New media	$32,945,000		**ASCAP Expenses**	
			Expenses	**Revenue**
Symphonic and concert	$6,041,000		Licensing services	$16,402,000
			Performing rights	$41,897,000
Membership application fees	$1,257,000		Membership services	$16,857,000
			Headquarters	$24,005,000
Interest and other income	$25,000		Legal	$15,632,000
Total domestic receipts	$614,779,000		**Distribution to Members**	
			Member	**Revenue**
Foreign receipts	$330,606,000		Domestic distributions	$461,854,000
			Foreign distributions	$323,054,000
Total receipts for 2013 (2014)	$945,285,000		Foreign societies	$66,076,000

Source: ASCAP Corporate Reports (2014).[18]

community. They have also worked diligently to reduce their operation expenses, which used to total above 25% many years ago, down to a very reasonable 10%–20% of the gross receipts. Can you imagine what it would cost for a music publisher and songwriter to actually go out to a radio station and collect 8 to 15 cents every time it played a song? Or how about chasing down a nightclub owner because a band he hired played one of your songs and he (because he hired them to play) owes you about a nickel? Wow.

Synchronization Licenses

Before public performance royalties can be collected from TV stations, TV networks, cable, websites, and so forth, the production company making the visual media product (film, TV show, etc.) has to negotiate a synchronization license for the use of the song in the production. Look at it this way: if you put a video you shoot yourself of your favorite dog doing tricks and add your favorite recording from, say, the group Walk the Moon, and then post it on YouTube, have you violated any copyrights? Here's the deal: no, yes, yes, and maybe!

You have every right to shoot your crazy, cute dog doing backflips, and you own the video as you created it yourself. However, if you take somebody else's song (copyright) and put it onto a video and you didn't get the correct sync license from the publisher or songwriter, then slap your hand, as you've violated his or her copyright. Thus, you are required to negotiate for a *synchronization license directly with the music publisher before* distributing it in your videos. That includes YouTube, except they are operating under a safe harbor (as the distributor). They didn't violate the songwriter's rights— you did. YouTube and other similar providers use filtering systems to detect illegal use of the music and when they discover it, they may take it off the service and send you a letter. However, if the crazy video with your cute dog is getting lots of hits or views, then the copyright owners may let it ride as they may perceive it as free promotion, and besides, they know you're broke and couldn't afford a lawsuit.

Live TV show producers have to license songs *before* they are used on a live TV broadcast. Several years ago, a good friend of mine was producing the number-one live cable show in the United States. He smiled as he paused his chain smoking for a moment and said, "Here, look at this." It was a several-page letter from an attorney, stating that if his band played even a couple of notes from any of the listed songs, they would quickly be sued. Isn't this fun! Let's say you decided to use the master recording from the group Walk the Moon with your video of your back-flipping dog. Is that legal? Well, who owns the actual recording of the song by Walk the Moon? The record label, so if you used the original recording, now you'll probably get both hands slapped. There are two copyrights on a recording, one owned by the music publisher and the other (the recording of the song) owned by the label. Master recordings have to be negotiated *directly with the label*, just as sync licenses have to be negotiated *directly with the music publisher*. What if you want to put on a stage play from a movie you saw or make a book into a movie or play? Get permission before you even consider it. And do not be surprised if they say no, as only music (audio) falls under the compulsory license, which by the way you still need to acquire a mechanical license and pay for when recording your own cover version of a major artist's recording.

Folio and Print Sheet Music

Print and folio sheet music and books used to drive the industry about 150 years ago. Not today. But we see musicians, students, arrangers, conductors, and others still use sheet music when performing and recording. Nashville still uses a numbers system, sometimes passing on the sheet music for sessions. However, even the Internet has added to sheet music sales as over the past 20 years software companies have developed websites that allow users to download sheet music, and staff paper to write music, directly to their own computers and printers. Sheet music royalties for songwriters and music publishers are usually a small percentage of the suggested retail list price, often 10%–20%. In addition, most publishers turn the work over to book printing and publishing companies, which then pay the publishers and writers their share. Table 6.3 shows the different types of licenses available and the representation agencies that issue the licenses and collect

Table 6.3 Licensing Music. These are the different types of licenses available and the representation agencies that issue the licenses and collect the fees (paid as royalties) for the use of music defined in the six exclusive rights.

Type of License	Exclusive Right Covered	Purchaser of License	Agency
Mechanicals	Reproduction Distribution	Labels, music companies	Harry Fox Agency or direct
Digital Mechanical	Downloading by transmission (DPD, digital phonorecord delivery)	Providers of digital streaming and limited download	Harry Fox Agency or direct
Sync	Reproduction Distribution with a visual presentation	Film and visual production companies	Direct
Master	Reproduction Distribution	Film and visual production companies	Direct
Public Performance Blanket	Public performance	Mass media radio, TV, clubs, restaurants, businesses, and venues	ASCAP, BMI, SESAC, CCLI, churches
Digital Public Performance Transmission	Performance by means of digital transmission, webcasting, and radio simulcasting	Satellite, digital providers, and broadcast radio (passively transmitted)	Sound Exchange
Print	Reproduction Distribution	Book publishing companies	Direct
Transcription	Reproduction Distribution Public performance	Audio productions (resale to chain stores) and in-store and elevator music	Direct
Dramatic or Grand	Stage performance	Stage production company (theater)	Direct
Foreign	All	Foreign mass and digital media, plus foreign production companies	Harry Fox, ASCAP, BMI, SESAC

the fees (paid as royalties) for the use of music defined in the six exclusive rights.

Future Trends

The world is quickly changing and just as the printing press ushered in new opportunities, so do our digital technologies and innovations. However, in this case billions of dollars in royalties owed to songwriters, music publishers, labels, and artists have been lost forever. According to the Recording Industry Association of America,

> *One credible analysis by the Institute for Policy Innovation concludes that global music piracy causes $12.5 billion of economic losses every year, 71,060 U.S. jobs lost, a loss of $2.7 billion in workers' earnings, and a loss of $422 million in tax revenues, $291 million in personal income tax and $131 million in lost corporate income and production taxes*

> *. . . -In the decade since peer-to-peer (p2p) file-sharing site Napster emerged in 1999, music sales in the U.S. have dropped 47 percent, from $14.6 billion to $7.7 billion . . .*

> *-From 2004 through 2009 alone, approximately 30 billion songs were illegally downloaded . . . only 37 percent of music acquired by U.S. consumers in 2009 was paid for . . .*

> *-Frontier Economics recently estimated that U.S. Internet users annually consume between $7 and $20 billion worth of digitally pirated recorded music.*[19]

And these are old figures. CNN's David Goldman stated in his article "Music's Lost Decade: Sales Cut in Half,"

> *Now just 44% of U.S. Internet users and 64% of Americans who buy digital music think that that music is worth paying for, according to Forrester. The volume of unauthorized downloads continues to represent about 90% of the market, according to online download tracker Big Champagne Media Measurement.*[20]

GMR

Many famous recording artists who are also songwriters and music publishers are moving their affiliation from one of the three American PROs to Global Music Rights, which is a new company that collects royalties worldwide. ASCAP, BMI, and SESAC have agreements with many other world agencies, yet they all take their share of the pie, so GMR is becoming an important industry development. Their business is structured toward niche writers and publishers, so all U.S. copyright law and license collection methods used by ASCAP, BMI, and SESAC are also offered and enforced by GMR.[21]

Catalog Valuation

Okay, let's have some fun. You are a little wild and crazy and over the past few years have written several songs that actually made a little money. You learned from the mistakes you've made and joined one of the performance rights organizations, pitched your material, and had them recorded by local, regional, and an over-the-hill rock star. And you've recorded the songs and placed them yourself on iTunes and Spotify. Let's say a friend of a friend of a friend knows the head of A&R at Universal and gets your songs to her. We'll call her Cindy, and after she listens to all the material she decides one of the songs is just what she needs for a newly signed recording artist. Cindy knows that you've formed your own music publishing company as she found all the information she needs to contact you from BMI. Instead of trying to buy one song, she decides to buy all of your songs.

You registered all material with the Copyright Office and formed your own music publishing company as a DBA, in order to receive the songwriter's and the music publisher's shares of any royalties generated. Cindy offers you $5,000 for your catalog of songs. What are you going to do? What does this really mean in dollars and cents? Five grand sounds a little cheap—maybe she'd give you ten. The problem is that you really do not know how much your publishing company is worth. You've decided to figure it out, so you look at all your old receipts and write the information into a simple chart. Or should you just take the money and run? Value estimates are subjective and difficult to determine, but there are also some basic principles to follow, so here we go! Businesses are sold all the time in capitalist society. The formula can be complex, but it's often based on the five-year average income (total profits of five years) divided by five. That figure is tied to a multiple between 5 and 15 years, and it assumes that if you sold the company and it continued to gross the same as it had in the pass, it'd just take 5–15 years to pay itself off.

In our example, your songs in your publishing catalog earn $18,675.17 over five years and an average "yearly" income of $3,735.03. The assumption is your songs will continue to generate income for many years and so the question is, should you take the $5,000 cash? So here's what you do. Take the average income and estimate the 5-, 10-, and 15-year payoff price. That means you would be fairly paid using $18,675.17 as the bottom figure, $37,350.30 as the middle or 10-year amount, and $56,025.50 as the high figure. Now, you know your catalog is worth between $18,675.17 and $56,025.59. How good of a negotiator are you? How badly does Cindy want the songs? By the way, the song she wants is the one you never released, or she could just use a compulsory mechanical license, but she wants to control first-use rights and if it becomes a hit, she and her music publishing company/label will make more money without you in the picture.

However, fools are born every day and I admit I've been one of them. Bad deals and trusting the wrong people are a common business mistake. Just make sure you get offers in writing and talk to an attorney. What, then, is an honest price? That depends on your negotiation skills and how much (or badly) Cindy wants it. The price you accept depends on how much you are willing to sell your company and at which price point. The deal also depends on how much the buyer wants your songs and how much she's willing to pay. Okay, let's assume you take the middle figure, or $37,350.20, and sell the songs and company.

The Big Surprise

But here's the really good news. Will you still get any royalties? Think about it: remember the different types of songwriter deals, and so when you sold your catalog, what did you really sell? You sold the ownership in the songs and company and the exclusive rights. But you will still receive your songwriter's share, which will probably be significant. Now, your songs will be recorded by major artists, and probably get tons of blanket license royalties and sync licenses royalties for placement in movies and films and other sources. How cool is this: you got a check for $37,350.20, plus what your songs earned over the last five years *and* all or a part of the songwriter's share on all future royalties. For how long? Your life plus 70 years until the songs fall into public domain. Rocking!

People Get Rich

Let's forget about all those future royalties and think about how to make money from the check you got. Let's invest the $30,000 into safe mutual funds in the market and use the $7,000+ to pay the taxes. On average those kinds of investments will double every seven years, so if you're 20 years old now and you invest over a long period of time here's what will probably happen. At 27 the $30,000 becomes $60,000. Another seven years and it's worth $120,000 and you're only 34. Seven more years and you're 41 with a cool $240,000 in the market. Forty-eight and you've got about half a million, 55 and you're ready to retire and live off the interest. That's how rich people often get rich. So can you. The money could be more than the million or a little less, and guess what—we didn't even count any of the additional songwriting royalties that will be coming in to your mailbox every three months. Now you know why this is such a cool business. Or we could work for 40 years delivering Pepsi and retire with whatever we could save and Social Security, if that's still around.

My point is simple: if you maintain some percentage or all of the copyright when you're making a deal, it may become valuable in the future. Of course, if you turn over your copyright ownership, usually done in a work-made-for-hire and in independent and staff writer deals, you'll receive your percentage of the songwriter's share.

Best Efforts

Publishers can only be proactive concerning a song's potential selection by an artist to record. Contracts customarily state that music publishers have satisfied their contractual obligations when they have, to the best of their ability, attempted to place a song with music users. It is very difficult for a songwriter (or his or her attorney) to define and prove that a music publisher did not attempt to place the song with various music users. In addition, publishers can usually provide a stack of invoices and paid statements as proof of their attempt to place the song with record labels, recording artists, and other types of music users.

Staff Writer Termination

Sometimes the marriage between the songwriter and music publisher fails. The quality of their agreement means very little unless both profit. Written contractual goals not fulfilled by you or the publisher may be terminated, but doing so, given the nature of the contract you have signed, might be difficult. Legal action is customarily required to prove nonperformance of a contractual agreement. Publishers may default on their contractual obligation if they fail to provide licenses to music users; pay royalties in a timely manner; place the song with a major recording artist or an artist signed with a record label; provide sheet music (of the song) to music stores and other retail outlets; place the song in radio and TV commercials, advertisements, and movie soundtracks; or protect the copyright against infringement.

Do the Math

Many songwriters wonder how much money they should receive in a royalty check. It's complex as it all depends on: (a) the type of deal you struck with a publisher, (b) how successful the song is at generating money due to its use in recordings, films, commercials, and so forth, (c) which performance rights organization you belong to (ASCAP, BMI, or SESAC in the United States or one of the hundreds around the world), (d) the exchange rate on the value of different currencies compared to the dollar, (e) and many other factors.

Now compare the different types of deals and determine how much money you made in each. Remember, three issues are being negotiated in every deal: (a) the songwriter's royalties, which are usually 50% of the total income but are called 100% of the songwriter's share, (b) the music publisher's royalties, which are the remaining 50% of the total income but are called 100% of the music publisher's share, and (c) the ownership of the copyright, which may be valuable and a sellable piece of personal property with the value based on popularity (how many companies may want to use it in the future and how many consumers may want to buy or listen to it in the future). Let's assume you've written a song by yourself that generates $100,000 in total royalties. In the work-for-hire deal you are paid $2,000 to write the song and the company that hires you owns the copyright. In each deal the copyright owner later sells the copyright for an additional $500,000. How much money did you make in each type of deal?

Sources of Revenue

It's amazing how many relatives come out of the woodwork when someone makes the top ten charts. The fact that an act is famous and known doesn't mean they are rich. As a matter of fact, they usually struggle financially to survive, unless they understand how the business works. Now that you have an understanding of the songwriting and music publishing business, let's look at labels.

Notes

1. "U.S. Copyright Office—Help: Author." U.S. Copyright Office. April 4, 2013. Accessed July 14, 2015.
2. "Works Made for Hire." Copyright.gov. September 1, 2012. Accessed July 14, 2015. http://www.copyright.gov/circs/circ09.pdf.
3. Ibid.
4. "Music Publishing 101." NMPA.org. 2014. Accessed July 19, 2015. http://nmpa.org/about/#music_publishing.
5. "Copyright and the Music Marketplace." Copyright.gov. February 2015. Accessed February 20, 2016. http://copyright.gov/policy/musiclicensingstudy/copyright-and-the-music-marketplace.pdf.
6. "Mission Statement." NMPA, National Music Publishers, About. 2014. Accessed July 19, 2015. NMPA.org.
7. Christman, Ed. "SESAC Buys the Harry Fox Agency." *Billboard.com*. July 7, 2015. Accessed July 26, 2015. http://www.billboard.com/articles/news/6620210/sesac-buys-the-harry-fox-agency.
8. Ibid.
9. Kowalke, Caitlin. "How Spotify Killed the Radio Star: An Analysis on How the Songwriter Equity Act Could Aid the Current Online Music Distribution Market in Failing Artists." *Cybaris* Volume 6, Number 1 (2015): 195–247. Accessed November 11, 2015.
10. Peters, Marybeth. "Statement of Marybeth Peters The Register of Copyrights before the Subcommittee on Courts, The Internet and Intellectual Property of the House Committee on the Judiciary." Section 115 Compulsory License. March 11, 2004. Accessed July 26, 2015. http://www.copyright.gov/docs/regstat031104.html.
11. "Notice of Intention to Obtain a Compulsory License for Making and Distributing Phonorecords of Nondramatic Musical Works." Copyright.com. July 23, 2015. Accessed July 27, 2015.
12. Mazziotti, Giuseppe. "The Politics of European Online Music Rights." *Music Business Journal*. December 1, 2011. Accessed July 27, 2015. http://www.thembj.org/2011/12/the-politics-of-european-online-music-rights/.
13. Kowalke, Caitlin. "How Spotify Killed the Radio Star."
14. "Royalty Policy Manual." BMI. 2015. Accessed July 27, 2015. https://www.citationmachine.net/chicago/cite-a-website/manual.

15. Williams, Paul, and John LoFrumento. "2013 Annual Report." Music, Made Possible. June 9, 2014. Accessed July 28, 2015. http://www.ascap.com/~/media/files/pdf/about/annual-reports/2013-annual-report.pdf.

16. Ibid.

17. O'Neill, Michael, Bruce Esworthy, Phillip Graham, James King, Stuart Rosen, Alison Smith, Ron Solleveld, and Michael Steinberg. "2013–2014 Annual Review." BMI, Press Credentials, 2014 Financial Results. 2014. Accessed July 28, 2015. http://www.bmi.com/pdfs/publications/2014/BMI_Annual_Review_2014.pdf.

18. "Annual Report." www.ascap.com. Accessed February 20, 2016. http://www.ascap.com/about/annualreport.aspx.

19. "For Students Doing Reports." Riaa.com. 2015. Accessed August 1, 2015. http://www.riaa.com/faq.php.

20. Goldman, David. "Music's Lost Decade: Sales Cut in Half." *CNN Money*. February 3, 2010. Accessed August 1, 2015. http://money.cnn.com/2010/02/02/news/companies/napster_music_industry/.

21. "Global Music Rights FAQs." Global Music Rights.com. 2014. Accessed July 27, 2015. http://globalmusicrights.com/faqs.

RECORDING LIGHTNING IN A BOTTLE 7

Edison the Inventor

Thomas Edison (1847–1931), what a guy! Little did any of us know that when he invented the telephone, record player, film camera, and projector, he'd start an entirely new industry. Sure, we'd had entertainment before—books and poems—but now with Edison's inventions some very bright businessmen turned those live performances and books into recordings, film, movies, and, eventually, computer games. His inventions, when used as the tools of production, started the record and film entertainment industry. In essence, Edison created how most of us receive our entertainment products delivered over radio, TV, and Internet, and in live theater.

Let's take a closer look at the full range of Edison's inventions, as described by Gerlinda Grimes on the website Howstuffworks.com. As you read, think about the kind of creative and entrepreneurial mind that made all these "everyday" things, some now fading from our use, possible.

> *The phonograph (1877)*—*Edison earned his nickname "The Wizard of Menlo Park" in November 1877 when he invented the world's first method of recording and playing back sound.*
>
> *The carbon microphone (1877–78)*—*Alexander Graham Bell may have invented the telephone, but it was Edison who invented a microphone that turned the telephone from a promising gadget into an indispensable machine with real, practical applications.*
>
> *The incandescent light bulb (1879)*—*Without a doubt, the light bulb is Edison's most famous invention. Scientists and*

inventors had been racing against each other for years, trying to invent artificial light. Edison cinched the win by creating an incandescent bulb with a carbon filament that could be practically reproduced.

The Brockton breakthrough (1883)—*Once the world had light, it needed a way to power that light. In the tiny town of Brockton, Massachusetts, Edison set out to construct one of the world's first three-wire electrical power plants as a way to show the world that electric power was safer and more efficient than gas power.*

The kinetoscope, kinetograph, and kinetophone (1888–1890s)—*Edison and his assistant William Dickson first invented the kinetoscope, a boxlike contraption that enabled a single viewer to watch a motion picture short through a peephole. Films were recorded with a motion picture camera called the kinetograph; later, the kinetophone attempted to add sound to moving pictures.*

Nickel-iron batteries (1901)—*Before steam- and gasoline-powered engines were popularized, some of the world's first automobiles were powered by batteries. Edison's nickel-iron batteries were an improvement, in terms of both ecological impact and charging time, over the more commonly used lead-acid batteries of the day.*[1]

We often think of creative people as the artists, painters, and symphony conductors, but Edison was also a brilliantly creative inventor of *devices.* The scientific discoveries he and others made were developed into innovative tools that by and large still sustain the majority of entertainment products we enjoy. The creation of these new entertainment devices and techniques really started with fresh ways to use electricity, with some important discoveries coming initially from the Serbian (now Croatian) American innovator Nikola Tesla (1856–1943), whose idea of alternating electrical current, as opposed to Edison's application of direct current, led to the foundation of wireless communication systems, the design of the modern generator, and other engineering-based inventions. Between Edison and Tesla, who came to America to work with Edison, the first light bulb, camera, projector, microphone, and sound recorder (a cylinder device) that led to the phonograph were all designed, tested, improved upon, and patented. Cool. Without any of that, we'd have a boring life, maybe even still candlelit, and we certainly would not have had the entertainment industry for so many decades and in the forms we do without these two scientists.

Edison the Entrepreneur

Edison was also an *entrepreneur*, having started the Edison General Electric Company in 1892, which specialized in communication inventions using his techniques for harnessing electricity for industrial and consumer uses. His company eventually became General Electric, one of the largest energy companies in the world, and he was smart enough to be the largest single shareholder. Edison's role in the development of the entertainment industry cannot be overstressed since he started the first movie production company in 1893, called *The Black Maria*, which was the nickname of his kinetographic theater.[2] Through this company, he improved on late 19th-century European efforts to design a movie projection system, among other things, and his inventions also led to the development of sound recording, motion capture techniques, and projection used in early films, notably in the groundbreaking 1903 film *The Great Train Robbery*, which was filmed by a cameraman who had worked with Edison and knew his devices firsthand. Edison, as the inventor of the building-block machinery of the modern entertainment industry, may have accomplished much of his amazing work motivated by the sheer joy of creating something totally new in ways that would improve the quality of life, but he also was fully aware the inventions had great value as industrial products and that others would pay for them. That's a wow! Yet, I wonder if he would have worked so hard if the law (at that time) did not protect his inventions, and if their acceptance by society couldn't have provided a living for his family. Can you see the correlations between Edison's inventions and property rights with the situation currently disrupting our entertainment industry's creative writers, artists, and companies?

Edison worked in a different sort of marketplace, and while he designed the tools we require in the modern entertainment industry, he did not have to deal with how others would use them. His ideas for scientific applications would enrich society, but he confined himself to discovering new things and making sure they worked for those who wanted to use them for a variety of purposes and needs. Naturally, he could not fully anticipate the full impact of his creative discoveries as derivatives and enhancements of his groundbreaking works came into being over time.

Today, the questions that face the entertainment industry, the "Edison tool users," if you will, are more about the law than the operation of the devices, but without the recording equipment Edison launched, we would not be having the discussion about copyrights, ownership, and distribution of products. Let's take a look in this chapter at one of the processes in the production of entertainment, an actual recording session of a ten-song album. You may be surprised by how much we need to know to actually accomplish the master recording. It is, after all, a combination of oil and water, creativity and business, pleasure and frustration, emotions and money. By reverse engineering to analyze a simple recording process, we should be able to get a glimpse of what it would take to create a feature-length movie and with this, we can see how far we have come, how the entertainment landscape has changed, and how much we still owe to Edison and his entertainment inventions.

The Creative Business Milieu

In the entertainment business, there is an underlying philosophical conflict between the creative and business sides of the industry. Successful industry professionals accept these philosophical differences and work together to interface the creative with the business. Nonetheless, when large sums of money are involved, opinions, sometimes called ideologies, often emerge around the relative value of art or the arts and the profits realized by monetization of the arts as business. Disagreements range from friendly discussions to lawsuits, which sometimes define the boundaries of the conversations and potential solutions to the personal and financial conflicts artists and creators experience.

Where does the rubber meet the road in these sorts of principled conversations? Within the discussion of how to make recordings is one good example, as recording session artists need to get their product to the consumer. In their experiences, we will see a meeting of labor of love with real labor that costs and makes money for the parties in the entertainment production process.

Most artists want to control their own sound and performances, but at the same time, the labels want to control the costs. The solution is to bring in a third player, the producer, who has to bridge both the creative process and the financial aspects of creating an album. They've got to create great sellable tracks *and* hold the line on budgets. Therefore, during the recording of an album, the creative and financial control of the session and budget is the producer's responsibility. Still, no matter what issues are involved, labels still fund and invest in recording artists and their creativity. Kennedy and Wenham overview the investment process in the International Federation of the Phonographic Industry's *Investing in Music*:

> New signings, leading to new releases, are the lifeblood of record companies. Based on data received from its members, IFPI estimates that one in four of all the artists on a typical international record company's roster were signed in the previous 12 months. Continually investing in new talent is a hugely risky business, as only a minority of the artists developed will be commercially successful in a highly competitive market.[3]

In the film production business, the executive producers invest millions in the production and the director has to deliver the wow movie. It is not easy and a very different knowledge base is required to be successful. Egos are involved in both audio and visual productions. However, smart producers in the recording studio and directors in the film industry know who's the boss. It's usually the people supplying the money and the person who's in charge of the quality of the production. And if it fails at the box office or at retail, then it's the producer of the album or the director of the movie who takes the blame. Thus, understand that producers and directors make a lot of money when the product is successful, but they are the ones who also have to find a job when it fails. The recording artists and movie stars are usually exempt from the blame for a failure—unless they are in a long series of money-losing products. Then they are gone also! Not because they do not have talent, but because consumers don't want to pay, use, watch, listen, or even steal their recordings and movies. That's a slow death that is almost impossible to recover from by starting one's career over. There are also many things that can go wrong, such as conflicts with the actors, lighting, audio, technical problems, location, weather, marketing, promotion, and the timing of publicity. It's a real crapshoot, but when it's right, when it's a wrap, the last note has been sung, and everyone gets goose bumps . . . Well, then, everyone on the set (or in the session) knows that something very special has been created that might touch the human spirit or even last a lifetime. And each person who worked on it or created it was part of the dynamic team effort that made the magic happen. Inspiring? Exciting? You bet!

The Ingenious Creative Technicians

It's the people behind the stars and public personalities who usually provide the essence for success in the creative and business systems. We hear and see only the finished products. But for every major star in the music industry and every major film star, there's as many as 100 unseen songwriters, scriptwriters, musicians, actors, producers, directors, set builders, lighting and sound engineers, marketing and social media types, labels, or film production companies *and* their teams of promoters, sales, distributors, and others who help create, promote, and manage the act. Today's wannabes have the computers and the ability to upload their self-made movies, videos, and music tracks to Internet providers, but what they rarely have is a wow song or story, the expert technicians required to create a competitive quality final master, and the money and business knowledge and skills to actually sell it. Are you starting to see where you might fit into the business?

The Pros

If we're trying to compete with the major leagues, the wannabes need competitive products at the same quality before consumers will even notice it. Mom, Dad, and a few friends might watch it on YouTube or download it from iTunes, but that's about it. It always starts with the wow. Next, the pros (experts at sound, lighting, recordings, filming, gaming, staging, etc.) have to come in to create high-quality entertainment products. British pop star Adele came to the attention of her label after one of her friends put up some of the singer's songs on MySpace when Adele was just 16 years old. Two years later, in 2006, Adele was signed to her first contract. Most of us wannabes simply do not have the experience and tools to accomplish the task of professional quality recording at a competitive level, and the chances of the big break are just that—a chance that someone will hear and be interested in the sound or the song or both. Do we have the talent and brains to hit a major league pitcher's 100-mile-per-hour fastball, throw a 60-yard touchdown pass, nail a three-pointer 30 feet out, or score a short-handed goal against the NHL's Pekka Rinne? Of course not! But it's really fun to take a shot at it!

The Brilliant Generation

This latest generation can often whip an older guy in a computer game blindfolded with one hand behind their backs. As an example, my two girls, Hannah and Chellie, at ten and eight years old, could always fix my computer problems. If

I couldn't figure out what was wrong they'd come over, give me the "Oh, Dad, that's easy" bit, hit a few keys, and gleefully walk away while I would be asking, "How'd you do that?" In a recording session, if the producer asks for a deeper, thicker tone (whatever that means) on the grand piano or a longer sizzle on the high hat, how are you going to do that? Try condenser mic a quarter of an inch off the sound board on the "dead" spot (which is only about the size of a silver dollar). Every piano is different due to the type and grain in the wood, so every sound is nuanced by the recording technique associated with the peculiarities of the particular instrument. You probably didn't know that trick, but my late friend Don Cary, who learned it from the guys at A&M Records, taught it to me. By the way, don't even consider touching the EQ when recording basic tracks! Learned that from another acquaintance who owned the most successful rock studios in Los Angeles and San Francisco, Wally Heider (1923–1989). The biggest rock stars in the industry loved the guy for his knowledge and skills even when he dressed like a farmer and was old enough to be their grandfather.

When the film director asks the writer to change the "meet cute" to establish a more subtle conflict between love and hate for a split second, how's it going to be done? A *meet cute* is the term used in the film industry that describes how two people (usually the starring couple) meet for the first time in an amusing or entertaining way. If the director wants the conflict to be nonverbal and asks the actress to show love in her eyes and hate on her face, how is she going to do that? How will the lighting and camera angles be changed to get the shot? How will the actress also give a hint of the future love/hate emotion that sets up the rest of the script? Well, you have to know or quickly improvise because they wanted it done yesterday and they're about to go over budget. These examples are a small snapshot of the genius and creative personalities who live behind the scenes or glass that make the artist or star look and sound great. And it shows us the who of the creative process, and, as you see, it takes a team to make a product and to shape an environment where success is possible.

Teams

As a part of a team of experts working together to create the entertainment products you'll need to be quick-witted, possess a great personality, and be consistently easy to work with to support the other team members. The computer whizzes may need to work with the math expert software writers to develop just the right tension between winning and losing the game. Can't have it too easy or too hard. How will the videogamer's choices be converted into skill levels that will provide just the right emotional response between winning and losing the game to encourage the player to continue? In a film production, how is the writer's vision turned into the conflicts of the storyline? How do you light the set, dub in the sound effects, select the perfect type of makeup that will not reflect the red light from the gels—and the list goes on! In the recording sessions, the musicians, background singers, audio engineer, and producer must work together to create magic; or, as Quincy Jones used to call it, the "lightning in a bottle" that emotionally supports the royalty artists. Of course, once the products have been created there are other teams of executives and experts who finance the productions, budget the expenses, manage the acts, publicize the people and the events, and market the music, images, and merchandise to consumers.

Lots of Hats

The perception recording artists and stars often have is they simply have to sing, play their instrument, act, and provide shameless self-promotion. But the smart ones also become entrepreneurs, business owners, and sometime even employers. As I have been in sessions with some of the most successful record producers, engineers, musicians, and artists in the world, I can tell you it is a significant head adjustment. The studio musicians have to have a connection with producers because that's who hires them, just as personal managers hire the on-the-road-musicians to back up the famous recording artists on the tour. Thus, the musicians and backup singers who cut the tracks in the studios are not usually the ones on the tour with the famous dudes.

There is a very strict "respect for authority" in the studios. Who you talk to and when are important to know. Usually it's very wise to keep you ears open and your mouth shut until asked a question or given an opportunity to introduce yourself. The talented technicians, who are the film camera people, lighting techs, and stage managers, all have their bosses, and in reality, the producer in the recording studio tells the lead musician, who tells the drummer and bass player what they need in their performance. The director in visual productions usually tells the techs running the switcher which camera to "take" for the live feed or the lighting director what to have the techs guys fix when the egotistical star has a slight shadow on the forehead. It has to be perfect! Even the stars (including the men) need to know what type of makeup to use for different types of lighting, camera shots, and angles. And that covers just the basics for getting started! The more you know, the more you can help, and the quicker you'll find employment by respecting the professional roles each member of the team plays.

Attitudes are really interesting. It's amazing to hear the superstar call the producer "sir" or "ma'am" or by their first name in a humble manner, respectful of their position as boss and insightful creative coach or teacher. It's also a blast when I've heard them say in a very direct manner to an act that's being a jerk to "just shut up and sing." That usually happens only once, and it's not the superstar who keeps the job if he or she continues to disregard the authority of the producer. Same thing in the film industry with the director, only even more so, as there are only a few really great directors. The ones who have the talent to make an actor give the performance of their lifetime are rare. The really smart stars know their position at that moment and always show respectfully how lucky they are to be cast in the role or to be working with the producer, director, and crew. It's much different from what you'd think listening to gossip, publicity agents, and TV hosts gushing over an act. But that's the reality of the business, and everyone knows how to play his or her role and to work together to make it happen! Everyone knows that the final recording, movie, or whatever product will be good or bad based on each person's passion, knowledge, personality, and communication skills as part of the team, contributing to anything that needs to be done.

Oil and Water

In the business, there is always a strange balance between the creative and business-minded men and women who own, operate, or work at industry-related companies. It's this crazy balance that determines the nature of the entertainment companies and industry's products and overall success. Realistically, industry success involves passion, a unique world vision of humanity and society, street and learned knowledge, talent, an understanding of business principles, connections, money, and an excellent team of experts who are constantly searching for profits, which are becoming increasing harder to assure now that most of the music in the world, for instance, is acquired without payment. The product's potential for success is based first on a great book, story, or song and then the artist's ability to contribute through his or her talent in some meaningful manner to a message of emotional importance to various segments of the consumer market.

Professional musicians, artists, producers, cameramen/women, and other industry insiders assertively earned their own success by having a positive attitude, meeting the right people, perfecting their artistic and business intellectual talents, obtaining a quality education, and being in the right place at the right time. Successful industry-related entrepreneurs provide products and start-up businesses that solve problems, create win-win situations, and provide better entertainment products to the world's population.

Luck is important, but personal goals, developing your creative and analytical talents, a positive attitude, finding your niche, and educating yourself to convert your passions into a career are important. If a major producer in the music industry or a film director asks you, "What do you want to do?" the answer "I don't know" isn't helpful. Entrepreneurship is becoming one of the main keys to success in the entertainment business. Industry professionals customarily associate with other colleagues for commonality and friendships, and to enhance career opportunities. Nobody can make it in this business by him- or herself. It's too big of an industry. Work in teams! There are many opportunities to be found by working in teams with others who have similar interests, goals, and connections. Let's look at some important teams in the industry more closely to see how they collaborate to bring quality products to consumers.

The Recording Team

Recording artists are idolized as superstars, but it is the creative team of talented studio musicians, backup singers, audio engineers, and record producers that makes the recordings that lead to the artist's fame. Having the ability to sing well is one thing; getting it recorded with magic is quite another. Major artists usually need to reach iconic status before consumers buy (or steal) the recordings that pay for the distribution, promotion, and publicity required to launch careers, especially long-term careers that have a specific brand that brings consumers back year after year.

Producers

The session producer is the captain of the ship, directing and stimulating the creative elements of the audio engineer, studio musicians, and recording artists. Industry-proven skills and knowledge include a passion for music, an understanding of music theory and sight-reading, and in some places knowledge of the "number system." Producers also need to know copyright laws, music publishing, business finance, and the music business subsystems. Musicianship skills are essential for music producers. Successful producers communicate well with others, are decisive leaders, understand basic accounting skills for controlling budgets, and have strategic management skills for administration. In addition, producers commonly have a consummate understanding of recording studio acoustics, basic electronics, equipment capabilities, and an almost magical ability to "marry" the right song to the right recording artist.

Producers know other industry professionals through networking, showcases, parties, meetings, professional organizations, industry events, award shows, and so forth, and *successful* producers know who the industry movers and shakers are (the insiders and personalities who manage and control the industry), as the majority of their work originates from those connections. Ever heard of Ray Charles, Wilson Pickett, Led Zeppelin, Bob Dylan, Carlos Santana, Dire Straits, or Aretha Franklin? Jerry Wexler (1917–2008), who started his career in the music business as a journalist (with a degree from Kansas State University) for *Billboard* in the 1940s, became the guy behind the scenes who launched their careers.[4] He helped black entertainers break into the business when others wouldn't. As Bruce Weber (2008) described Wexler at the time of his death,

> *Mr. Wexler was something of a paradox. A businessman with tireless energy, a ruthless streak and a volatile temper, he was also a hopeless music fan. A New York Jew and a vehement atheist, he found his musical home in the Deep South, in studios in Memphis and Muscle Shoals, Ala., among Baptists and Methodists, blacks and good old boys.[5]*

He was a living contradiction who used societal challenge and personal tensions to create pure magic.

Independent producers typically begin their careers learning the trade by producing demo sessions for publishers and custom or vanity albums that are commonly nonlabel productions paid for by the recording artist or band. Some are demo recordings that showcase the recording artists or songs, which are then "pitched" to a record label. Great demo recordings often lead to the next step in the career of a producer: working for a label in the A&R department, recording development deals.

Record label staff producers are hired by labels to produce development deals for acts. The label pays for all expenses, including the studio, musicians, and vocalists. The producer gains additional opportunities to work with better musicians, artists, and studio equipment. If the tracks are perceived as sellable, exciting, or quality, labels may sign the act and release a single. If a label passes on the act, the cost of the development deal is commonly used as a tax write-off.

Major independent producers are at the top of their profession and are paid handsomely. Extremely successful producers earn $1,000–$6,000 per side act they are producing. In addition, they receive producer points (usually 1%–3%) of the artist's royalties on each unit sold. Major producers are highly respected for their abilities to produce hit records. Their success makes them behind-the-scenes stars within their own industry. Because major producers gain so much power and respect in the industry, they control their destiny as clever entrepreneurs by letting labels and artists bid for their services. Major labels often hire very successful producers as executives to run a label and use and protect their talents instead of competing with them. Makes sense, doesn't it?

Musicians

Musicians change our world; life would not be the same without them. All musicians are artists, as it doesn't matter where they play or how much money they make. Although the recording artists usually get the fame, it is the musicians who create the tracks and "feel" in the studio who are the real heroes. Without them, singers are poets—so who are these people who might make someone a lucky star?

Garage/jam band musicians are usually the high school students who jam to learn cover tunes of popular acts. This often becomes their first step toward a professional career in the industry. Most musicians volunteer to perform in small clubs, coffeehouses, churches, and other types of local gigs to learn the business and improve their performance skills. Many continue to improve their performance skills and knowledge during college by jamming with others who share their interests and taking private lessons from classically trained musicians. Others simply learn how to make recordings on their computers using GarageBand, Pro Tools, and Logic Pro software programs.

Working musicians receive money for their music performances. Serious musicians join unions and audition to become members of various levels of performance groups, including orchestras, bands, and other types of entertainment ensembles. Many use their musical skills to supplement other types of occupations. Others continue to struggle either part-time or full-time to develop a "sound" and a "following" that may lead to greater financial opportunities through their creative musical performances.

Road musicians and event artists are the musicians and vocalists who are on tour as an opening act or as the musicians hired to perform with a recording artist. Others are often classically trained musicians who form smaller groups to perform at special social, political, or other types of events. It is very rare for the road musicians of a major recording artist to be on the actual hit recordings they are playing live. That is the job of studio musicians. Road musicians' salaries are related to the stature of the act they're backing. Major star, more bucks. They can make $1,000 per show, plus per diems that cover hotel, food, and travel. Known acts or over-the-hill artists generate rates closer to a few hundred dollars per show. What road musicians have to consider is what the cost of going on the road will be relative to how much other work—that is, studio or solo work—might be missed and how to establish a pay rate that reflects what their talent and time are worth to the act. They have to put on their entrepreneurship hat to make the deal and not be swayed by the potential excitement of going on the road with the act. Basic union (AFM) scale is $200 a day plus $20.00 for the pension fund. Rates vary depending on the "local" you're a member of. As an example, in Nashville, the road scale is $200 plus the $20 pension for a total of $220. In Los Angeles it is not listed, and thus it varies depending on the local the musician is a member of. However, the union has a lengthy pack of bylaws, which sets rates and scales on all types of performances in various types of live and recorded media.

Studio musicians are most often heard on major recording artists' recordings, yet very few consumers know them by name. Great studio musicians may make hundreds of dollars in a three-hour recording session and more than a couple thousand a day, charging double or triple. They often become true stars known only by the best producers and recording artists. They are recording artists themselves, but are rarely known by the public. Studio musicians play their instruments to highlight the marriage of the song to the vocal characteristics of the recording artist; plus they somehow emotionally play the instrument in a unique expression, reflecting their emotions, personalities, and life experiences. Some of the best musicians in the world are studio session musicians, so of course, most record labels and recording artists use master session studio musicians for their album projects. Recording opportunities include music soundtracks for TV shows, movies, commercials, jingles, and major artists' recordings. Possessing musical talent is imperative if you want to be a recording artist or studio musician. In addition, a music education and an understanding of copyright law, music publishing, networking, and the music business are important for success. As should be clear by now, there are times and places to go "solo"; at other times, you have to be part of a team, and when it comes to the business of entertainment, you must also be affiliated with professional groups who work to protect your interests as an artist and an industry professional.

Musicians Union—The American Federation of Musicians

The American Federation of Musicians represents about 90,000 musicians in the United States and Canada for recordings and live performances in negotiated agreements for studio and other recordings, TV, films, jingles, concerts, stage shows, symphony, opera, and ballet. Major record labels are signatories to the American Federation of Musicians' Sound Recording Labor Agreement (formerly the Phonograph Record Label Agreement), which governs the wages, benefits, and working conditions of all its members, including studio musicians. According to information posted on their homepage, www.afm.org,

In the mid-1800s musicians in the United States began exploring ways to improve their professional lives. They formed Mutual Aid Societies to provide members with loans, financial assistance during illness or extended unemployment and death benefits. A number of these organizations became early unions serving various constituencies, but problems arose between them due to competition. In 1896, delegates from these organizations gathered at the invitation of American Federation of Labor (AFL) President Samuel Gompers to organize and charter a musicians' trade union. A majority of the delegates voted to form the American Federation of Musicians (AFM), representing 3,000 musicians nationally. They resolved: "That any musician who receives pay for his musical services, shall be considered a professional musician." Within its first ten years, the AFM expanded to serve both the US and Canada, organized 424 Locals, and represented 45,000 musicians throughout North America.[6]

Membership advantages include a retirement fund, health insurance, and AFL-CIO benefits. The AFM uses franchised booking agents, who usually work for 10% commission to connect union musicians with people who want to hire them. Most of the agents represent musicians and recording artists locally and nationally. If you are a musician playing gigs and do not want to get stiffed (play the gig and not get paid), representation by an AFM licensed agent will usually prevent that type of problem.

Audio Engineers

Audio engineers mix the creativity of music and the artistic capabilities of the artists, musicians, and background vocalists to the logical, physical, and technological potential of the studio's acoustics and equipment to capture (record) the best performances. Suggested skills and knowledge include a basic understanding of the business of music and entertainment, music theory, copyright laws, music publishing, marketing, management, and business finance. Knowledge of computer software programs Pro Tools, Logic Pro, and even GarageBand is important as the industry moves toward less expensive modes of recording to save money and improve the profit and loss statement. Basic electronic, acoustics, and recording equipment varies by studio, and having a continuous desire to learn the latest technical advances in recording equipment and audio production is fundamental. Communication skills and the ability to work with highly creative individuals are crucial. Business courses are also helpful, as most audio engineers often become producers, and both are entrepreneurs. They also need to be able to have fun, experiment, be a little crazy, bend the rules, and do whatever it takes to get the best creative performance possible out of the artists. In their book *Come Together: The Business Wisdom of the Beatles*, Courtney and Cassidy state,

> George Martin (the Beatles' producer) brought a new engineer, Geoff Emerick, into the control room. . . . Emerick had raised thought the EMI ranks quickly and was young enough to have a healthy disregard for the "rule," a fact that endeared him to the Beatles. Emerick was not fazed when John Lennon told him that his voice should sound like the Dalai Lama singing from a mountaintop. Run the voice though a rotating speaker cabinet (normally used for an electronic organ). Problem solved. He thought nothing of flipping the tape around and recording a guitar solo with the

tape reversed, so that when the song is played forward, the gui-tar solo is played backwards (this is a distinctive and now easily recognizable sound, in which each notes fades in and ends with the "attach" of the plectrum). He was willing to cut a tape into little pieces, glue it all back together in random pieces, and add it to the final mix if that was what the Beatles wanted.[7]

Most lease a recording studio, computer, and software, and hire a music publishing company, and *some* own their own independent record labels. The careers of the most successful audio engineers start with basic lessons in recording or on their own computer, recording themselves and others. Careers often follow a similar trajectory from entry level, perhaps in an internship in a music business or business program that leads in time to becoming the head engineer or an entrepreneurial opportunity to own your own studio. The typical job title, experiential path, and requisite skills are these:

- *Entry-level audio engineers* are often college students who are working in a studio as an intern. At the same time, they are recording local musicians on the latest computer recording software. They are rarely paid for their efforts; however, they are given opportunities to learn the basics and to meet members of the creative team.
- *Second engineers* are a step up in the process from entry-level engineers. They set up microphones, cables, headsets, the console, and alignment of the tape machines. Second engineers also run the tape machines, keep the log sheets during the sessions, and occasionally travel with artists on the road to hone audio and mixing skills.
- *Staff engineers* are sometimes employed at a recording studio depending on the type of the work in which the studio specializes. They accomplish any level of session, from demo to master, with major recording artists and triple-scale musicians. Staff engineer positions are commonly found in nonmusic recording centers, where great engineers are difficult to find. Annual salaries range from $50,000+ to huge sums for engineers with great track records.
- *Major-artist/independent engineers* are at the top of their profession, with a long list of hit records they helped create.

They are "on-call" for and most often work on master sessions for label artists and known superstars. Their pay is negotiable, ranging between $1,000 a day, plus expenses, to huge per-side (whatever they want) salaries based on their track record. It is very rare for audio engineers to also receive points from the artist's points; however, in very special situations, it has happened. Most studios located in major music recording centers (Los Angeles, New York, and Nashville) have replaced staff audio engineers with independent engineers. The location for the production has become less important as many artists have studios in their homes or vacation homes, and many use computers and software to record all their tracks instead of renting a major studio.

Recording Artists (Royalty Artists)

Royalty artists are usually signed by record labels in work-for-hire deals for the specific purpose of creating recordings. The artists are paid a percentage (a royalty) of the profits from the sales of the product. The royalties are called points and a point is a percentage of the suggested retail list price. I know this sounds a little crazy, but the points are the percentage of your income as determined in your deal. Everybody wants a piece of the royalty artist's royalties, so the artist needs to be careful in signing the initial deal as the producer may want a percentage or points. The amount will not come out of the label's profits; rather, it will come out of the royalty artist's points. If you have an artist manager, they may want a percentage of how much you are being paid, and the same with an audio engineer, to whom you should never pay a percentage. Royalty artists usually sign work-for-hire deals for specific recordings. However, in 360 deals, the artist is also responsible for other revenue-generating work, such as concert tours, promotions, and endorsements. The label retains the copyright of the artist's musical and vocal performances and, thus, the artist gives up ownership (the copyright) of his or her performances.

Record deals are just weird when compared to a typical bank loan, where you have to prove you can repay the debt or put up something of value to guarantee it. On the front end of a recording deal, the label acts as a bank, providing financial security to pay for the artist's recordings, tours, and advances. However, the artist never owes the label any money! It is the label that has to pay off the loan (to itself) by applying the artist's royalties from each unit sold (in the old days before peer-to-peer [P2P]), and, now, a combination of royalties and percentages of total net income from corporate sponsorships, merchandise, and touring. Therefore, it is the label that signs the artist, finances the recordings, and markets and sells the act's products and shows to recoup expenses and split profits. What does it cost the artist? It depends on the deal, percentages, and support by all label employees to enhance and aggressively sell the act through their image and branding. How well does it work? That depends on how excited the fans get and how much money they throw at the act's products and shows. And nobody knows what that might be until it actually happens. Welcome to the emotion business! Spend millions to create, promote, market, and sell the act and their products and within a week or two, everyone knows if they'll hit or not. Welcome to the odds and the financial risks.

Virtuosos are celebrated as the best in the world based on their musical and vocal talents and fame. Examples of virtuosos are opera singer Luciano Pavarotti (1935–2007), violin virtuoso Itzhak Perlman (b. 1945), and the classically trained singer Andrea Bocelli (b. 1958). All these artists have been highly sought after to make recordings, both in live performances and through studio recordings that sell to niche markets around the world.

BGVs (background vocalists) provide the harmony in recording sessions and live performances. It is amazing how often background vocalists are used to blend their voices with the main star to improve or "carry" the superstar act.

Professional singers are found everywhere: on recordings, in choirs, advertisements, jingles, and movie soundtracks;

you name it—they are there. Professional singers (vocalists) frequently net more money than label artists, as they do not have to support an entourage of road musicians, managers, producers, and label marketing and promotion executives.

Demo singers find entry-level work at publishing companies, singing demo recordings. Nonunion members and college interns may earn $10–$50 a song, which is not much money, but singing in a recording studio is different from performing in a choir or on stage. Demo sessions provide the novice studio singer with some valuable studio experience. The acoustics, lighting, and monitoring are often quite complex. So, the more experience gained in the studio, the quicker an artist may become a professional recording artist. There is also another advantage, as producers, A&R, and artist managers often listen to the demo tapes, hoping to find a great song. Sometimes, they also find a great voice! Therefore, singing on demos is a great way to gain studio experience and be heard by many industry insiders who are seeking the "next big thing."

Singers Union—The Screen Actors Guild/ American Federation of Television and Radio Artists

As we saw earlier, it is useful to be part of a larger entertainment professional organization that will enable you to follow major changes in the industry and connect you to others within your field. The Screen Actors Guild merged with the American Federation of Television and Radio Artists on March 30, 2012, when more than 80% of both unions membership voted in favor of the merger.[8] According to the SAG-AFTRA website,

> *This formed the most powerful union ever for more than 160,000 actors, announcers, broadcasters, journalists, dancers, DJs, news writers, editors, hosts, puppeteers, recording artist, singers, stunt performers, voiceover artists and other media professionals.*[9]

The American Federation of Television and Radio Artists has represented recording artists in studio sessions, television productions, and live performances since the late 1930s.[10] It also represented actors, announcers, sound effects artists, and other personalities working in radio and television. Representation was focused in six areas: sound recordings, broadcast news, commercials, industries, and new technologies (non-broadcast industry films) and entertainment programs (television).[11] About 600 companies (including all the major labels and music companies) signed The Code of Fair Practice for Phonograph Records.[12] Accordingly, the SAG-AFTRA webpage states,

> SAG-AFTRA represents approximately 160,000 actors, announcers, broadcast journalists, dancers, DJs, news writers, news editors, program hosts, puppeteers, recording artists, singers, stunt performers, voiceover artists and other media professionals. SAG-AFTRA members are the faces and voices that entertain and inform America and the world. With national offices in Los Angeles and New York, and local offices nationwide, SAG-AFTRA members work together to secure the strongest protections for media artists into the 21st century and beyond . . . actors, announcers, broadcast journalists, dancers, DJs, news writers, news directors, program hosts, puppeteers, recording artists, singers, stunt performers, voiceover artists and other media professionals. Our work is seen and heard in theaters, on television and radio, sound recordings, the internet, games, mobile devices, home video: you see us and hear us on all media distribution platforms. We are the faces and the voices that entertain and inform America and the world . . . SAG-AFTRA is committed to organizing all work done under our jurisdictions; negotiating the best wages, working conditions, and health and pension benefits; preserving and expanding members' work opportunities; vigorously enforcing our contracts; and protecting members against unauthorized use of their work.[13]

As such, SAG-AFTRA works to protect all aspects of the entertainment industry, both seen and unseen by the public. One of the major environments for such a group is found in the studios, which are key to all entertainment production.

Studios

To make money, labels and music companies invest money into their signed artists and hire a recording team to create a sellable recording. But they've also got to find a place to do the work, a recording studio for music, a film production studio and set for a movie, and a live stage for theater. Traditional studios tend to be earmarked by their acoustics (sound), equipment, cost, and reputation for making hits. In the past, these studios were an integral part of the music business: the place where dreams were made and sometimes broken. In general, traditional studios fall into six categories:

- *Master studios*—most master studios are solidly packed with the latest acoustic audio equipment and special effects toys producers use to create great tracks.
- *Project/mini studios*—these are mainly for a low-budget, limited pressing master tape, or a movie soundtrack; as such project studios cover the gap between the very expensive master studios and the basic demo studios. They are often filled with dated equipment and minor acoustics.
- *Postproduction studios*—after overdubbing sessions, producers and artists often move to the special equipment in the postproduction studio to mix the master tapes to computer flash drives.
- *Demo studios*—these studios are often the older analog studios with minor special effects equipment. They are used for demos as well as vanity and limited budget recordings.
- *Garage/basement studios*—these are the bottom-line basic studios often found in homes of young musicians and producers.
- *Dorm rooms and bedrooms*—it's hard to believe that you can use a dorm room or bedroom as a "studio," but with Logic Pro, Pro Tools, and other computer software programs, the need for the acoustic and equipment of a major studio are diminished.

Acoustics

Most music companies still use traditional studios on some or all of their artists' albums, probably because the talents of the recording artists, musicians, audio engineer, and producers combined are still only as good as the studio's equipment and acoustics. Poor acoustics can destroy even the best creative efforts of world-class artists. Floating walls, ceilings, and floors are used to isolated sound from room to room in performance studios and control rooms. Grooves are cut into the floors, walls are mounted on rubber tubes, and ceilings are often spring-loaded. Windows between the performance rooms and control room are often made of two thick double- or triple-pane pieces of glass separated by several inches of air space. Soundproof doors complete the package. Studios are usually divided into the following:

- *Performance rooms* are where the royalty artists and musicians perform.
- *The control room* is where the producer and audio engineers supervise and control the quality and amount of the recorded signal.
- *The equipment room* isolates the noise and heat generated by tape recorders, amps, computers, room equalizers, and most of the other electrical equipment needed in the traditional recording process.
- *The storage area* is where microphones, music stands, headsets, cords, direct boxes, and other recording gear are stored.

Aesthetics

While musicians and vocalists create the music in the performance studio, the audio engineer and producer enhance the quality of the sound in the control room. They may use tone controls (equalizers) on the console to improve the quality of the signal. Just as colors of paint help create emotions in a piece of art, engineers use acoustics, microphones, and consoles to emphasize and match the "feel" of the song to the image and vocal characteristics of the recording artist.

Microphones

Sir Charles Wheatstone (1802–1875) first used the term "microphone" in 1827, but it wasn't actually invented until 1876 by Emile Berline (1851–1929) and used as a telephone voice transmitter at the U.S. Centennial Exposition.[14] Mics convert sound waves into electronic signals. They all sound different, have different axes for picking up sounds, and vary in sound quality. Thus, it is extremely important that engineers know which ones to select, where to place them in the performance room, and what constitutes the correct proximity between the talent and the instruments. Proper selection and usage are often the difference between a "great" sounding session and something less. Basic differences, as controlled by the microphones, include the following:

- *Dynamic mics* have a coil or wire in the element (top) of the mic. Sound waves vibrate a plastic diaphragm connected to a coil of wire, moving the coil in and out of a permanent magnetic field. Electricity is created anytime you pass a wire through a magnetic force field. Thus, the movement converts the sound waves (acoustic waves) into an electronic signal.
- *Ribbon mics* operate on the same principle as dynamic microphones, except they have flexible, metallic ribbons (or foils) on which sound waves vibrate in and out of the permanent magnetic force field. Because the ribbon is flexible, the quality of the sound generated is often considered smoother and warmer.
- *Condenser mics* have two plates that hold a static electronic charge. One is a permanent plate and the other is moveable. Sound waves move the moveable plate, changing the distance between the two plates, which generates an electrical output. Condenser microphones need batteries or a phantom power source supplied through the microphone cable.

Pickup Patterns

Microphones have different pickup patterns or areas where they are most sensitive to sound. The patterns are used to acoustically isolate one instrument or vocal, excluding the leakage of sounds from other instruments or vocals into the microphone. Engineers also use pickup patterns to induce an acoustic mixture of "live" sound (omnidirectional) from a very isolated single sound source. Consumers often feel the quality of the mixture of the sound as an emotion when they hear the recording, but are largely unaware of the technical skills and training required to produce the sound that appeals to their emotions. Here are examples of sound pickup patterns:

- *Omnidirectional pickup patterns* allow sound waves to enter from all directions into the mic at approximately the same loudness level (measured in dBs).
- *Cardioid pickup patterns* cancel the sound coming from the sides and rear of the mic. Most engineers use mics with cardioid patterns to limit and avoid leakage problems that usually occur when using omnidirectional mics.
- *Bidirectional pickup patterns* receive sound waves from the front and back of the mic and cancel out sound waves coming from the sides of the mic.

Consoles

Consoles often appear to be straight from the bridge of the starship *Enterprise*. Consoles act as traffic cops, dividing and directing the microphone signals and connecting instruments to various destinations, all at about 186,000 miles per second. Each plugged-in microphone is one signal controlled through one module in the console.

Recorders/Software

Analog tape machines record a magnetic copy of both the amplitude (loudness) and frequency (number of vibrations per second) sent through the console to the tape recorder. The advantages of analog tape recordings are a "warmer sound" at a lower price. However, distortion of the signal and unwanted noises like tape hiss may occur with analog tapes. Digital machines, on the other hand, convert the microphone electronic signals into binary codes of zeros and ones (pulse or no pulse). They use recording tape only to store the binary code. When the codes are played back, the computer inside the tape machine converts the binary code into the music previously recorded. The advantages of digital recording include the lack of noise and distortion that are usually found in analog recordings. However, digital machines cost much more than analog machines (although great analog machines are becoming a rare commodity). Nowadays, computer software programs, such as Pro Tools, Logic Pro, and even GarageBand, are replacing analog and even older digital machines. These software programs sound great and cost very little compared to an actual physical studio (Garage-Band is often free when you buy a Mac computer).

Speakers

In the control room, the electronic signals (from the microphones to the console) are converted back into acoustic sound waves by monitors (speakers). To make sure they are providing a correct sound (flat frequency response), room equalizers are used to compensate for the "hype of the sound" created by the speakers and control room acoustics. As an example, control rooms are shot with "pink noise," which measures the frequency response to determine how to use the room equalizers to compensate for the acoustical problems.[15] The electronic signals coming into the console from the microphones, tape machines, effects, and computer software programs are amplified and divided into different frequencies (bass, midrange, and treble) by a crossover network (often in the speakers) and then sent to the corresponding bass, midrange, and treble speakers. Large speaker cones vibrate slowly and move a significant amount of air to reproduce bass frequencies. Smaller cones vibrating faster generate higher tones (frequencies). Tweeters reproduce the highest frequencies, adding clarity to the quality of the music or sounds you're hearing. The bass or lower frequencies have longer and more powerful wavelengths, while the highest tones are shorter and much more directional. Thus, room acoustics need to be correct in order for the audio engineer and producer to hear the signal correctly as they mix the tracks.

Effects

Effects equipment called "outboard gear" are devices used to improve, change, or fake the quality of the signal and overall "feel" consumers might perceive from listening to the signal (recordings). Of course, most of these "effects" are now generated within software programs. It is important for you to familiarize yourself with these tools as they help create the "desirability" of the product for the consumer. Among the commonly used tools and techniques to generate effects we find the following:

- *Compressions* that are used to reduce the amplitude (loudness) of a signal to record a proper signal on analog tape. Compression is also used to change the quality of the sound of instruments and vocalists. Settings include compression ratios and attack and release times used to fatten or tighten the quality of sound (the sound of a kick drum or bass guitar perhaps).
- *Echo* is used to make the singer or instruments sound as if they were in a different size room than the actual recording studio. Long reflections of 50 milliseconds or greater are perceived as a repetition of a direct sound wave. Think about singing in the shower: the "echo" sound created makes everybody a singer . . . except me!
- *Gates* are switches (set on a threshold) used to stop signals based on amplitude (loudness). Gates are used to block noise (other instruments) from leaking into microphones during the actual session. They are often used on drums and overdub vocals to "tighten" the quality of the sound.
- *Harmonizers* offer hope even for me when it comes to singing! Along with changing or correcting the pitch, they can electronically double (with a slight delay) one vocalist or instrument to make it sound as if there were two. The delay settings can be varied between echo and reverberation, which differ in time settings.
- *Limiters* compress the dynamic range of a signal to its maximum compression. The amplitude is "limited" to the

maximum settings, which tend to make the signal sound "thinner."

- *Slap* is a delayed sound perceived as a distinct echo, usually set for a delay of 35–250 milliseconds. Think of the early "rockabilly" recordings of Elvis (1935–1977), Carl Perkins (1932–1998), and Johnny Cash (1932–2003), and you'll know how producers used it in the 1950s. It used to be created by recording a new vocal out of sync with some previously recorded vocals (the delay caused by the physical distance between the tracks coming off the playback heads and the new track being recorded on the record head).
- *Reverberations* are a delay of ten milliseconds or more after a direct sound wave. Engineers use it to make the vocalists or musicians sound as if they are in a large auditorium instead of a small, acoustically controlled recording studio.

Studio Schedules

In major recording cities like New York, Los Angeles, Nashville, Memphis, Miami, or Seattle, sessions are "booked" in three-hour blocks scheduled from 10 a.m. to 1 p.m., 2 to 5 p.m., 6 to 9 p.m., and 10 p.m. to 1 a.m. This allows producers, musicians, and vocalists to move from project to project during the day by going from one session to another as all the studios are working on the same hourly schedule. The hour between sessions is for tearing down the current session and setting up the next session.

The Recording Process

It's time to create the magic in the studio that allows consumers to feel goose bumps. Producers are in charge of the recording processes, budget, and material—the songs the artist sings. They use their years of experience, musical knowledge, and opinions about "what" the public wants to hear to produce acts. In addition, producers often have a personal communicative talent that helps them encourage a maximum creative performance out of the singers and musicians individually and at the same time, together, to create the magic or goose bumps we get when we hear the final recording. Not an easy thing to do! Producers have to select the perfect song for the perfect artist and somehow encourage all the musicians and singers to create a final master recording that is superior to the original demo. Creative people can be weird and are often very difficult to work with, so the producer also has to maintain control of the session or it may turn into something resembling a circus, costing the label thousands of dollars per hour. According to record producer Jimmy Bowen,

> *If you're going to be a record producer, you must have an overall knowledge of the music industry. From a business standpoint you must, at all times, know where the industry stands. . . . for example, if you're about to meet with a label head to discuss producing an act... if you know what's going on in the state of the industry, you're not going to blow yourself out of the deal by coming in too high (asking for too much money).*[16]

Preproduction

Bowen claims there are five basic steps in preproduction for recording an act: (a) finding the artist, (b) researching the artist, (c) finding the material (songs), (d) hiring the musicians, and (e) setting up the session.

> *You've got to have an artist to work with; that is the first thing. When you're first getting started you will probably work with artists who you wouldn't work with later on, but more than likely, you'll both be at the same stage.... You'll produce some artists you probably shouldn't. You'll cause some marriages to happen that aren't the ideal marriages, but you must get some experience. When this happens, make sure you always do the very best job you can. If you realize that it's a super bad marriage, get out of it as gracefully as you can. Save the artist's time, save the artist's money, save your own reputation. You always want to maintain a business approach. If it doesn't work you're sorry it didn't work, but you'd rather not waste any more money.*[17]

Marrying the right song to the right artist is essential. The two must fit together in a way that allows the melody and lyrics to reinforce the perceptions consumers have of the artist's image. Bowen continues:

> *The first thing I suggest you do with an artist, when you first get involved with producing a specific artist, is to get a discography of that artist (a copy of all their recordings over their entire career, even the bad stuff). The artist has known themselves all of their life, you haven't! Don't try to come in after the artist has been singing for 12–14 years, working on their craft and hope you're going to jump on the train and ride a winner. Go at it seriously. . . . There's so much you learn from researching the artist. You learn what not to do and you learn what to do. It's like anything else. A doctor wants an x-ray to try and figure out what's wrong before shooting medicine into you, at least a good doctor does. Some of them write prescriptions over the telephone. Beware of the doctor who does that. Beware of a producer who's ready to go to the studio a week or two after he's met the artist or gets the deal. He's looking for a miracle to happen; he's hoping it will all work out.*
>
> *. . . Being a producer, you've got to be a psychologist, not a psychiatrist, but a psychologist. You've got to dig into the artist's head. You've got to find out what makes an artist tick, what makes him give 100% sometimes, 80% sometimes, and 20% sometimes, or you're going to be looking for luck, hoping the night you booked the studio you're going to get 100% and you may not get it. . . . Have a deep relaxed discussion with the artist, read your artist and find out what environment will let them open up, relax, and have the best place to have a rap with you. If it's in a bar over a beer or at his house, go do it. If it's at your house, do it. If it's at the studio late at night go do it. Wherever. Find a place where that artist is more comfortable, where he'll open up to you, so he or she will talk to you, so you can start to read and understand them.[18]*

Producers are required to find and select the right songs for the artist to record. According to Bowen,

> *You take a hit song if it comes from an ex-wife. You take a hit song if it comes from your biggest enemy. You owe that to the artist. Your job is to make hit records. He's paying you. It's his life and his money. Go out and hunt those songs. Anywhere you get a hit song is great. The obvious place to look is publishers and you*

need to have a good rapport with them. We look for material all the time. . . . When I say the hunt is on, that's just what I mean. It's not going to come to you very often. You've got to create the relationships before the great songs are brought to you. You've got to go out and make yourself known. . . . Thirty days before you're going to record an artist, put the word out. Then the songs flood in. The problem is that when you're going though songs (the last few weeks before a session) there's a tendency to take the best of the worst. So, have a big, long listening session for songs, cull out of 50 about 6 or 7. Get away from them until the next day and then go back and listen to those 6 or 7. You'll probably throw away 3 or 4 of them. . . . Then play the songs for the artist. . . . With the great songs, you'll both just light up. You'll light up, the artist will light up. The great ones, you don't miss them. If you miss the great ones, then, I'd suggest plumbing or electrical work, because you aren't going to make it. . . . Also, when the hunt is on research old songs. Don't eliminate the possibility of 20 or 30 years of success.[19]

The quality of the studio musicians, the recording artist's performances, the production, and the engineering all contribute to the final *sound and feel* of the recording. Bowen claims that

It's like casting a movie . . . if you put together a bad cast you're going to have a dog movie. . . . Talk to the artist about the musicians that he or she likes or doesn't like. Never cast a session and have a musician on it the artist has a bad vibe with. You're just putting a negative into your music that you don't need. . . . The whole trick is to figure out everything around the artist so when you go to tape, to make your record, you've got everything at max. You want the musicians at max; you want the artist at max. You want everything at its best. The ones that break through are the magical records that happen that way.[20]

Basic Tracks

Are you ready to lay down your first track in the studio? Do you have your artistic and technical team in place? Let's get to it! It is going to take a lot of time and effort, so will it be worth it in the end? That is up to you!

The first stage of a recording session in the studio is called *basic tracks* or *tracking*. The rhythm instruments and scratch (rough or practice) vocals lay the foundation for the rest of the instruments and vocals, which are cut later in their final versions. Session instruments can include drums, bass guitar, piano, and electric and acoustic guitars. The recording artist adds a reference or scratch vocal for the musicians to listen to while they play. It is resung and rerecorded later as a finished vocal. The session process involves the musicians and vocalists listening to the demo recording of the song, tuning up, and practicing the song at least a few times to prepare for the recording. Many of the studio musicians in the music centers of Los Angeles, New York, and Nashville learn, adapt, and contribute their musical creative talents to the essence of the song in just a few minutes. Basic tracks are recorded once the musicians and vocalists are ready, the audio engineer has set the microphones, monitors, headphones, and equalization and effects levels, and the producer is satisfied with the quality of the sound and musical performance during practice.

Figure 7.1 The first step in the recording process is the performance of the musicians and vocalist in the studio converted by microphones or directly into electronic signals that are sent to the board (or console). At the console the signal is split into five feeds or "sends" to (1) the 24-track tape recorder or computer hard disk drive, (2) the effects (outboard gear), such as compressors, limiters, reverb, and pitch controllers, (3) the monitor mix of the tracks sent to the amp and control room speakers, (4) a safety copy backup machine, and (5) back into the headset of the musicians and vocalist in the studio.

Basic Tracks Signal Flow

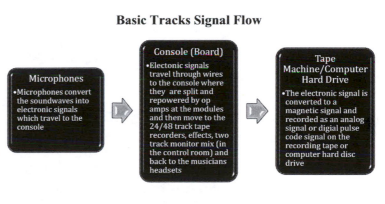

Microphones
•Microphones convert the soundwaves into electronic signals which travel to the console

Console (Board)
•Electonic signals travel through wires to the console where they are split and repowered by op amps at the modules and then move to the 24/48 track tape recorders, effects, two track monitor mix (in the control room) and back to the musicians headsets

Tape Machine/Computer Hard Drive
•The electronic signal is converted to a magnetic signal and recorded as an analog signal or digial pulse code signal on the recording tape or computer hard disc drive

Recording the right amount of signal on the tape is difficult. Too much signal will distort the tape. Too little signal will cause the playback to be noisy. Digital tape recorders and computer programs record a greater dynamic range, which solves some of the engineer's distortion and noise problems. If most of the musicians play poorly, the recording process is repeated. When the producer approves the basic musical tracks, the minor musical problems are usually fixed later. Logs are completed, which are written notes about the songs being recorded. Notes include length of the song, problems that need to be corrected in the overdub stage, and equalization and effects settings. Logs are stored in the box with the master tape. They are used in later recording sessions to alert the engineer and producers of necessary issues. Once the song has been recorded, everyone listens during tape playback for mistakes and opportunities to improve their performances. If recorded properly, the "take" becomes a "master."

Signal flow in basic tracks from the mics in the studio into the console is split and sent to the 24-track tape machine or hard disk drive (computer); the monitors in the control room (speakers); the musicians' headsets (earphones); the effects (add echo, reverb, pitch control to signals); and the two-track master tape machine or computer. Playback (to listen to the recording) reverses the signal flow from the tape machines or computer hard disk drive and allows the producer, audio engineer, musicians, and vocalists a chance to check their creative work.

Overdubbing

Adding instrumental parts and vocals to the previously recorded master tape is called overdubbing. The prior tracks are saved and the tape machine is placed into sel-sync (short for selective synchronization). Invented by guitarist Les Paul (1915–2009) in the late 1940s, sel-sync turns the record heads on the previously recorded tracks into playback heads. This allows the new instruments and vocals to be recorded in time with the instruments and vocals that were first recorded.

Signal flow in overdubbing is a combination of basic tracks and playback. The signal is played off the record head in playback and sent to the console, where it is split and sent to the monitors, effects, two-track/DAT, and through the cue system to the musicians' and recording artists' headsets (the signal is converted from an analog or binary magnet signal into an electronic signal by the playback heads and then sent to the console). Once the musicians and singers hear the signal (their previously recorded tracks), they play or sing their new tracks. The originals recorded in basic tracks are saved and the new tracks are added in sync with the previous tracks

Figure 7.2 Once the producer is satisfied with the "take" it is time to add additional musical instruments and vocalists onto the master tape or master Pro Tools tracks. The playback signal comes from the record heads (in the record head stack) turned into playback heads to produce a playback signal that is sent to the console. Once the musicians and vocalist hear the tracks being played back in their headsets, they play or sing along with the tracks in their headsets. So what is really happening is the signal from the previously recorded tracks is generated from the source (tape machine or computer) and sent to the console and then on to the musicians, who play along with what they are hearing. At almost the same time the new signals being performed by the musicians and vocalists are sent to the console and then onto the effects, new tape tracks, or computer control room monitors, and mixed back into the artists' headsets. The delay between the previously recorded tracks and the new tracks being recorded is so fast our ears never hear it.

Recording Overdubs

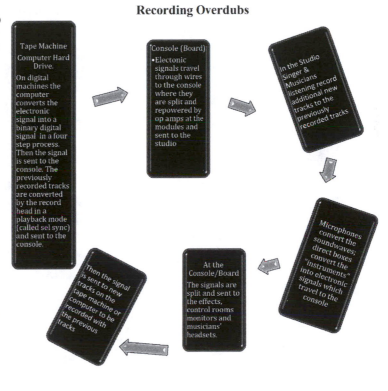

(microphones convert the artist's music into electrical signals, which travel back to the console to be split and sent to the monitors, effects, and two-track/DAT and are recorded as new tracks on the 24- or 32-track tape machine). "In sync" means that the record heads (on tape machines) are turned into playback heads, so that all tracks are played back, and the new tracks are recorded at the same time. Thus, the overdub stage is a combination of playback and recording at the same time.

Mixdown

The last stage in the recording process is mixdown. After all the instruments and vocals have been recorded and overdubbed, the 24, 32, or whatever number tracks are mixed down to 2 tracks (stereo): left and right. Producers and audio engineers mix the tracks according to the "style" of the music genre (rock, hip-hop, blues, jazz), create stereo images (using pan pots, which are switches and faders on the console that move the signal from left to right or anywhere in between), and create 3-D depth (by acoustic positioning and pre-echo signals correlated to a dry signal). Of course, the mixdown process is highly creative, and often depends on the perspective of the producers. Some labels use different producers to mix the same track, and then select the one they want to release. Album cuts and dance mixes are sometimes different from the version of the song released to radio. Producers may even change the amount of compression on vocals and instruments for an AM radio release compared to an FM or digital release, as AM radio is primarily midrange, missing much of the bass and higher treble. Digital radio solves these problems because the signal is a binary code, and the music is generated by the computer chip in the receiver. The mixdown process is shown in Figure 7.3.

"Signal flow" is the converting of the magnetic signal stored on the tape of the analog or digital tape machine or hard

Figure 7.3 The final mix is the process of mixing the 24 or 48 tracks to a two-track master.

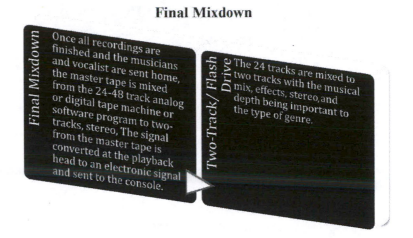

Final Mixdown

Final Mixdown

Once all recordings are finished and the musicians and vocalist are sent home, the master tape is mixed from the 24-48 track analog or digital tape machine or software program to two-tracks, stereo, The signal from the master tape is converted at the playback head to an electronic signal and sent to the console.

Two-Track/Flash Drive

The 24 tracks are mixed to two tracks with the musical mix, effects, stereo, and depth being important to the type of genre.

drive into an electrical signal. It is sent (op amps) to the console, where the producers and audio engineers may redirect it to effects equipment (reverb, echo, pitch, slap) and will finish by equalizing the signal (e.g., changing the frequency response, bass, treble), panning (stereo positioning), editing, and mixing.

Mastering

If the recording team of musicians, recording artists, audio engineers, and producers has done its job, the creativity of the recording session has been "captured" on tape or computers. The final mix, stored as an email attachment, two-track analog or digital tape (DAT), or on a flash drive, is sent to the mastering lab to be processed, and then pressed into CDs or released for digital downloads. The Moving Picture Experts Group developed compression systems to move digital video (movies and video) and digital audio data (the audio for the movies and video) quickly by reducing (compressing) the amount of digital information. The music industry adopted this format to digitally download recorded songs as MP3s.

Computer Programs

To the detriment of the recording studio business, many artists now forgo the cost of major recording studio sessions in favor of computer-based software. Great tracks are great tracks, and most consumers do not care if they were created in traditional studio sessions, costing hundreds of thousands of dollars, or in a dorm room for a couple of pizzas. Years ago, the only way to record great artists and end up with goose bump–inducing tracks was in recording studios. Not today, as computers and software technology have provided almost anyone with the tools they need to create great tracks without the previously required studio sessions. Some major artists are recording only basic tracks in studios, reserving the overdub and mixing stages to be edited with computer programs at their leisure. They can even email their vocal recording to producers, who remix the final tracks.

Technology has caused financial problems for many studios, musicians, and recording artists. However, it has also launched many creative opportunities for artists who, before this technology, could not afford the major studios, musicians, and singers. Software has leveled the playing field, and almost anyone can create tracks consumers might enjoy. However, once great recordings have been created (in a studio or on a computer), consumers still have to "discover" them before they know if they like them. Who can accomplish that? The answer is *you* if you know how the industry works, and of course, the experts at the music companies and labels whose business it is to make recordings and acts profitable. Just because you can create great-sounding tracks on a computer does not mean that you can sell them. Consumers usually want more—an artist with an image, live shows, merchandise, physical copies, and things they can hear, see, dream about, associate with, get emotionally excited about, and hopefully spend money on. Therefore, creating tracks in a studio or on a computer is only one of the ingredients required for success in the business.

Super Producers

Because of the financial problems caused by P2P and consumer behavior, some of the artists and their labels are hiring *superproducers*. These are the guys who wear about five hats at once as they produce the tracks using software programs in their own home studios, play the instruments, sing background harmonies, audio engineer, mix, and sometimes even develop the business plan and budgets. Labels save the cost of renting studios and hiring musicians and singers, and they often find that the tracks sound great and are delivered faster and at significant financial savings. Now that you understand a little more about the recording process and the business, let's take a look at what it might cost to actually record in a studio using union musicians.

Types of Recording Sessions (Scales and Rates)

There are several types of actual recording sessions, with the differences being based on union or nonunion agreements and if you live in a right-to-work state. The amount of money the recording team members receive is based on their fame, track record, contacts, and the kind of sessions the label authorizes. Record budgets are usually determined in advance by the label or whoever will own the copyright of the recording, just as music publisher own the copyrights of the songs. Budgets are based on many factors, such as the cost of the royalty artists, harmony singers, union agreements, the purpose of the recordings, and the producer's and engineers' fees. However, the types of sessions are usually based on the union scale. These are distinguished as follows:

- *Master scale* is the highest level of union recording session payment for the musicians, with the final master recording (owned by the label) having no restriction on the number of units they can press, sell, and offer for digital download.

However, in all types of sessions, the labels also acquire a mechanical license for the use of the song from the music publishers. Remember the © is the notice the songs being recorded are copyrighted, and the ℗ is the notice the recording of the songs is also copyrighted. A ™ by the name of the label indicates the name of the label is trademarked, and the ® by the label or artist's name is the notice the names are trademarked and registered with the U.S. Patent and Trademark Office.

- *Limited pressing* means what the title claims, limited sales of units. The label may use the master recording to press, press and sell, or digitally download (or any combination thereof), or give away the limited number of only 10,000 copies (units). In addition, limited pressing session recordings may not be used as movie soundtracks, original cast shows or theater presentations, or TV productions. Why would anyone want to pay a limited pressing scale? Easy—the musicians cost less and if we're not sure how well the album might sell, then the label still gets the great musicians at a lower rate. What happens if the album takes off and sells 10,001 copies? Then the label or the producer has to pay the difference in the LP scales and master scale fees. Also, remember the word "master" here means two different things. There are "master" recording sessions, which mean high-scale payments to musicians listed as a "master scale." Then there are the "master recordings," which are the raw and mixed-down versions of the studio recordings owned by the label used to make CDs and digital download copies for sale to the public.

- *Low-budget* union recordings are also available for labels and producers. In this case, the producer has to sign a *side letter* stating the total cost of the production will not exceed $99,000. There is also a "concept" album, a type of low-budget album with a production limit of $40,000. Producer fees are excluded in low-budget recordings like these. Once again, the reason for the type of session is to keep the artist's and musicians' costs low. If the cost of the project exceeds the budget limits, then the difference between master scale and low-budget scale must be paid, as with the limited pressing recording option described earlier.

- *Demo session* recordings and their copies can't be sold as they are used to demonstrate (demo) a song or vocalist being pitched to industry executives. Demo sessions, or calls, are typically scheduled for either a two- or three-hour time frame. As an example, the music publisher uses demo sessions to quickly record 3–5 songs an hour with "unknown" singers and very basic instrumentation. Those recordings may then be pitched to major labels, artists, managers, and film supervisors. Everyone knows that if the demo is picked up, then the song will be recorded again with major studio musicians and a signed royalty artist. The key is the "feel" of the demo to capture the essence of the energy in the song or abilities of an aspiring singer, or to showcase the fit of the song with the label, and/or to generate enough interest to get financial backing from the label to proceed with the recording.

- *Indie* or *nonunion recording sessions* are often accomplished in right-to-work states, sidestepping union-generated labor requirements. Anybody can record talent without using union members. Yet, most nonunion sessions are by non-professional artists, engineers, and producers who are in the process of developing their creative abilities. In addition, many states have right-to-work laws that allow for nonunion recordings. The best musicians and vocalists are usually union members and most likely (but are not supposed to) work for less than union scale. Unions help protect their members from unscrupulous producers, scam artists, and financially shaky record labels that promise payments but rarely deliver. Additionally, most nonunion recordings are used only for pitching songs and showcasing potentially new recording artists. If a label signs a nonunion act, it will usually (as it is a signatory to union agreements) recut the tracks with union musicians or pay the difference between what the musicians were paid and union scale. Additional nonunion sessions by bands and other acts that create their own labels are fine if uploading the works to the Internet or using subdistributors, such as CD Baby.

The rate scales fluctuate, but Table 7.1 shows you what they were for 2015–16. Over the past several years, the changes have been minimal.

Employer Terms

AFM musicians are hired "by the session call" (three hours, except for demo sessions, which can be either two or three hours), and SAG-AFTRA vocalists are hired by "the hour or number of songs recorded per hour" with a two-hour minimum (whichever is greater). Producers are usually hired "by the side" (one song) and paid half up front (when they agree to do the project) and the other half when the label accepts the final master tape. Some also receive "producer points" (often 1%–3%) out of the royalty artist's points. Audio engineers may be hired by the hour, session, day, or project. Studios may be rented by the hour, session, day, or project.

Typically, only the very best studio musicians and vocalists receive master session calls. A traditional master recording session is a three-hour call with a maximum length of 15 minutes of actual saved recorded time, four sides (songs), and a limited number of overdubs for corrections and sweetening. However, there are also overtime rates that, of course, cost more. One of the musicians must be declared "the leader" for the musicians and there is a correspondingly "a contractor" for the singers. Union musicians (AFM) must have a leader (one musician), who is paid double scale and acts as the main communication link between the other musicians and producer. The leader also provides the "card" that is signed by the musicians and turned into the union, who then bills the label for the musicians' work. All AFM session players are paid scale (set rates) depending on the type of session, plus $24 for health and welfare and an additional 11.99% of the gross scale session as scale pension (EP) funds. A typical recording budget often appears as follows.

Recording Budget Expenses
SAG-AFTRA Scales

The SAG-AFTRA contractor (leader), distinct from the AFM contractor, is paid a range of $50+ per hour or song as the leader of the singers on union calls. This person is paid a

Table 7.1 Rate Scales. Scale rates for professional musicians and singers.

Type of Session	Master (3-Hour Call)	Low-Budget (3-Hour Call)	Limited Pressing (3-Hour Call)	Demo (3-Hour Call)	Indie
Producer	Negotiated range $2,000–$5,000 per side	Negotiated range $500–$2,500 per side	Negotiated range $500–$2,500 per side	Negotiated range $50–$200 per side	Negotiated
Audio Engineer	Negotiated range $100–$150 per hour	Negotiated range $50–$100 per hour	Negotiated range $50–$100 per hour	Negotiated range Intern: $50 per hour	Negotiated
Musicians	AFM Side Musician	AFM Side Musician	AFM Side Musician	AFM Side musician	Nonunion Side Musician
	$397.36 per 3-hour session	$223.23 per 3-hour session	$196.50 per 3-hour session	$156.00 per 3-hour session	Negotiated
	Plus H&W $24.00 per session	Plus H&W $24.00 per session	Plus H&W $24.00 per session	Plus H&W $24.00 per session	None
	Plus pension $47.64 (11.99%) per session	Plus pension $26.76 (11.99%) per session	Plus pension $23.56 (11.99%) per session	Plus pension $18.70 (11.99%) per session	None
	Total $469.00 per 3-hour session	Total $273.99 per 3-hour session	Total $244.06 per 3-hour session	Total $198.70 per 3-hour session	Negotiated
Lead Musician (Double Scale)	AFM Lead Musician	AFM Lead Musician	AFM Lead Musician	AFM Lead Musician	Nonunion Side Musician
	$794.74 per 3-hour session	$446.46 per 3-hour session	$393.00 per 3-hour session	$312.00 per 3-hour session	Negotiated
	Plus H&W $24.00 per session	Plus H&W $24.00 per session	Plus H&W $24.00 per session	Plus H&W $24.00 per session	None
	Plus pension $95.28 (11.99%) per session	Plus pension $53.53 (11.99%) per session	Plus pension $47.12 (11.99%) per session	Plus pension $37.40 (11.99%) per session	None

(Continued)

Table 7.1 Continued

Type of Session	Master (3-Hour Call)	Low-Budget (3-Hour Call)	Limited Pressing (3-Hour Call)	Demo (3-Hour Call)	Indie
	Total $914.02 per 3-hour session	Total $523.99 per 3-hour session	Total $464.12 per 3-hour session	Total $373.40 per 3-hour session	Negotiated
Royalty Artists (Soloists and Duos)	SAG-AFTRA Royalty Artist (lead vocalist)	SAG-AFTRA Royalty Artist (lead vocalist)	SAG-AFTRA Royalty Artist (lead vocalist)	SAG-AFTRA Royalty Artist (lead vocalist)	Nonunion Side Artist (lead vocalist)
	$227.00 per side or hour	$227.00 per side or hour	$227.00 per side or hour	$227.00 per side or hour	Negotiated
	Health & retirement $28.38 (12.50%) per side	Health & retirement $28.38 (12.50%) per side	Health & retirement $28.38 (12.50%) per side	Health & retirement $28.38 (12.50%) per side	None
	Total $255.38 (per side/hour)	Total $255.38 (per side/hour)	Total $255.38 (per side/hour)	Total $255.38 (per side/hour)	Negotiated
Singers (3–8)	SAG-AFTRA Royalty Artist (lead vocalist)	SAG-AFTRA Royalty Artist (lead vocalist)	SAG-AFTRA Royalty Artist (lead vocalist)	SAG-AFTRA Royalty Artist (lead vocalist)	Nonunion Vocalists
	$103.50 per side or hour	$103.50 per side or hour	$103.50 per side or hour	$103.50 per side or hour	Negotiated
	Health & retirement $12.94 (12.50%) per side	Health & retirement $12.94 (12.50%) per side	Health & retirement $12.94 (12.50%) per side	Health & retirement $12.94 (12.50%) per side	None
	Total $116.44 (per side/hour)	Total $116.44 (per side/hour)	Total $116.44 (per side/hour)	Total $116.44 (per side/hour)	Negotiated

Table 7.1 Continued

Singer Contractor (on 3–8)	SAG-AFTRA contractor $50.00 (per side/hour)	SAG-AFTRA contractor $50.00 (per side/hour)	SAG-AFTRA contractor $50.00 (per side/hour)	SAG-AFTRA contractor $50.00 (per side/hour)	None Required or Negotiated
Studio Rental	Master Studio		Project/Mini/Post-Production Studio	Demo Studio	Demo/Garage Studio
	$150.00 (per hour)		$125.00 (per hour)	$50.00 (per hour)	$50.00 (per hour)
	$350.00 (per session)		$300 (per session)	$125 (per session)	$125 (per session)
Cartage (Handling of Instruments)	Cartage $12.00 (per session)	Cartage $12.00 (per session)	Cartage $12.00 (per session)	Cartage $12.00 (per session)	None
Per Diem (Royalty Artist Only)	Per diem $150.00 (per day)	Per diem $75.00 (per day)	Per diem $75.00 (per day)	Per diem $75.00 (per day)	None

Source: AFM[21] and SAG-AFTRA (2015).[22]

per-hour rate or a number of songs rate, whichever is higher. In addition, stacking or adding vocalists—say, having three singers record it twice to sound like six—is an additional call. The SAG-AFTRA contractor provides the "card" and other paperwork submitted to the union for payment. SAG-AFTRA's payment scale is determined differently than the AFM scale, which can drive anyone crazy until you've experienced it. Payments depend on the "number of sides" or "actual master takes," the number of overdubs, also called a side, and the length of the actual recording in minutes and seconds, which is used as a multiplier. Thus, if we add all the "sides" together and the length of each is a total of less than four and one-half minutes, the producer or label is billed for one side at scale. If, however, the takes add up to a figure between 4:31 and 9:00 minutes then the bill is doubled to two sides. Overdubbing adds sides to the total. A new recording or side in each multitrack is an additional side. So, to calculate the number of sides, take the original track, add any overdubs or additional tracks recorded, and multiply by the corresponding multiplier based on the actual length of the recording. Whew!

Contingent Scale

The musicians and vocalist may receive postrecording payments based on units sold. Of course, with the streaming and free downloads contingent payment checks have almost stopped hitting the mailboxes. Still, if you were a musician or vocalist on the last Taylor Swift album, you're probably in good shape—more money is headed toward your mailbox—because she refused to make it available for streaming. As an example, in the SAG-AFTRA agreement, vocalists receive between 50% of their original check to 75% based on the number of unit sales over 157,500 units. If the album or recording is sold at gold (500,000 units), then it's 60%, and anything over 1,000,000 (which is called a platinum record) receives 75% of the original recording scale.

Just think—we had to know all of this stuff or hire someone who gets it before we completed a simple professional album. Wow, take a shot at what else you need to know if you're going to produce a movie, TV show, or computer game. It's really a great, wonderful, yet complicated business, isn't it? Okay, now that we have seen the rules of the game, let's determine a typical recording budget.

Notes

1. Grimes, Garland. "What Did Thomas Edison Invent?" How-StuffWorks. 2015. Accessed September 21, 2015. http://science. howstuffworks.com/innovation/famous-inventors/what-did-thomas-edison-invent1.htm.
2. "Thomas Edison Most Famous Inventions." Thomasedison.org. 2015. Accessed September 21, 2015. http://www.thomasedison. org/index.php/education/inventions/.
3. John Kennedy and Alison Wenham. "Investing in Music: How Music Companies Discover, Develop and Promote Talent," IFPI. org, 6. Accessed March 19, 2011. http://www.ifpi.org/content/ library/investing_in_music_2010.pdf.
4. Weber, Bruce. "Jerry Wexler, a Behind-the-Scenes Force in Black Music, Is Dead at 91." *The New York Times*. August 15, 2008. Accessed October 20, 2015. http://www.nytimes.com/2008/08/16/arts/ music/16wexler.html?ref=obituaries&_r=0.
5. Ibid.
6. "History of the AFM." Official Website of the American Federation of Musicians. 2015. Accessed October 20, 2015. http://www.afm. org/about/history.
7. Courtney, Richard, and George Cassidy. "The Corporate Connection." In *Come Together: The Business Wisdom of the Beatles*, 101. Vol. 1. New York, NY: Turner, 2011.
8. Finke, Nikki. "SAG-AFTRA MERGER APPROVED! 81.9% of SAG Ballots Returned Voted Yes; 86.1% of SAG-AFTRA; Single Union Effective Immediately." Deadline.com. March 20, 2012. Accessed September 30, 2015. http://deadline.com/2012/03/sag-aftra-merger-approved-screen-actors-guild-american-federation-television-radio-arts-251114/.
9. "History." 2015. Accessed October 1, 2015. http://www.sagaftra. org/history.
10. "AFTRA History." 2015. Accessed October 1, 2015. http://www. sagaftra.org/history/aftra-history/aftra-history.

11. "About Us." 2015. Accessed October 1, 2015. http://www.sagaftra.org/content/about-us.
12. "AFTRA Sound Recording Code." 2015. Accessed October 1, 2015. http://www.sagaftra.org/files/soundrecordings_onesheet.pdf.
13. "About Us." 2015. Accessed October 20, 2015. http://www.sagaftra.org/content/about-us.
14. Bellis, Mary. "The History of Microphones." Inventors.about.com/money. 2015. Accessed September 30, 2015. http://inventors.about.com/od/mstartinventions/a/microphone.htm.
15. Personal interview with industry insiders and audio engineers, including Jim Williamson, producer and audio engineer of Sound Emporium Studio, Nashville, TN (1978–2004).
16. Wacholtz, Larry. "Record Production." In *Inside Country Music*, 184. Vol. 1. New York, NY: Billboard Books: Watson-Guptil, 1986.
17. Ibid.
18. Ibid.
19. Ibid.
20. Ibid.
21. "Scales, Forms & Agreements." Nashville Musicians Association. 2013. Accessed February 20, 2016. http://www.nashvillemusicians.org/scales-forms-agreements.
22. "Sound Recordings." Accessed February 20, 2016. http://www.sagaftra.org/sound-recordings/sound-recordings.

THE LABEL BUSINESS— RECORDING BUDGETS 8

The Invention

The Parisian Édouard-Léon Scott de Martinville (1817–1879) was the first to make a sound recording, in 1857. He developed a visual way to convert sound waves into physical lines on a rotating cylinder, applying for a patent and his invention of the phonautograph.[1] Scott de Martinville had invented a process to record a soundwave, but not how to play it back. As a matter of fact, when we pause and think about it, the entertainment and music industry has always been device-driven, so let's start this chapter by taking a look into the history of essential processes and devices we now take for granted that allowed recordings to happen, which led to the development of record labels and fostered other entrepreneurial activities that still support the modern industry.

In 1877, when Thomas Alva Edison invented the playback process enabling him to record "Mary Had a Little Lamb," he simply reversed Scott de Martinville's process. Edison combined a little entrepreneurial wisdom, added his own innovative thinking, and bingo, his transformation led to the birth of the music business. But here's the amazing part of the story. Édouard-Léon Scott de Martinville had also recorded (probably using his own voice) the French song "Au Clair de la Lune," never thinking it could be reproduced. His thing was creating a visual record of a sound, making it doubtful he thought about how it might be played back since that had never been done. In 2007, some of the original phonautograph recordings made on cylinders were found and converted digitally to their original states as sound recordings.[2] Thus, it appears that Thomas Edison was probably the second person to actually record a song, following de Martinville's recording. Because they were inventors, interested in physics and audio engineering, they likely had no idea the impact of their inventions on our music industry and the availability of entertainment products in the years to come.

Recording Labels

Several of the most successful labels in the early stages of the business were started by entrepreneurs who were either creative artists themselves or executives who visualized the profits of marketing live performances of artists into "recording artists." Columbia Records was formed in 1886 around the earliest version of a "talking machine," with sound quality that is really awful by today's standards. But their concept was to use the invention for the creation of a new type of business model by selling recorded copies of live acts. Since a machine had been invented that could play recordings, they thought why not sign live artists and record them so that consumers will buy tickets and use the recordings to create a memory of the concert or event, which we know as building a fan base. The Recording Corporation of America, better known as RCA, was created as a support platform to sell the Victor Talking Machines in 1901. Capitol Records was created by the three-person team of songwriter Johnny Mercer (1909–1976), movie producer Buddy Desylva (1895–1950), and retail store owner Glenn Wallicks (1910–1971) in 1942. The store was called Music City, located at Hollywood and Vine, and it had private listening booths. The three formed Capitol Records by combining the elements of a modern label, including the creative side, film side for soundtracks, and even the retail stores to sell products. Jack Warner (1892–1978) jumped into the business in 1958 by forming Warner Bros. Records to increase another revenue stream for his movie soundtracks. The entrepreneurs had developed a way to monetize Édouard-Léon Scott de Martinville's and Thomas Edison's inventions that would grow into a terrific multibillion-dollar business. They provided a high-quality form of entertainment at a very reasonable price to the consumer. In addition, the new recording industry, like the movie industry, united a nation with songs by great artists that had only been available live before the recording and playback process increased the size of the consumer market and created new ones across the socioeconomic spectrum.

Magnetic Recording

Samuel F.B. Morse (1791–1872) invented the telegraph in 1844 that sent an electronic signal over a wire in the form of short or long pauses, leading to Alexander Graham Bell's (1847–1922) device that was the first to convert acoustic sound waves into electronic waves and then back again into acoustic sound waves in 1876. The theory of passing a wire through a magnetic force field showed the means of creating electricity (electromagnetic frequency; EMF) for consumer use. This led to magnetic recordings and playback functions (on a wire), invented by Valdemar Poulsen (1869–1942) of Denmark in 1893.[3] He also described how paper with a coat of magnetic powder could be used to record a signal that, in turn, inspired German Fritz Pfleumer (1881–1945) to develop the first magnetic recording tape in 1929.[4] This series of sequential inventions illustrates how one discovery leads to another opening up entrepreneurial opportunities to create entertainment products, such as phonograph players, vinyl singles, albums, and, God help us, those lousy-sounding audio cassettes. The same innovative process seems to have taken place with films, and projectors in theaters, leading to cheap consumer versions of film cameras, and then video cameras and those old, blurry TV screens of 525 scanning lines, consisting of tiny dots or "pixels," blinking faster than our eyes blinked, so we saw only what appeared to us as a constant moving picture.

The Star System

The film industry discovered it could increase profits if it revived the practice of making some of its movie actors into personalities good for fan magazines and into screen stars consumers wanted to follow. The star system, rooted in notions of fame and celebrity, was a great business decision to enrich company profits. Actors were signed to film companies/studios (just as many athletes are currently with sport teams) and given new names, acting lessons, homes, money, and so forth. However, the seven-year deals placed them under strong studio controls in both their careers and personal lives. Actresses Betty Davis (1908–1989) and Olivia de Havilland (1916-) brought lawsuits that Davis lost and de Havilland won, which changed the contract system. In 1950, Jimmy Stewart (1908–1997) was the first actor who negotiated his own deal and won a percentage of the profits his movies made. That shifted the power in the movie business to the actors who were most in demand.[5]

During the early stages of the recording industry, the artists who signed with labels and positioned themselves in more than one media had a better chance of becoming household names. Thus, we find stars worked as recording artists, and performed on live tours, in stage shows and in the movies. The vocalists and musicians who drove recording profits in the early years of the industry include Al Jolson, Tommy Dorsey, Glenn Miller, Perry Como, Dinah Shore, David Bowie, Elvis Presley, Frank Sinatra, Dean Martin, The Beach Boys, The Everly Brothers, The Beatles, The Grateful Dead, Joni Mitchell, The Doobie Brothers, Fleetwood Mac, Smokey Robinson and The Miracles, The Temptations, Diana Ross, Marvin Gaye, The Commodores, The Jackson 5, Michael Jackson as a solo act, Stevie Wonder, Dolly Parton, Hank Williams, Sr., Buck Owens, the Carpenters, Joe Cocker, Cat Stevens, The Police, Janet Jackson, and many others. You probably know that some of these recording stars also made movies, and one notable trio of singers—Sinatra, Martin, and Davis, Jr.—were part of the Rat Pack (which also included Joey Bishop and Peter Lawford). Sinatra, Martin, and Davis, Jr. doubled their value (at the very least) in the entertainment industry as they were solo music acts and film stars with notable followings in each media. The Rat Pack was reinvented in 2001 with a remake of the 1960 *Oceans 11*, which has become the Oceans 11 trilogy of movies, starring George Clooney, Brad Pitt, and Julia Roberts.

History of Computers

Researching who developed the first computer is a little like the Magical Mystery Tour, which is the title of one of the Beatles' songs. There are so many different answers to this whodunit that a better way to answer it may be to ask a different question: Who contributed what to the devices we now call a computer? The best review of the history of the development of the computer is probably an online article titled "Computer Hope" by Nathan Emberton of Salt Lake City. According to Emberton,

> The word "computer" was first used in 1613 to describe a human who performed calculations ... In 1837, Charles Babbage proposed the first general mechanical computer, the Analytical Engine. ... Babbage's youngest son . . . was able to perform basic calculations. . . . The Z1 was created by German Konrad Zuse in his parents' living room between 1936 and 1938. It is considered to be the first electro-mechanical binary programmable computer . . . The Colossus was the first electric programmable computer, developed by Tommy Flowers, and first demonstrated in December 1943. The Colossus was created to help the British code breakers read encrypted German messages. . . . Atanasoff-Berry Computer, the ABC began development by Professor John Vincent Atanasoff and graduate student Cliff Berry in 1937. . . . It was an electrical computer that used vacuum tubes for digital computations . . . The ENIAC was invented by J. Presper Eckert and John Mauchly at the University of Pennsylvania and began construction in 1943 and was not completed until 1946. . . . It occupied about 1,800 square feet and used about 18,000 vacuum tubes, weighing almost 50 tons. Although the Judge ruled that the ABC computer was the first digital computer . . . The early British computer known as the EDSAC is considered to be the first stored program electronic computer. The computer performed its first calculation on May 6, 1949 and was the computer that ran the first graphical computer game,

> *nicknamed "Baby." . . . The TX-O (Transistorized Experimental computer) is the first transistorized computer to be demonstrated at the Massachusetts Institute of Technology in 1956.*[6]

And that's just the very short version of the history of the computer. Most of us do not think about our iPhones and other household devices as computers, yet it's often the microelectronic chips developed in computers that make them function. Everything from our TVs to rice cookers has computer chips in it, which convert the binary codes of the signal into a high-definition TV monitor or cook the sticky rice at just the right temperature and moisture. Computers have also improved the quality of the music and video recordings, from solving math problems and breaking war codes with mechanical and vacuum tube devices to producing the crystal clear quality of a recording, film, or video with a digital camera, laptop computers, or "smart" cell phones. There are other economic advantages to using computers in the entertainment industry as broadcast transmissions require far less bandwidth and are less expensive to use and maintain than the corresponding analog systems. As an example, one fiber-optic line the size of a human hair can handle a hundred thousand phone calls and a cable of fiber-optic lines can carry over several million calls at the same time.

Digital Recording

In the early 1950s, the Whirlwind machine was introduced, using a magnetic core RAM, and later the transistor replaced vacuum tubes.[7] These inventions reduced the size of computers from huge, clunky, room-swallowing machines into personal computers with workstations. I was in Los Angeles in the late 1970s visiting my friend the late Tom May, manager of the A&M recording studios, where the old Charlie Chaplin film studios were housed. Tom took me into one of the studios being set up for a new album by jazz horn player Herb Alpert (b. 1935) and his band. Spread across the control room floor were the modules from the second digital tape recorder in the world, developed by 3M. The engineer was frantically trying to figure out how to make the machine work, as he was having a problem keeping the modules electronically isolated from each other. The damn thing didn't work and he was freaking out, trying to figure a way to correct the problem. He was under the gun as Herb Alpert and his band, The Tijuana Brass, were walking into the session in about 20 minutes. What he was struggling with can be explained this way. Digital recordings take the incoming signal from the microphones and mixing boards and convert them into a digital binary code made up of 32 or so zeros and ones. The code is then recorded on a tape or hard disk drive as a magnetic pulse code, which represents no recording (zero) or a one (a present pulse of magnetism). When a digital machine is played back the pulses—or lack of them—are converted by the computer back into the binary codes of zeros and ones. The computer inside the tape machine then uses a string of the numbers (the binary code) to convert it into an audio sound, or frame of a picture if using a digital camera. The quality is amazing as the computer is making the sounds we call music, just as the computer chips in our TVs are creating and coloring the TV monitor. To better understand why the engineer was stressed, read on to understand more about how the code works.

Binary Codes' Unexpected Effect

Gottfried Wilhelm Leibniz (1647–1716) invented the binary code number system around 1685 in Germany.[8] It's what makes computers work, yet it took us a few hundred years more to invent the rest of the stuff we needed to build an actual computer. Binary code, as we saw in Chapter 1, uses the different combinations of the zeros and ones (zero being off and one being on) found in many logic circuits. It is a very cool concept when converted into ASCII, where the binary code numbers may be programmed to represent a letter. As an example, the bit code (zeros or ones) that represents the word *entertainment* is 0110010101101110011101000110010101110010011 101000110000101101001011011100110110101100101011011100 1110100.[9] And converted into ASCII, the word *entertainment* appears as 101 110 116 101 114 116 97 105 110 109 101 110 116.[10]

Binary codes, for all the good they do in improving quality by eliminating various types of noise from audio recordings, also make it possible for anyone with a copy of the code to replicate the product, and thus steal it.

The Advantages of Digital Recording

In analog recordings, when a copy is made it's called a dub. The dub is one generation (or recording effort) beyond the original, which is called the "master." The problem is that each generation or copy decreases the quality of the sound as the distortion from the recording process and the hiss on the tape are usually doubled. After about ten or more dubs, the copies start to sound quite inferior to the master and thus are worthless if you're trying to sell them. Take a pure glass of water and start adding salt. A little salt, no big deal and, in fact, it might help heal a sore throat. But ten spoons of salt make the water worthless. By the way, do not drink saltwater, as all it will do is make you thirstier!

Some young people like the distortion and noise found on vinyl recordings. They hear it as a "warmer" quality of sound. But for as much as they have acquired a new fan base, vinyl recordings simply do not sell well; the total revenue is significantly less than 1% of all music recordings each year. Usually a hit recording on vinyl is lucky to sell over 50,000 copies. When you tack on breakage problems, the vinyl recording business model's economies of scale are too risky, meaning there is not enough financial return on the company's investment. The front-end expenses of recording, mastering, pressings, shipping tied to marketing, promotion, and publicity require a large investment that is not typically offset by significant profits.

Advances in recording technology should have an impact, maybe even a positive one, on the recording industry, so let's take a dive into what a record label really is, how it functions, and how that entertainment is monetized for profit by the team that creates it. As you read, let's ponder: with all the changes in the music and entertainment industry today, do we really need labels and what functions have they served and do they serve? How might they need to transform themselves to stay viable? What are your questions?

Major Labels: Traditionally Surviving or in Need of Transformation?

What is a major record label? The current business models are much different from the labels of the past, where many subsidiary brands were used but actually owned and distributed by one of six major labels. Warner Music, for instance, sold recordings under labels including: Asylum, Atlantic, East West, Elektra, Fueled by Ramen, Nonesuch, Reprise, Rhino, Roadrunner, Rykodisc, Sire, Warner Bros., and Word, as well as Warner/Chappell Music Publishing, which controls over 1 million titles.[11] Many of those old label brands continue to exist currently, but the labels have dramatically reduced their roster of acts, which simply means they have let go many of the former famous artists for economic reasons.

Traditional Business Model

Labels tend to have a five-step process as part of their business model to find, sign, create, promote, and sell an artist's recordings. Traditional labels are usually divided into several departments, all of which have specific tasks to accomplish as they attempt to profit from the recordings of their artists. The departments have their own autonomous but interlinked responsibilities. They must accomplish their individual missions while working together with the other label departments to accomplish the overall strategic mission of the label, which is, of course, to create and sell recorded music. Of course, due to the changes in the marketplace caused by digital innovation and the lack of copyright enforcement, labels have been forced to morph their recording business models into a branding business model. If we can't sell enough recordings to pay the bills and make a profit, what should we do? Still, they have to make the recordings that are becoming more difficult to sell for promotion and publicity purposes. In addition, they have to recoup their expenses and, with luck, profit from new revenue streams. It is essential to know the departments found in a major label and how they interact, whether you plan to be an artist or an industry executive. And not all the intersections at a label are self-evident, as the legal department and the royalties department, which is the label's equivalent of a payroll department, are part of the administration directly serving and reporting to the president of the label.

Legal Department

Legal is responsible for writing and negotiating contracts, as well as signing

- Songwriters or their songs
- Copublishing deals
- New and returning recording artists
- Label employees' government forms

They are also responsible for negotiating mechanical, performance, sync, and other licenses with domestic and foreign agencies, and representing the label in legal actions. Accordingly, attorneys negotiate the "terms of agreement" between the creative and business subsystem players of this creative business. Again, it is creativity informing business principles. Most labels have one or more attorney on retainer, which means they are paid a monthly stipend for their part-time employment. As you may have noticed, labels usually have their own music publishing companies, which means they want to have their acts recording their published songs to increase overall total revenues. In the digital age, this has been very important as some labels may now pass on an artist if they can't also get their music publishing to be as profitable as they want or need. The economics of the game have changed dramatically as streaming increases and record sales drop.

Royalty/Payroll Department

Payroll or the royalty department pays the bills and the royalties, accepts sales payments, and reviews artists' recoupment. Financial experts control the flow of money in the label's and the artists' accounts. Most expenses are billed to the artists' accounts and recouped through album sales. That's how the old business model worked, yet we'll soon understand why that model simply does not work anymore. "Recoupment" is a standard practice in the business that is often seen as unfair for acts as the label may make millions before the average artist makes their first dollar in royalties. However, more labels have switched their deals to 360s and now negotiate various types of deals from what were revenue streams owned by the acts. Accountants add the recording costs to the rest of the label expenses (artist development, photography, tour support, and most of the marketing, promotion, and so forth) to the recording artist's ledger. In traditional recoupment deals artist royalties are paid only *after* all the label's expenses for the recording, marketing of the album, and all additional expenses (tour support, and so forth) have been repaid from unit sales. These expenses are called the "all-ins" and are usually 100% recouped by the label before acts are paid. Artists are regularly supplied with quarterly or semi-annual profit and loss (P&L) statements, often through their personal managers.

The Department of Artists and Repertoire (A&R)

Better known as A&R, the department of *artists and repertoire* is the communicative link between the label's creative and business executives. They are the front door or filter for acts and songs the label may want to sign and then develop into (shall we say) money machines. It drives artists crazy to think this way, but labels are in the business of making money, not satisfying egos. The CEO, president, and vice president of each label department meet at least once a week to review the label's acts and marketing strategies. They review units sold and analyze marketing, promotion, and publicity that might increase sales. This is a serious business meeting as decisions are often made about which acts to resign (pick up their option) and which acts to drop—that is, electing to pass on an act's option. Decisions are made by the group as a whole and based on the profit and loss projections of each act's recordings (in the prestreaming days). The department is sometimes divided into two sections, creative and administrative, as in the case at Sony Music. Creative A&R acts as the "ears" of the label, always searching for new artist and songs that will help make the label become more profitable.

A&R is more than the people in the office. It is also the key process that determines what the label has to sell, as Figure 8.1 shows you. Spend some time with this figure to learn the mechanics of one of the central departments in the sustainability of a recording label's business.

The folks in A&R do not have an easy job. Between a half million to 1.2 million dollars are spent on "breaking" an individual recording artist; however, the actual cost generally depends on the type of artist, the genre of music, and the type and size of the target audience. Different players, such as executives, producers, and directors, make the same types of decisions in the movie business. The "stakes" or risk of business is just a lot higher. Instead of a million-dollar investment in a new recording artist, the investment may be $10–$100 million for the production and marketing of a new film project. Their budget usually starts at about $10 million for a low budget and may quickly hit $100 million when the marketing, promotion, and publicity kick in.

The department of A&R also acts as an administrative facilitator between the creative and business subsystems. Songwriters create songs, musicians and vocalists record demos, and writers pitch their demos to music publishers. Publishers in turn pitch the songs they have acquired to A&R. Most labels simply leapfrog the steps by having their own music publishing companies under the label banner. That keeps everything "in house" and improves profit statements at the end of

The A&R Process

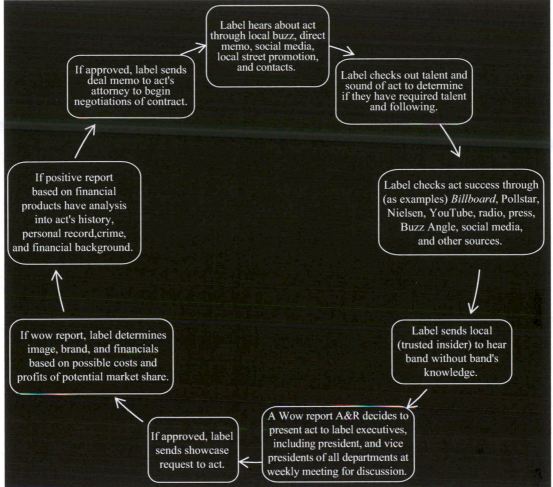

Figure 8.1 The A&R process is more detailed than most think. After all, if the label is going to drop a million or so dollars into an act's career, it assuredly wants to reduce the risks of doing business by confirming all the financials, projects, personal investigation, and everyone at the label agreeing to sign the act.

the year as the music publishing royalties are now added to the labels'. Songs selected by A&R are evaluated for current album projects and pitched to the recording production team, consisting of the producer, artist, and sometimes the artist's

325

personal manager. In preproduction, the recording artist, producer, artist's manager, and A&R representative determine which songs, musicians, and recording studio to use. However, the producer makes the final decisions about which songs to record, studios to rent, musicians and singers to hire, and audio engineer to use. A&R is often required to establish a reasonable recording budget and time frame for the project. After the final master recordings are completed, if the master tapes are accepted they are then converted (mastered) into consumer products, such as CDs and digital downloads, and distributed, marketed, and promoted through the mass media, cell phones, and Internet sites to generate profits.

Labels/music companies create sellable products (recordings) through major recording deals with expenses in the recording budgets including renting great studios; contracting AFM musicians and SAG-AFTRA vocalists; hiring major producers and audio engineers; and factoring the cost of the mechanical license at statutory rate (or at the CCC rate) multiplied by the estimated number of units pressed and sold or digitally downloaded. Budgets for the actual multitrack recording range from $50,000 to $200,000, plus mechanicals are established and budgeted by A&R administration. The recording costs plus mechanicals and advances are usually defined in the deal as 100% recoupable and are considered all-ins.

Artist Development

Once an act has been signed to a recording deal, labels have to teach them how to walk and chew gum at the same time. Only kidding, of course, yet most new acts have to be developed into a sellable image that will "catch" the consumers' attention long enough for them to get hooked. Labels seek recording artists who are able to capture the imagination of the human spirit and of the current popular culture, themes, and beliefs. New artists who have established a local image, a fan base, and a successful number of units sold (even in small numbers) are seen as having potential, as they may have already developed a meaningful image that communicates some emotional message to consumers. Labels view small successes in local markets as an indication that if the act is developed, recorded, and exploited (marketed), they may become very profitable. According to record producer Jimmy Bowen (whose artists included Frank Sinatra, Dean Martin, and Hank Williams, Jr.),

> *You should always strive to work with an artist who is a one-of-a-kind. They are the only ones that really, really break through. You can record an artist who's a copy of an artist, you can have success with them, but it's not really worth the effort. Anytime you have an artist who's like somebody else, you're starting off in second place, no matter how well you produce it. If you're producing a one-of-a-kind artist, you're doing something new. . . . Sometimes a one-of-a-kind is not obvious when you first start with them. Usually, there will be a little something, you'll hear, just four bars somewhere, you'll see them in person, you'll hear them with a guitar or piano someplace, and you'll catch that little moment of magic. You've got to hunt it, look for it. You're always hunting that gold, that one-of-a-kind. And when you see it, it may be 4 years away; 6 months away; that's your problem. You've got to figure out when it is, see if you can get it into a point where it's valuable before you lose the opportunity to work with it.*[12]

Artist development may be its own department or part of A&R or even sales and marketing within a label. However, marketing is usually defined in the music business as distribution and unit sales, whereas development is often divided into creative, artist, and product development in the music and entertainment industry. As these are areas of psychographic research, let's pause for you to see the richness of what goes into the idea of artist development:

- *Image development* considers the artist's image, including acting, speech, appearance, demeanor, and social manners. The artist's image frequently has to be enhanced. Trainers are hired to help the artist control a weight and/or drug problem, dentists to improve the smile, and stylists to suggest various types of cosmetics, clothing, and hairstyles for concert, stage, TV, and/or movie appearances. Image consultants are hired to suggest methods to help the act appear comfortable and confident on stage and with the media.

- *Creative development* focuses on songwriting, music publishing, and the ability to promote the artist through shameless self-promotion. They may also supervise video script treatments, budgets, and production and provide the finished videos to the marketing department for promotion.
- *Branding and product development* supervises and coordinates the development of the label's actual album projects through each department at each stage of the marketing, promotion, publicity, and distribution process. Branding has become very important, as album sales have decreased. However, in the traditional business model, branding was not as important as it currently is, as labels have had to find new revenue streams. The marketing plan, promotion, publicity, and street date launch are usually based on the act's image tied to consumer preferences. Remember this is the emotion business and it's the consumer's reactions to the act's image and recordings that open up the merchandise and touring opportunities. More will be said about this in Chapter 10 within the contexts of media use and risk management for the label.

Promotion

Promotion or "exposure" of the artists and their recordings is vital to the financial success of the artists *and* their label. It is usually handled as a "campaign," with most of the label's efforts being focused on breaking a new release to radio stations and Internet sites. In addition, the labels often hire independent promoters to motivate radio station airplay as well as retail sales. Publicity, which helps drive promotion, is often thought of as free promotion because the artists do not have to pay for the stories, pictures, and articles used in the mass media. Nevertheless, professional publicists who *are* paid by the label or artist's manager supply most stories to the mass media. Budgets for promotion are expensive, usually in a range of 100% to 200% of the all-ins. If the recording, pressings, and mechanical licenses total $200,000, then it's not uncommon to set a promotion budget of $200,000 to $400,000. Major superstars actually require less in the promotional budgets, as they are already famous, yet they usually want more money in their advances or recording budgets.

The three levels of promotion are *national*, *regional*, and *local*, all of which have their advantages and disadvantages. National ads are expensive and targeted toward several demographic and psychographic types of consumers. Regional ads build support in only a section of the country, and the label may therefore miss its true market. Local ads (from free press to posters on local telephone poles) are usually poorly produced and sometimes do not generate an increase in consumer buys; however, they can start word-of-mouth conversations leading to a buzz about a new act. World promotion is often contoured to local and regional consumer preferences, as a marketing campaign or advertisement in France may often be very different from one in Japan.

Tour Support Tied to Marketing Plans

Labels also provide tour support through financial advances to place the act on the road. These shows are considered events that provide primary opportunities to create radio airplay, fan hype, and word-of-mouth publicity, all of which are tied to the artist's self-promotion. Of course, this is the label's effort to increase consumer excitement and bottom-line product profits by raising the level of consumer awareness and motivation to buy. Before 360 deals, labels would rarely receive any money from the merchandise bought by fans at the concert. Now, it is usually part of the recording deal and will probably soon include other event income sources, like ticket sale profits and corporate sponsorship endorsements. Note that product distribution, promotion, and publicity are designed by the label's marketing plan and usually occur before or at the same time as the street release of the recording or album. While a show in one part of the world might be seen as very cool, it may be offensive in another part of the planet. It is important for the long-term security of the act's career to be aware of the local customers and norms.

Publicity

Publicity is about telling consumers the story behind the artist and his or her image; plus it's very powerful when combined with a corresponding promotional campaign. While promotion provides a *sample* of what the labels or music companies have for sale (recordings in traditional deals, plus concert tickets, merchandise, and corporate sponsorships in 360s), publicity tells consumers *why* they might enjoy the artist and their recordings. Effective promotion is an attempt to have the artists perceived in ways that match or create consumers' interest in the act. Remember that the emotional connection between act and consumer is what counts in the entertainment and music industry. The three variables in consumer responses to an entertainment product can be easily summarized as like it, hate it, or don't even notice—and you certainly don't want to be in that last group!

Publicity from the label's perspective includes stories and interviews planted in the trade magazines, popular press, and through social media. It also includes appearances on cable and network TV talk shows and the human-interest stories, pictures, and feature articles publicists provide the media. All of these publicity efforts, known as impression opportunities, are designed to communicate messages consumers can use to form an emotional connection to better understand themselves consciously or subconsciously. The best entertainment products are linked to situations in the consumer's life or a specific worldview. If consumers get a warm and fuzzy feeling, connecting that with the artist and the music they've recorded is great for the label. Labels hope we get excited enough about our emotional connection that we decide to buy a CD, steal it, catch the act on tour, and buy lots of merchandise.

Shameless Self-Promotion

Once the album is completed and they are on tour, the artist's job becomes shameless self-promotion on talk shows and through interviews and/or performances. The act has to reflect the "image" that matches how they are perceived by consumers who need to buzz over a new album, book, movie, and tour. Keep it simple, as it's always better to enhance "who the artist is" than to fake an image. Faking it and making it appear believable or real are what actors and great crooks do. Labels use psychographics in the form of consumer lifestyle research to determine the placement of stories (about the artist's life or daily events) in trade magazines, the popular press, in-store endcaps (those main aisles facing shelves in the stores), images in social media, ads for concert tours, and even paparazzi photos in digital media. Of course, all the publicity releases are used to enhance the image of the act being a sexy dude, bad guy, or amazingly attractive girl band. The labels deliver the publicity hits to the media fans are using, especially the websites, where they get their news and share their enthusiasm for entertainment products through social media. These materials and placements cost money as well, and budgets are usually 50%–100% of the all-ins. However, it is much less expensive to advertise directly to the fans on the devices they use than it is to buy national radio and TV spots that won't be noticed by most of the market. The good news is that great publicity tied to glitzy and well-thought-out promotion will provide results in as little as a two- to four-week campaign. It either works and the act is on its way toward success or it fails and the label will move on to the next act or project. A lifetime to get a shot and only two to four weeks to decide if it's going to work. That's another kind of wow!

Marketing and Sales Department

In business, marketing is defined as the four Ps of *product*, *placement*, *price*, and *promotion*. In the music business, marketing is defined as the distribution and sales of recordings in the traditional sense, and as branding opportunities associated with the acts in the 360 business models. Consumers' illegal downloading of entertainment products and innovation in the industry have changed that—remember the perfect storm and the principle of creative destruction are at play. Correspondingly, recordings are quickly becoming the buggy whips of our time, as consumers no longer need to buy the recordings. Though sales are not totally dead, even a blind mouse can see that the cheese (album sales) is going away. Previously, weekly sales figures related to the marketing plan indicated the need or lack of promotional radio and tour support to increase profits. The marketing and sales department makes sure that CDs (product) are distributed to retail and iTunes (place), at the right price (price), and that radio and YouTube have consumers buzzing (promotion). Retail stores' customer base, location, and co-op advertisement are considered as well. Highly trafficked retail websites, such as iTunes, are always considered first for CD and digital sales.

Retail

Delivery of the recordings to wholesale and retail outlets using worldwide mega-entertainment distribution channels is vital. Timing is imperative, as labels want consumers to be able to easily buy, stream, or, in some cases, steal the recordings they have just "discovered." The idea is steal them if you don't buy them so that we know where to set up touring opportunities and profit from the live shows and merchandise. Sounds odd, right? But reaching the consumer market in any way they can is what the labels need to do.

Traditional distribution of recordings (one of the subjects of Chapter 10) used to be through one-stops (record chain stores, most of which are now out of business), mass merchandisers (Best Buy, Target, Walmart, and Kmart), record clubs and per inquiry (PI) advertisements through local late-night TV ads. However, the old business model of brick-and-mortar stores is quickly reducing their inventory of CD and DVD products, and it appears that within three to five years, they will stop stocking and selling them. Digital sales usually follow the same street release date schedule, with buys available for digital download through Internet websites and other providers. The problem is that people just do not buy many recordings anymore and the labels can't make money on such a few sales unless you're a trendy artist.

Streaming Royalties

Streaming has become one of the new revenue sources for labels, but not for the recording artists. Spotify and other streaming services must negotiate license deals with labels to play the recording of music over their systems. About 70% of all the money collected by streaming services, after expenses and taxes, is now sent to the three labels—Warner Music Group, Universal, and Sony—and any independents that acquire enough streaming time to receive royalties. Artists have been understandably upset, because of the lack of payments from streaming, or the minimal payments they may receive to make up for the royalties lost from album sales. Of course, some of the artists never negotiated a percentage of streaming royalties for themselves, so labels do not have to pay them. There has been such an uproar from artists that Sony has agreed to drop its package fee deductions from royalties (described in Chapter 9, but here, think of it as advances charged against records broken in shipping) and Warner Music Group has declared it will pay approximately 3% of revenues, if and when they decide to sell their shares of ownership in Spotify. At present, the three major labels also appear to own up to 18% of Spotify, so in addition to the 70% of the profits the labels receive for licensing fees, they receive 18% of Spotify profits and are struggling for some programming power.

YouTube Video Royalties

YouTube videos tied to master recordings of major stars or wannabes' recordings (a cover) of major hit songs are illegal. The wannabes rarely acquire the correct licenses—master licenses from the label (copyright holders of the recordings) for the actual recordings and sync licenses (copyright owners of the song) from the music publisher. However, the music publisher and labels know the wannabes don't have any money and it is often a waste of time to pursue legal action against them. So, what are the owners of the copyrights to do? Demand a percentage of advertisement from YouTube in order for them to be allowed to remain in business on the Internet. YouTube is a safe harbor, meaning that it's outside of the law, as long as it scans all content and takes down the content when it is notified of or recognizes a copyright infringement. The labels turn a blind eye and, instead of shutting it down, simply demand money. At the same time, the labels may take the position that the music played on YouTube is at least "heard" as free promotion.

Mash-Ups

In 2014, two students at Belmont University recorded a very cool mash-up of Taylor Swift's songs "Blank Space" and "Style." Louisa Wendorff was a School of Music major and Devin Dawson, who was my student at the time; he graduated in December 2015 as a music business major. A female student in the very next class after Devin's had seen the mash-up on YouTube and she sent an email to Taylor, with whom she is acquainted. Next thing you know, 27 million hits plus, Taylor Swift tweeted her compliments about the cover to her fans. At the time, Louisa Wendorff was a School of Music major, and Devin Dawson, was a student of mine, who graduated in December 2015 as a music business major. Of course, it was their great vocals and student recordings, video, edits, and other efforts that made it all happen.

Good for them! The video can be seen at https://www.youtube.com/watch?v=7m3o5LuFKxg.

Free Goods and Promotional Copies

Free goods to the retail outlets are used to stimulate rack stocking and potential sales of new artists and releases. Buy 100 and labels may give you another 25 free. If the retail outlet sells 30 copies, they return 95 copies and keep all the money paid for the 25 "free copies" and the net money received from the customer minus the cost of buying the other five albums at wholesale. To better understand the concept, if you own a store, would you buy at wholesale a new act nobody knows exists and then try to mark up the price to make a profit? Of course not, as nobody has ever heard of the act and does not even know you've got it in the store. The labels, therefore, create a win-win situation where the store owner won't lose any money if the albums don't catch on with the public, but it is in the inventory if (and that's a big if) radio, YouTube, and streaming start to play it and people want to buy it. Of course now labels have to ask themselves, would anyone actually buy it or would they rather illegally download it? How are the labels going to make any profits on their financial investments in new talented artists?

Promotional copies can't be sold to customers. Promotional copies are used by the marketing and promotion departments to provide samples of the releases to broadcast radio stations and TV talk shows to encourage websites, local papers, and magazines to print stories about the artist and his or her career, recordings, and tour dates. Likewise, broadcast radio receives free promotional copies to play on air, which can't be sold, and retail outlets get free goods (over the units purchased from distributors at about 60% of the suggest retail list price) to sell for whatever price they want. Interactive streaming is different, so Spotify must pay labels for the rights to transmit their recordings. Remember labels own the copyright to the recordings. They licensed the right to record, distribute, and sell the recordings from the music publisher, who owns the copyright of the song. When Spotify and other streamers pay the labels, do they pay the recording artist some of the monies collected? Usually not at this point, as it depends on what's in the artist's contract with the label! It does appear that labels are starting to realize the difference in the sums of money they would have paid artists for album sales is not being made by SoundExchange royalties (which we'll talk about later) and streaming royalties collected by the labels.

Cost of Doing Business

Labels have already run the numbers before they sign an act. They use current and past recording artists' financial reports (both successful and failures) to determine their financial risks of investing in the artist. They want to estimate *as closely as possible* their risks and potential profit on every dollar spent. Part of it is just running the numbers on the expenses of recording, development, marketing, promotion, and publicity to estimate a reasonable figure, which will usually be between $800,000 and $1.2 million. Movie companies, Broadway theaters, computer game companies, and book publishers all do the same thing. It takes money to launch a book and millions more to create and distribute a major film. Small independent films may be budgeted at a minimum of $10 million. Welcome to the business of entertainment. It really does take money to make money. Ask yourself, why? Let's run through some of the same concepts the label executives have to before they decide to sign any artist.

The Recording Budget

Understand that artists, labels, music companies, and producers have many options when establishing recording budgets. Artists usually want everything, including many musicians (horn sections, strings, etc.), famous producers, and more time. With egos often huge to even outrageous with new acts, labels have to both persuade and negotiate with them to concentrate on the business of entertainment and keep them focused on what is monetizable.

Superstars know the power they have is in their loyal fan base and how well their last recording did in making a profit. Acts often think they are the next "big thing" (which is good) as it's a difficult business based on rejection. So having a reasonable ego is understood. However, don't push it and always remember label personnel work harder for the artists who are decent and kind and even harder for the ones who remember their first names. Budgets put realistic constraints on the creative process.

Figure 8.2 It's amazing how much knowledge is required to answer the simple question "What does it cost to record a ten-song album?" To find the answer, know the various types of sessions available, the cost of each stage of a recording session and the costs of the producer, recording artists, musicians, audio engineer, and background singers, plus studio rent and per diem rates.

The Three Variables of Recording

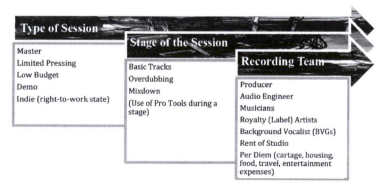

Type of Session
Master
Limited Pressing
Low Budget
Demo
Indie (right-to-work state)

Stage of the Session
Basic Tracks
Overdubbing
Mixdown
(Use of Pro Tools during a stage)

Recording Team
Producer
Audio Engineer
Musicians
Royalty (Label) Artists
Background Vocalist (BVGs)
Rent of Studio
Per Diem (cartage, housing, food, travel, entertainment expenses)

Too high a budget and the artist and label may never make any profits; too little money and the quality of the album may suffer. Although the artist is often responsible for submitting the budget, it's usually the manager or producer who writes it. Still, the process gets everyone in the same ballpark and establishes realistic and solid financial figures. It also puts the artist "on notice" and provides a measure of the financial limits to which the label will respond when hearing the artist's "creative" demands: more horns, maybe—an aromatherapy infusion system to help them reach the right mood, probably not.

So, now, you need to role-play the part of the label, and Table 8.1 is something you, as a business, live with every day. Think of it as the 30,000-foot view of all you and the act need to agree on and do so you both can be in that win-win-win situation.

Table 8.1 Recording Budget Table. Recording sessions can be very expensive. Producers are usually in charge of the recording budgets instead of the royalty artists as they maintain control over the spending. Artists may tend to party more and blow the money, so the labels place financial decisions in the more responsible hands of the producer. Many labels now require the royalty artist to submit a full budget to the vice president of A&R, and then once approved, it's sent to the producer for any updates. It's a great way to make sure the artist is aware of all the recording costs and also helps the artist to understand that in the recording sessions the producer is the boss.

Recording Team	Responsibilities	Union	Basic Tracks 10 a.m.–1 p.m.	Overdubbing 2–5 p.m. 6 p.m.–9 p.m.	Mixdown 10 p.m.–1 a.m.	Totals
Producer						
Audio Engineer						
Musicians						
Royalty Artist (Vocal)						
Background Singers						
Studio						
Cartage per Diem						
Totals						

Recording Team Member Scales and Session Budget

With the cost sheet, we can roughly determine the cost of one song recorded in a studio and then multiply that figure by the number of songs on the album you, as the label, want to record. By figuring the cost of an average song, we may calculate and, thus, predict the cost of a typical 12-song album at master, low-budget, limited pressing, or demo rates (remember what you learned in Chapter 7 and apply that here). Let's look at the cost of hiring a producer first.

We want to start by recording one song using a wonderful modern studio (the best), SAG-AFTRA singers, AFM musicians, royalty artist payments (at triple scale), a known producer, and an audio engineer for each stage of a recording session (basic tracks, overdubbing, and mixdown). The full recording budget (shown earlier) combines the three main factors labels use in determining the recording costs—the type of session, the needed personnel-producer, number of musicians, studio—and the per diem expenses to be in the studio. What is the total cost for the album and at what restrictions? Background vocalists are usually the last to be recorded, so no need to price them out until the third session. Also cartage and per diem are paid for the entire day to the royalty artists and musicians, sometimes even if not all are required on all sessions.

Producers—Types of Session Scales/Payments

The producer is paid a per-side rate (one song), which we're assuming will take four sessions to complete. We've got four sessions scheduled for the day as we plan to complete one song a day over the next two weeks.

Audio Engineers—Types of Session Scales/Payments

Audio engineers make or break all types of producers. Once a major producer finds someone who "gets it" and is quick,

Table 8.2 Recording Budget Expenses. Here are the recording cost sheet figures, as you saw in Chapter 7, for determining an actual recording budget (2015). First determine what type of recording session you want to pay for, master being unlimited pressings and sales; limited pressing restricted to 10,000; low-budget restricted to about $100,000 total; and demo, which is not allowed to be sold if recorded by union musicians.13 Then determine the cost of the recording team members and studio rental, plus per diem for each stage of the session, including basic tracks, overdubbing vocalists, and overdubbing additional musical instruments, plus mixdown. Then once the total is determined, simply multiply that figure by 10 for a 10-song album or 12 for a 12-song album. Bingo, the recording budget is finished.14

Type of Session	Master (3-Hour Call)	Low Budget (3-Hour Call)	Limited Pressing (3-Hour Call)	Demo (3-Hour Call)	Indie
Producer	Negotiated range $2,000–$5,000 per side	Negotiated range $500–$2,500 per side	Negotiated range $500–$2,500 per side	Negotiated range $50–$200 per side	Negotiated
Audio Engineer	Negotiated range $100–$150 per hour	Negotiated range $50–$100 per hour	Negotiated range $50–$100 per hour	Negotiated range Intern: $50 per hour	Negotiated
Musicians	AFM Side Musician	AFM Side Musician	AFM Side Musician	AFM Side Musician	Nonunion Side Musician
	$397.36 per 3-hour session	$223.23 per 3-hour session	$196.50 per 3-hour session	$156.00 per 3-hoursession	Negotiated
	Plus H&W $24.00 per session	Plus H&W $24.00 per session	Plus H&W $24.00 per session	Plus H&W $24.00 per session	None
	Plus Pension $47.64 (11.99%) per session	Plus Pension $26.76 (11.99%) per session	Plus Pension $23.56 (11.99%) per session	Plus Pension $18.70 (11.99%) per session	None
	Total $469.00 per 3-hour session	Total $273.99 per 3-hour session	Total $244.06 per 3-hour session	Total $198.70 per 3-hour session	Negotiated
Lead Musician (Double Scale)	AFM Lead Musician	AFM Lead Musician	AFM Lead Musician	AFM Lead Musician	Nonunion Side Musician
	$794.74 per 3-hour session	$446.46 per 3-hour session	$393.00 per 3-hour session	$312.00 per 3-hour session	Negotiated
	Plus H&W $24.00 per session	Plus H&W $24.00 per session	Plus H&W $24.00 per session	Plus H&W $24.00 per session	None
	Plus Pension $95.28 (11.99%) per session	Plus Pension $53.53 (11.99%) per session	Plus Pension $47.12 (11.99%) per session	Plus Pension $37.40 (11.99%) per session	None
	Total $914.02 per 3-hour session	Total $523.99 per 3-hour session	Total $464.12 per 3-hour session	Total $373.40 per 3-hour session	Negotiated

(Continued)

Table 8.2 Continued

Type of Session	Master (3-Hour Call)	Low Budget (3-Hour Call)	Limited Pressing (3-Hour Call)	Demo (3-Hour Call)	Indie
Royalty Artists (Soloists and Duos)	SAG-AFTRA Royalty Artist (lead vocalist)	SAG-AFTRA Royalty Artist (lead vocalist)	SAG-AFTRA Royalty Artist (lead vocalist)	SAG-AFTRA Royalty Artist (lead vocalist)	Nonunion Side Artist (lead vocalist)
	$227.00 per side or hour	$227.00 per side or hour	$227.00 per side or hour	$227.00 per side or hour	Negotiated
	Health & retirement $28.38 (12.50%) per side	Health & retirement $28.38 (12.50%) per side	Health & retirement $28.38 (12.50%) per side	Health & retirement $28.38 (12.50%) per side	None
	Total $255.38 (per side/hour)	Total $255.38 (per side/hour)	Total $255.38 (per side/hour)	Total $255.38 (per side/hour)	Negotiated
Singers (3-8)	SAG-AFTRA Royalty Artist (lead vocalist)	SAG-AFTRA Royalty Artist (lead vocalist)	SAG-AFTRA Royalty Artist (lead vocalist)	SAG-AFTRA Royalty Artist (lead vocalist)	Nonunion Vocalists
	$103.50 per side or hour	$103.50 per side or hour	$103.50 per side or hour	$103.50 per side or hour	Negotiated
	Health & retirement $12.94 (12.50%) per side	Health & retirement $12.94 (12.50%) per side	Health & retirement $12.94 (12.50%) per side	Health & retirement $12.94 (12.50%) per side	None
	Total $116.44 (per side/hour)	Total $116.44 (per side/hour)	Total $116.44 (per side/hour)	Total $116.44 (per side/hour)	Negotiated
Singer Contractor (on 3-8)	SAG-AFTRA Contractor $50.00 (per side/hour)	SAG-AFTRA Contractor $50.00 (per side/hour)	SAG-AFTRA Contractor $50.00 (per side/hour)	SAG-AFTRA Contractor $50.00 (per side/hour)	None Required or Negotiated

Table 8.2 Continued

Studio Rental	Master Studio	Project/Mini Postproduction Studios	Project/Mini Postproduction Studios	Demo Studio	Demo/Garage Studio
	$150.00 (per hour)	$125.00 (per hour)	$125.00 (per hour)	$50.00 (per hour)	$50.00 (per hour)
	$350 (per session)	$300 (per session)	$300 (per session)	$125 (per session)	$125 (per session)
Cartage (Handling of Instruments)	**Cartage** $12.00 (per session)	**Cartage** $12.00 (per session)	**Cartage** $12.00 (per session)	**Cartage** $12.00 (per session)	None
Per Diem (Royalty Artist Only)	**Per diem** $150.00 (per day)	**Per diem** $75.00 (per day)	**Per diem** $75.00 (per day)	**Per diem** $75.00 (per day)	None

respectful, knows the software, music, acoustics, board, and mics, and works well with everyone, they usually use them all the time. Audio engineers are hired on reputation and the hit records they've cut, plus location and networking with major producers and A&R personnel at labels. Take from this that success is often as much defined by your skills with the tools of audio engineering as it is by your personality, flexibility, and willingness to listen and respond to what artists and labels say they want in the final mixdown. Also remember that your expertise as someone trained in sound and sound environments makes you a creative person, so bring that and a great attitude to the session and you may have steady work for a long time!

Lead Musician (Instrument)—Types of Session Scales/Payments

We'll hire AFM union musicians, who are the best in the world. They are expensive, yet save money in the long run as they are comfortable in studios and quick, and the final quality of the recording will be much better than if we thought we saved money by hiring nonunion and nonstudio musicians. However, the union musicians are paid on a three-hour call or one "session call," as each session is three hours long. The producer usually calls the lead musician, who then calls the other side musicians. The lead musician is paid double scale and holds the signature card for everyone to sign after the session for payment.

Side Musicians—Types of Session Scales/Payments

Let's return to Table 8.2 for the cost of hiring an AFM side musician as the 2015 scale rates are listed. Once you decide on the type of session, then you'll have to multiply the cost of one musician times the number of musicians required for the session for the basic tracks or overdubs. The way most musicians, as we saw with the engineers, are hired is based on networking and reputation, mostly based on word of mouth, so the work done in a session becomes part of the musicians' "business card."

Major Royalty Artists (Vocals)—Types of Session Scales/Payments

The purpose of this "party" is to help our label's royalty artist hit the big time. The studio is where the magic happens first. The type of session you book depends on the purpose of the recording. For example, the master scale rates give the labels the right to sell as many units as possible, whereas the lower demo rates for union musicians create recordings that can only be given away as demos.

Background (Vocals) Contractor—Types of Session Scales/Payments

The SAG-AFTRA contractor is the leader of the background vocalists and is required anytime three or more BGVs perform on the session. Rates are the same based on the number of hours to record one song or the number of songs recorded in one hour. The contractor is the leader of the other singers, sometimes helping with arrangements and the one who carries the signature card for proof of work. Contractor scale is $103 per hour/song, plus 12.5% for health and welfare. The scale is the same for all types of union sessions, and thus the number of vocalists on low-budget and limited pressings is often very low. It's really rare to find backup singers on demos, because of the cost of the singers and the fact that union demos are not sellable.

Background Singers (Vocals)—Types of Session Scales/Payments

The background vocalists receive a scale of $103 per song or recorded song (whichever is greatest) plus 12.5% for health and welfare, for a total $131.38.

Studio Rentals

Recording studios have sliding scales of rental costs correlated to the purpose of the session. Master studios are usually the most expensive with the newest equipment, plus they can charge based on their outstanding reputations established by the names of the acts, producers, and world smash hits recorded in their facilities. Studios are usually rented on a three-hour sessions call (which is the most expensive) to all-day rate, which is usually 25%–40% less if booked for a week at a time.

First Overdub Session (One Song)

The second session is usually recorded after the basic tracks were recorded in the first session. Usually the bass, drums, two guitars, keyboards, and a scratch vocal are recorded on the basic tracks session. Then, what instruments do you want to add to the basic track instruments? Now is the time to add different instruments, such as horn sections and strings. If we're doing the one song in four sessions over one day this session is scheduled from 2 p.m. to 5 p.m.

Second Overdub Session (One Song)

Now that we've got all the instruments recorded, it's time for the royalty artists to "do their thing" and prove they're worth the big bucks. Royalty artists may want to record the vocal by themselves in the studio with only the producer and engineer, while others like to record the vocals with a live band or orchestra during the basic tracks. Some can nail the required vocal easily; for others it's note by note, or, as the engineers say, punching and pulling second by second. If all else fails, there's always auto-tune for the famous who can't sing anymore. Some royalty artists tend to tense up in the situation, so their scratch vocal (they recorded on basic tracks to make sure the rhythm and timing were correct) may be used instead. As an act ages, phase adding mics and acoustics may be used to "capture" whatever small volume of the voice is left. The harmonizing singers are last. Their job is to provide the ah's and lyrics, but also to cover mistakes and fill in the gaps. Also if the BGVs are recorded before the royalty artist's track is laid, it's difficult for the star to sing around them, so the BGVs usually record last. Musicians should be finished with all the tracks, so they will not be at this session. Royalty artists (signed with a label) are usually paid triple scale for their singing. Take the hour scale times three hours for the session times three again for the triple rate and figure what this will cost you. If we are doing one song at master scale in our four sessions over one day, this session is scheduled from 6 p.m. to 9 p.m.

Mixdown Session (One Song)

The night session from 10 p.m. to 1 a.m. is usually for mixdown of the recorded tracks into a two -track master that will then be "mastered" into the matrix to press CDs or uploaded to various sites for sale and streaming. Notice that all the recordings are finished, so no musicians and vocalists are present, just the engineer (sometimes) and the producer. Mixdown involves three issues that need to be considered by the producer and the engineer: the musical mix based on the genre; the stereo effect balancing out what is on the left- and the right-hand side; and the depth of the actual recording, giving it the 3D quality of sound. There is also an increase in the musicians' scale for the 10 p.m. to 1 a.m. session and, therefore, we are finished with them before this session. As an example, the total cost for a side musician is $469.00 per three-hour session on the first three sessions (including scale at $379.36, plus health and welfare at $24, and the pension fund contribution of 11.99% of scale, in the case of master scale, is an additional $47.64). However, during the 10 p.m. to 1 a.m. session, the basic scale is increased to $463.12 plus the $24.00 for health and welfare, and the pension fund of 11.99% of scale ($55.59), for a total of $542.71. That's an increase of $73.71 per musician and the lead musician is also paid an additional amount.

Cartage and Per Diem

Additional session expenses include cartage, which is the cost of instrument delivery and tuning, and the per diem expenses budgeted for the royalty artist, such as the hotel or label's condo expenses, plus limo and food.

Done, Done, and Done!

It is now time for me to tell you that most labels require their recording artist to actually submit what we've just done. So, while you may have questioned why we did this and why it is important, here is the reason. If you are the artist, or you are the label, you both have to know how much your recording is going to cost. It helps the artist understand the business and the label grasp its financial investments in their creativity before they step into the studio. Can we all say wakeup call! It is your role to be both the entrepreneur and the executive to be sure you are getting the best deal. Then the very last thing is to multiply the cost of that one average song by the number of songs on your album, and bingo, you've got your recording budget ready to give to the head of A&R at the label for approval. This discussion was only about the cost of the recording. Think about what it would cost if we were shooting a feature-length movie, staging a Broadway play, or developing a new computer game.

In the next chapter, we will consider in more detail how the major labels operate, especially how they organize their budgets and their operations to maximize their chances of staying in business in the changing environment that surrounds them.

Notes

1. MacKinnon, Eli. "Edison Voice Recording Is Old, but Not Oldest." Live Science. October 26, 2012. Accessed October 14, 2015. http://www.livescience.com/24317-earliest-audio-recording.html.
2. Ibid.
3. Schooner, Steven. "The History of Magnetic Recording." November 5, 2002. Accessed October 15, 2015. http://www.aes.org/aeshc/docs/recording.technology.history/magnetic4.html.
4. "Valdemar Poulsen Biography (1869–1942)." 2015. Accessed October 21, 2015. http://www.madehow.com/inventorbios/94/Valdemar-Poulsen.html.
5. Orsatti, Ken, and Melanie Webber. "How SAG Was Founded." 1995. Accessed October 13, 2015. http://www.sagaftra.org/history/how-sag-was-founded/how-sag-was-founded.

6. "When Was the First Computer Invented?" 2015. Accessed October 15, 2015. http://www.computerhope.com/issues/ch000984. htm.

7. Ibid.

8. "Gottfried Leibniz." 2015. Accessed October 16, 2015. http://www. computerhope.com/people/gottfried_leibniz.htm.

9. "Free Online Modify and Convert Text Tool." 2015. Accessed October 16, 2015. http://www.computerhope.com/cgi-bin/convert.pl.

10. Ibid.

11. "Warner Music Group, Investors Relations." News Release. March 31, 2013. Accessed October 15, 2015. http://investors.wmg.com/ phoenix.zhtml?c=182480&p=irol-newsArticle&ID=1819485&high-light=.

12. Bowen, Jimmy (as quoted by Larry Wacholtz), *Inside Country Music*, 136, Billboard Books, New York: Watson-Guptil, 1986; Wacholtz, Larry. *Lights, Glitter and Business Sense*, Thumbs Up Publishing, Washington: Cheney, 1984, 184–198.

13. "Scales, Forms & Agreements." Nashville Musicians Association. 2013. Accessed February 20, 2016. http://www.nashvillemusicians. org/scales-forms-agreements.

14. "Sound Recordings." Accessed February 20, 2016. http://www. sagaftra.org/sound-recordings/sound-recordings.

ODDS OF THE GAME 9

Getting Noticed

Acts and bands are still often discovered through local personalities: radio station DJs, local promoters, and record storeowners who inform regional label distribution and promoters about the "local buzz" an act is creating. Without the band's knowledge, labels have personalities, street promotion, or regional employees attend a live performance to determine the act's potential. Often the response of the audience is seen as just as important as the band's actual quality of performance. The band's originality, uniqueness, age, industry knowledge, and sellable image are also evaluated to determine if they can be molded or developed into a major recording act.

At the same time, wannabes write their song, compose it on Pro Tools, and upload it to YouTube, thinking they are about to become rich and famous. It takes two to three times of doing this until they often realize nothing happened, and it's best not to expect the phone to start ringing or a wonderful message to pop up in email.

What happened? Some may realize that either no one ever saw or heard the upload, or nobody liked it. However, once in a while, someone will dig it and tell someone about why they enjoyed it, and then they email someone and the buzz, the word-of-mouth wow, starts. Great! The odds are less than 1% of becoming rich and famous: however, there may still be a decent living in it *if* you know how the industry works. From the previous chapters you know you'll have to join one of the industry professional representation companies—ASCAP, BMI, and SESAC—if you want to be paid as a songwriter or music publisher. I hope you also understand that either you need a music publisher to represent your material (songs) or you'll have to start your own publishing company if you want to be paid for the copyright (back to Chapter 3 if memory serves). And if you want your songs pitched to major acts, then you'll need professional song pluggers, who in the majority of cases deliver songs to A&R at the labels, producers, artist managers, artists themselves, and music supervisors for film placement sync licenses and other opportunities to get your work before the buying public.

The Level of the Game

Record labels have to filter out the wannabes, invest in the most talented, and use the best producers, musicians, promoters, and marketers in the world to launch an act. It's the wannabes vs. the experienced professionals—no different from a local high school football team vs. the NFL's New England Patriots. Who do you think knows the game better and who do you think is going to win? Of course, it's wonderful that any of us can play the game and throw a self-composed computer recording up on the Internet. However, most of us will fail to gain a yard in the entertainment *field of play* and that may be *the best thing* that could happen, because a lack of consumer interest is a form of knowledge that may help you or an act you are interested in seeing develop spend more time on their business plan and business channel development, as discussed in Chapter 4.

Good isn't good enough! If the song isn't a wow or the consumers do not feel the recording's emotional sensation, then the act's name, image, and brand are ignored like the plague. Your goal as a creative person is to have your works speak for you. To do so, your knowledge as an entrepreneur needs to be rock solid, the marketing and audience for your song clearly defined, the song promoted professionally in various areas of the media, the social networks buzzing, and then magic *might* be accomplished. Labels, even though they can't make any serious money selling albums and singles anymore (unless the act is a major heartthrob superstar), are still vital

349

to the success of the creative writers, musicians, and artists as they have the money, connections, experience, and business knowledge to succeed. They've just had to invent a new way to cash in once the act has been discovered as the talk of the town or region. The labels are the creative and executive decision makers who can turn local success into national and, oh yeah, world success. Let's take a quick look at what the three major labels have accomplished.

Warner Music

In the late 1990s and early 2000s, consumers' use of P2P, the AOL merger, and a few questionable management decisions caused serious financial problems for the huge media corporation Time Warner. As record sales went into the toilet, labels became less attractive financial investments. In 2003, Time Warner sold Warner Music to Edgar Bronfman, Jr., and an investment firm. Bronfman had previously owned Universal Music and sold it to Vivendi. Bronfman, despite having very little successful music business experience in his past, seems to be one of those rare executives who just *get it*. In the book *Fortune's Fool* (2010), he admits, "The days of the record industry dictating formats to the consumers are gone, and they're not coming back."[1] Bronfman knew the old business model was broken and gone forever unless Congress restored some copyright protections against illegal downloads, which we all know has not yet occurred.

The Catch-22 Economies of Scale

One of the ways Bronfman changed the industry was by making all his new acts sign 360 deals. In the system he stepped into as a new record label owner, and as we saw in the previous chapter, labels would frontload the artists and pay for all the recording, development, promotional work, publicity, and marketing expenses, which for a new act averaged $800,000 to $1.2 million and between $3 million and $5 million for established superstars in much of the United States. Those figures are re-arranged in global terms to $500,000 to $2 million based on the location and economics of the local territories or markets. Let's face it: it costs money to take raw talents and develop them into desirable recording artists. And once they become famous, then their egos and corresponding advances also tend to rise. The labels recouped 100% of the all-ins and 50% of the marketing expenses out of the artist royalties on each unit (recording) sold until the entire debt was paid. In addition, if the marketing costs (including promotion and publicity) were tied to tour support (and they usually were) then the 50% recoupments were raised to 100%. Some would say, what a rip, but honestly, the label is financing an act and their creativity to the tune of millions while the acts have to just show up and sing.

Labels made or lost money to the way the deals were structured with their artists and how many units were sold. At least that's the way it was. Thus, the deals were often perceived as favoring the label, which would often break even between 75,000 and 150,000 units sold, and against the act, as they wouldn't start receiving any royalties until the recordings often sold two or three times as much. However, what if the act took off and became famous? How much money did it cost them? Zero or close to it. And once famous, they received 100% of the touring, merchandise, and corporate sponsorship royalties and other sources of revenues that the label made possible by investing about a million into their careers. But times have changed!

The Catch-22 Solution—The 360 Business Model

We can all see the handwriting on the wall, as the labels can't sell enough recordings any more to make a profit on their investments. Though Bronfman changed things by offering the 360 deal, it is usually the same deal as was previously made for albums and music publishing. But the 360-deal concept comes from the added percentage of the artist revenues from the touring, merchandise, corporate sponsorship, and other revenue required to be paid to the label. That's the only way labels could stay in business. They had to start taking a percentage of the artist's share of the income generated by their fame. At the same time, new acts realized that playing the Pro Tools and YouTube self-promotion game wouldn't get them anywhere. They still needed the financial support, knowledge, experience, industry, and media contacts of the labels if they were to have a shot at a serious career. New acts still need the professional producers, songwriters, musicians, and business executives to make it in the business. Sure, there have been a handful of acts that have made niche careers for themselves on YouTube, but after we count our fingers on both hands, it's hard to think of others. The economies of scale had to be rebalanced in a manner that brought the balance between financial gain and loss back to a reasonable level. The 360 deal accomplishes that, but it is not the final answer.

Professional Experience

It appears the professional creatives and executives in the industry really are special and know what they are doing to create the hits. Some of the superstar acts, once they have become famous, have started their own labels to save money; yet, they usually still need a major label (Warner Music, Sony, or Universal) to distribute and promote the product. So, what do major entertainment companies really do to make a profit in the industry? What do their business models tell you about how to succeed in the business as a business? Let's take some time and analyze Sony's business model as an example of a company trying to evolve with the larger entertainment and music industry.

Sony Music

The music business for Sony was directed through Sony Music Entertainment (Japan), and Sony Music Entertainment International (SMEI) in the rest of the world, until the Sony/BMG merger (Bertelsmann Music Group, formerly owned by Bertelsmann AG of Germany). Sony has moved its world headquarters for music publishing to Nashville,

Tennessee, and has explored distribution of its movies, shows, and music directly through TVs and other devices consumers own. Buy a Sony TV and you are connected directly to Sony Entertainment over the Internet. Sony has struggled with its entertainment production side of the industry compared to the product manufacturing side. The business of entertainment can be a risky business, no matter how thoughtful the leadership is in generating new models of entertainment distribution. According to their Form 20-F filed as an annual report with the U.S. Securities and Exchange Commission (SEC) in 2015, Sony expressed its concerns about the impact of copyright infringement on the future of their business.

These reports and filings help investors analyze the potential financial risk and benefits when buying different types of stock. Once again, we have a chance to look at the difference capitalist and other types of government make in entrepreneurial opportunities, and not exclusively in the entertainment and music industry. Smart people who want to become rich let their money make more money. Yes, that's what I said: let your money make you rich instead of working day to day to pay bills to survive. We have that opportunity only in capitalist governments, as any of us can take a few bucks each month and place it into a stocks-based money market account. In a few years, we'll actually have more money! Love it. Now investing in a single company's stock means that you are buying into a business where you may make or lose money based on how successful the business is. A word of caution, as I've blown this several times. I have invested in single company stocks and lost money. When I was a lot younger I could have and should have bought Apple stock for less than $5 a share. Today, I'd be a millionaire! Stupid me! However, I have also invested in 401k plans and IRAs for retirement, so if I ever retire I'll have a little money to live on besides Social Security (if it still exists). Sony has been a wonderful investment for most, but is it still? Let's take a look at the whole corporation and then the label.

Sony is a global entertainment company that has many "divisions" that produce entertainment products, such as movies, TV shows, musical recordings, and computer games, plus the devices to enjoy them. That costs a lot of investment money on the front end, and there are many risks involved, including competition, competitive introduction of new products, research and development, restructuring due to market changes, acquisitions, joint ventures and investments, and price structuring to wholesalers and retailers. The reliance on external business partners, inventory of parts and products, foreign exchange rates, world economic conditions and financial markets, laws and regulations in different countries, hiring and signing the most talented workers and creative artists, cost of operation indexes, increasing cost of equipment and talent, labor interruption, value of equity, securities, interest rates, the destruction, loss, alteration, or unauthorized access to data, litigation, and brand devaluation are just a few of the issues entertainment corporations such as Sony must consider.[2] Breathtaking, isn't it, and so diverse—it is like its own little world!

Sony's Entertainment Business

Continuing to read the 2015 SEC filing, we learn that Sony defines its content business as entertainment productions and products, including movies, music, and computer games, and they state these "are subject to digital theft and illegal downloading." Ouch! Sony claims,

> *software and technologies that enable the duplication, transfer or downloading of digital media files from the Internet and other sources without authorization from the owners of the rights to such content have adversely impacted and continue to threaten the conventional copyright-based business model by making it easier to create, transmit, and redistribute high quality, unauthorized digital media files.[3]*

Sony is also concerned about sales of physical entertainment platforms as the industry slides toward streaming.

Sony appears to be saying that if the decline accelerates and if streaming cannot increase enough (by generating paying consumers, perhaps?) to offset the decline in sales, then the music industry may be negatively impacted. That is simply a way of saying bye-bye, if the decline in potential profits continues. To counter the P&L statement losses, Sony has over the last three years completed restructuring "in the amount of 77.5 billion yen, 80.6 billion yen, and 98.0 billion yen in the fiscal years 2013, 2014 and 2015."[4] About 20,000 employees' jobs were terminated. In addition, it's clear that if the decline continues to increase then it may no longer be worth the investment in the music business. As the authors of the filing explain, "While alternative models for selling entertainment content have emerged, such as digital downloads and subscription streaming services, these revenue streams may not be sufficient to offset the decline in physical media sales."[5] All of the entertainment companies are facing the same situation. They also worry about their economies of scale, distribution processes, and profit and loss statements tied to the lack of legal copyright enforcement and shifts in consumer behavior. But they also are very concerned about licenses and intellectual property rights enforcement on which their business depends. Here, the filing directly speaks to issues of ownership:

> *such licenses may not be available at all or on acceptable terms and Sony may need to redesign or discontinue marketing or selling such products as a result. Additionally, Sony's intellectual property rights may be challenged or invalidated, or such intellectual property rights may not be sufficient to provide Sony with competitive advantages. Such events may adversely impact Sony's operating results and financial condition.[6]*

Sony Pictures is a signatory to union contracts, including new three-year agreements with each of the following:

> *The Screen Actors Guild-American Federation of Television and Radio Artists ("SAG-AFTRA") ... American Federation of Television and Radio Artists ("AFTRA") . . . the Union of British*

355

> *Columbia Performers; and the Directors Guild of Canada . . . the International Alliance of Theatrical and Stage Employees ("IATSE") and the Teamsters.*[7]

All of those agreements and the jobs they protect disappear if Sony ever decides to drop its entertainment production businesses. As you can see from these excerpts, the label executives are arming the federal regulatory agencies with precise information on the impact of streaming and illegal downloading, which could be used by legislators to explore refinements and amendments to the copyright, fair use, and safe harbor regulations and practices.

Legal Issues

In 2011, and again in 2014, Sony made the news when it suffered cyber-attacks that disclosed private business information, credit card misuse, and identity theft issues. As a result Sony now faces legal action costing millions to defend. According to the 2015 F-20 Form, "In the fall of 2014, Sony Corporation's U.S. subsidiary, Sony Pictures Entertainment Inc. ("SPE"), was subject to a cyber-attack that resulted in unauthorized access to, and theft and disclosure of SPE business information, including employee information and other information."[8] It is not unusual for large corporations to face legal actions. Most are irrelevant cases that cost more to defend than they are worth. However, the people who pulled off the cyber-attacks may or may not be caught for their illegal stunts. They may live in a third-world country not allowing Sony or nations to seek legal actions. However, cyber-attacks can cause serious havoc for "brand value," personnel, employees, stockholders' confidence in investments, and consumer relations. They cost additional money, stock value, and, sadly, even as happened here, financial market devaluation of the business. Looking deeper into the report, we also see additional types of businesses Sony owns, including electronics, entertainment devices, banking, and life insurance. In 2015, sales grossed 8,215,880 yen (in millions). You can calculate the conversion rate into dollars with 109.0 yen equal to 1 dollar and 138.8 yen against 1 euro. Pretty amazing, right?

Cost of Business—Sales, General, Administration, and Operating Expenses

As of March 2015, Sony employs 131,700 people worldwide, of which 7,600 work in the Sony Pictures division and 6,900 work in the music segments of the company. The majority of workers—93,900—are found in the electronics division, making devices. Some employees lost their jobs due to restructuring based on economic conditions and consumer sales. Approximately 50,000 are employed in Japan and 81,700 are located outside of Japan.[9] Operations expenses are supported by stockholder share values in Japan and in the United States. The value of Sony stock has in the last year ranged from $28.65 per share to $15.93. Sony's contractual obligations and commitments total about 19,435 billion yen over the next five years. Long-term contracts with recording artists, songwriters, and entertainment companies total about 6.3 billion yen and the expected cost of film and TV production and programming rights is budgeted as 12.6 billion yen. However, the long-term commitments over the next five years are reduced to about 5% of those figures.[10]

Profit and Loss Statement

When looking at Sony's expenses in mobile communications, game and network services, imaging products, and home entertainment, plus device development and production, adding in wages, administrative costs, and operational expenses, including office spaces, the company profits were flat at 39,729 million yen before taxes. Total operating income was actually a loss of 220,436 million yen, primarily in financial services and corporate job elimination.[11] It didn't make much in profits except (and here's where we want to pay close attention) in the exchange rate between the yen, dollar, and euro. In other words, Sony's real profits for financial year 2015 were not tied to product sales and so forth, but to the fluctuating exchange value of the yen—less valuable compared to the dollar, more valuable compared to the euro. Thus, Sony was able to continue to pay its Japanese employees their same salaries (but it cost them less due to the devaluation of the yen) and make money from the sale of their products and services paid in (higher-value) dollars and euros. Amazing! What a crazy business, and my guess is that most of us had no idea about the cost of doing business as a major entertainment corporation. Now that we've looked at one of the huge mother ship corporations, let's continue our look at the last of the three remaining major record labels, Universal.

Universal Music Group

Music Corporations of America was renamed Universal Entertainment Group as part of Edgar Bronfman's revitalization of MCA, which he bought from Matsushita Electrical Industrial of Japan. He and stockholders used revenues from their sole ownership of Seagram Liquor of Canada, which at the time was posting about $75 million in profits per year.[12] He also bought Polygram in 1998 from Philips Electronics of Holland and merged the two entertainment giants under the umbrella name Universal Entertainment Group.

Then he sold all of his holdings of Seagram Liquor Company, plus Universal, to Vivendi of Paris, France. Vivendi also bought 51% of Canal Plus, a film production and pay-TV company for a grand total of around $34 billion in stock and cash. What the heck, a billion here, a billion there—soon we'll be talking about real money. It also assumed the $6.6 billion debt Bronfman created by buying the various entertainment companies to form Universal.[13] Then NBC-TV bought most of the old film masters from Vivendi and EMI (Electrical Musical Instruments) of London, England (the owner of the Beatles' masters and many others), went bankrupt. The EMI publishing business (the © business) was sold to Sony and the master recording business (℗) was sold to Vivendi and placed under the Universal Music Group banner. Universal is now considered the most successful major label in the world, owning about 50 labels, including Decca, Deutsche Gramophone, Disa, ECM, Interscope, Geffen, A&M Records, Island Def Jam Music Group, Mercury Records, Polydor Records, Show Dog–Universal Music, Universal Motown Republic Group, Universal Music Latino, and Universal Music Group–Nashville. It also owns Universal Music Publishing Group, containing more than 3 million titles and a merchandising business with it subsidiary Bravado, which designs branded products for tours and online specialty stores. It also has deals (as do Sony and Warner Music) with streaming providers, such as Spotify, iTunes, and others. It owns 90% of Dailymotion, an aggregation and distribution platform for promotion, film trailers, entrepreneurs, independent acts, and others who want to provide services and products to specific targeted consumers.[14]

Other Types of Labels

Other types of labels besides the majors are still around and a few may become powerful by the success of the acts they have signed. The keys to being a successful label include (a) the artists you sign, (b) world distribution (before the Internet), (c) promotion, and (d) publicity (which only the majors can usually afford).

Disney

It's hard to think of Disney Music Group when we're considering major labels unless you're a teenager or maybe a little younger. Both Hollywood Records and Walt Disney Records primarily support the other parts of the Walt Disney Company, which are mostly associated with the film, television, and cable systems, ESPN, the ABC Network, and book publishing. It owns 35 radio station networks and, of course, has its theme parks, stores, and cruises. These include names you know: Touchstone Pictures, Lucas Film, Walt Disney Pictures, DisneyToon Studios, Walt Disney Animation, and Pixar Animation Studios.[15] Disney is a great company that develops young talent acts and then places them on their labels, radio networks, cable and TV show, movies, and, if they become teen idols, on the road in concert tours (discussed more fully as a revenue generator in Chapter 12).

Affiliate Labels

Affiliate labels are usually partially owned by one of the three major labels, an artist or group of artists, or another type of industry insider. Because there is shared ownership, some of the money may come from the major label or from the industry personality who shares ownership. It all depends on the power of the acts, personalities, and potential profits of the major and affiliate labels. Album releases are usually distributed to the local promoters with the major label distributor, which provides unit sales through traditional retail outlets, and promotion and publicity through the label's established media connections. They are different from independent labels as the deals are limited to acts and recordings the major labels want to work with and release. A good example is Madonna and her former affiliate label, Maverick Records, which was with the same company and the same label she was signed with. However, if she had her own affiliate label, then she could sign acts and have them distributed. If they made money, great, and if they lost money, they became a tax write-off.

Independent Labels

Indies are usually controlled, owned, and established by major industry personalities, such as record producers, artists, or label executives. Sam Phillips of Sun Records discovered Elvis and many other acts. Producers use their own labels to land distribution deals with major labels, and established artists may form a label to distribute artists with an assumption the business won't succeed as they need tax write-offs. The most successful independent labels include Curb Records, which over the last 30 years has launched many superstars, and Big Machine, which has Taylor Swift and is now signing many new and independent acts, such as Stephen Tyler, White Lightning, Reba McEntire, Florida Georgia Line, Martina McBride, Ronnie Dunn, and Tim McGraw. Independent labels need the cash to cover all the front-end expenses themselves and, therefore, they own the copyright of the recording, or, as we are starting to say, the ℗. Then they can negotiate with any of the three majors for distribution, promotion, publicity, and other services. As an example, Taylor Swift is an act on Big Machine Records, which is distributed by Universal Music Group, owned by Vivendi of Paris, France.

The range of business processes, no matter what the type of label, is largely similar to that shown in Figure 9.1.

Figure 9.1 Labels use a new business model that requires them to still find and sign wow talent, create wow recordings, and then activate a marketing plan based on promotion, publicity, price, and placement (distribution). The difference is that the price is not free or a low monthly fee, which provides access to the recorded music, such as Spotify instead of the sales of CDs. Thus, revenue is generated on access to the sound, live shows, and point of purchase of merchandise, as well as companies that want their products to be associated with the act's brand.

Administrative Processes & Revenue Streams-Common Across Label Types

Find & Sign Talent

- Labels request showcases, investigates acts, analyze market cost of imaging and branding & assess the size of the potential market

Creates Recordings

- Labels know most "units" will not sell, yet they still need to have a wow song, tied to a wow recordings to market, brand & sell the act. Remember the steps in the recording budget process we covered in the previous chapter-(1) types of sessions, (2) stages of the recording sessions, & (3) cost of the recording team and studio.

Promotion

- Provides a "sample" of the recordings through YouTube, radio, media, and streaming. The exposure gained by promotion is to increase sales of albums, tickets, merchandise, corporate sponsorships, and branding product tie-ins.

Publicity

- The artist's backstory based on their image and brand when seen on TV talk shows, award shows, stories planted in the popular press and social media. Most acts have seven images based on different psychographic lifestyles, which enhance the consumer's emotional connection to the act, which improves profits.

Distribution

- In the old days this was easy, simply sell units of recording as CDs at Mass Merchandisers, (Best Buy), Rack Jobbers (Kmart, Walmart and Target), One-Stops (Mom & pop stores and chain stores), Record Clubs & PI Advertisements (per inquiry) on late night TV. Now instead of album sales which has lost up to 80% of its market, labels revenues are generated by access to the music through streaming such as Spotify, Apple, & Amazon's Prime, plus master recordings placed in films, branding, concert

Vanity Labels

Almost anyone can create music on a computer with digital technologies. Small, vanity, and one-man operations can get noticed by uploading materials (songs, recordings, videos) to the Internet, yet to reach the level of an independent company, they usually still need the revenue generated by album sales. Budgets are usually dramatically reduced to match a small, niche consumer base, which is willing to buy into unknown acts. Examples include fundraising on Internet sites, commercial endorsements, or selling unique types of pressings (e.g., vinyl and extended albums) to a small group of fewer than 50,000. It's a great place to start. But if you've signed a long-term seven-year deal with one of these labels and your career exploded, you're stuck with the small label. They may then (and usually will) sell your deal and sometimes that's not in your best interest if your deal points are weak. Then you're subject to the terms and conditions of the new label unless they happen to be nice guys and tear up the old deal and give you a better offer. But don't count on it. Vanity labels created by unsigned artists to sell their own recordings on iTunes or off the stage can't sell much of the recorded product anymore, even after a live performance. That is why you need a business behind you to make your career a success.

Specialty and Virtual Reality Labels

Specialty labels have small shares and a small presence because they are focused on a genre or a niche market, such as children's music or comedy albums, or a type of unique distribution, such as digital downloads, cell phones, or websites like MySpace and Facebook. The virtual reality labels do not often press CDs; they keep expenses low by providing only digital downloads to the Internet users, consumers' cell phones, and PDAs.

Promotion Branding Labels (PBLs)

Having recently emerged as a marketing tool for other products, Starbucks formed its own label and signed Sir Paul

McCartney. Many companies are signing artists in an attempt to sell other products. Others form labels to support manufactured artist products or related merchandise, such as dolls or depictions of artists in cartoons, or sign children and teenage stars to TV shows and movies. Now that you have read about how some of the leading labels were founded, how they have sustained themselves, and how they keep pace with changes in the entertainment and music industry, how about we spend some time considering how to get you into that world as an artist? What do you do to get "noticed" and get your career on the road?

Labels' Competitive Advantage

Unlike you now, most recording artists rarely enter into a recording deal with any real understanding of how labels or music companies work. They are often on an ego trip, dreaming of becoming rich and famous from the deal. Instead, artists need to consider what the label can and will do for them to launch their career and what they can do for the label to make a profit on their investment. Most recording artists never received any royalties from their recordings. The truth is simple: the record labels are not in the business of making artists rich—they are in the business of selling discs and digital downloads that will make *them* rich. However, we know that selling enough discs to make a profit does not work often anymore, so welcome to the 360 deals. Thus talent, as defined by the labels, is related to the amount of money they can make from an artist's recordings, tours, merchandise sales, and corporate sponsorships.

This is the merger of the creative system and business system, where neither can accomplish much in the worldwide market without the other, as we have seen. The creative system provides the songs, vocalists, musicians, audio engineers, and producers. The business system, consisting of labels, provides the money to fund the recordings, promotion, publicity, and distribution. It's a simple formula: creativity plus business administration and marketing equals potential profits for both. How much money each makes depends on how successfully the act sells recordings, tickets, merchandise, and corporate sponsorship, plus the almost irrelevant royalties paid on streaming. On the front end, labels treat artists with great respect. On the back end, in closed meetings, it is the number of units moved, merchandise sold, sponsorships, touring revenues, streams, music publishing, synchronization royalties, branding partnerships, plus break-even points achieved, that determines the respect the label has for an artist. This sounds harsh of course, but this is a business and who else would spend millions on an unknown artist for the promotion, publicity, and distribution of artistic products?

Investing in Talent

Labels have invested more than $20 billion into talent production and marketing in the last five years.[16] And still many acts complain about how the labels are screwing them. Well, what kind of contract did you sign? The labels have stepped up and put the money where their mouths are and strongly supported creative artists in the most difficult times of copyright infringement. Why are they doing it? Simple—they are building the future stars of tomorrow who will generate profits for all. Seventy percent of unsigned acts know they need a record deal to launch their careers as they do not know how to do or have the money for the production, promotion, and marketing of their brand.[17]

Getting a Deal

Here is where the business model has really changed from in the old days, when you just had to convince label executives of your talents. Now, we also need a very successful fan base, tour, buzz, and record sales, and hopefully some idea of your image and potential branding. Start locally, and if the consumers start talking about your act, the labels will find out (ears to the ground) quickly as they are always looking for the "next big thing." However, getting signed now means more than selling albums, so branding, merchandise sales, and touring potentially play into the decision. How can you get noticed once you make the buzz happen?

Websites

Acts are rarely first noticed on websites such as Facebook, MySpace, YouTube, or some other related site. Sadly, the odds are millions to one of being discovered, as most of the artists posted do not have the potential to develop into major recording acts. In addition, getting noticed on a social network is not enough to interest a label. However, it is a good first step to see if you can build a real fan base for your act. Label executives know that recordings made with the use of Pro Tools and other acoustic improvement software do not give an act the necessary talent and personality required to record for major labels and to perform live before huge audiences. Again, take a different approach. Develop a huge following on the Internet. Tie it to a local or a regional buzz based on your live performances and units sold. Get some local airplay and create your own fan base, and you have significantly put the odds of being "noticed" by a label in your favor.

Showcases

Live performances before a small invited group of guests (including music industry A&R personnel, managers, media, and concert promoters) can be a positive way to introduce your act to powerfully established industry personalities. Of course, we are assuming a certain level of success before you offer a showcase to the labels. However, if you've achieved a following on the Internet, a local or a regional buzz, airplay, and retail sales, it may be time to invite the industry to a performance.

Pitch Memo

If a showcase or tour is well received and you've built a significant fan base, it may be time to send a pitch memo to the heads of A&R at various labels. At this point, you're just letting the label know of your success and desire to connect. The first thing they do is check you out on the Internet to see what you sound and look like. It better be damn great or you're done. If they like what they see, then they'll usually make a few phone calls to radio stations, stores, and personalities they know in your local area to see if they have heard of you or if there's any buzz about your act. There better be! Clearly, timing is important, and a website before you are ready is a killer to your future career. If you have no buzz, that means you haven't arrived in your local community, so why should the label care about you? If they can't find you easily the first time they look for you, they probably will not look again—this is an overwhelmingly competitive market, and there is no time for searching for the elusive "you." If there is nobody excited about you and talking about how cool you are, there is no point in trying to go big. Without any local buzz and a great Internet presence your "cold turkey" pitch (first-time contact) will rarely be answered unless investment money is also included, as in a master tape lease deal with a "no financial loss" guarantee for the label. However, they will still not be interested unless they are wowed, and besides, the pitch memo is really just an introduction. As difficult as this sounds also remember that the three major labels always have one ear to the ground and want to be the first to discover the next big thing. So, when you're ready, let them know about you and what you do with a well-crafted entrepreneurial and creative plan of attack.

Financing Your Own Deal

Most of those out-of-nowhere deals have to be rejected by the label because of the risk associated with self-financed acts (unless you're a proven success, like Garth Brooks). However, a pitch memo is a notice to the label that you are serious about pursuing a deal with them. Be very careful as there are lots of independent labels that will promise you the moon, take your money, sign you, and then nothing happens except you get the shaft. They may even put out a few albums and then just say they did their best, but it failed. Isn't it funny how some people will take our money if we're dumb enough to offer it? So don't unless it's tied to a legit major label, with a business plan and ironclad contract, and executives with a very successful track record, and then you're still usually throwing your money away.

Communication With the Label

In some cases, communicating with the label is as simple as using legal representation to contact them. Your legal rep can contact other industry personalities to invite them to listen to demos or attend concerts. However, the legal representatives need to be known to the industry executive before they'll get a return phone call. Don't be afraid to contact the labels directly. You should prepare an electronic press kit, consisting of a bio, recordings, videos of concerts, and copies of any press articles and advertisements. With that in hand, you can use the pitch as a cover or memorandum of introduction to the label execs. If the label is interested, they will send either someone they know (who lives nearby) or someone from the label to check you out by calling or catching a show unannounced. And if the label representative really likes what they hear, then they'll be watching the audience (not you) to see if the crowd is also into your show or if they're just talking and drinking (which is a killer). Let's say the magic happens—then what?

The Label Team

As we saw in the last chapter, label decisions about which acts and songs to sign are often based on research, experiences, gut instincts, and consumer reaction. The profitability and survival of the label depend on its decisions. To avoid being a casualty of a poor decision, many labels make their "final" signing decision by committee consensus. The committee consists of the label's vice presidents of each department, an A&R person representing the act, and the label's president and/or chairperson of the board. The purpose is to involve everyone as a team to campaign for the signing of the artist, and then to work together for the successful selling of their respective recordings. Once the label has committed itself, the team to help you launch your career is formed and the deal-making process begins. To be sure you make the best decisions possible for yourself, both now and in the future, be sure you understand well what I describe ahead about that deal. I even recommend you find a way to commit this to memory as on the day of that phone call, you will be so excited that you may jump at the chance and forget that this is a business and that you have one chance, right then, to take the actions with your attorney that will be significant to your future success in the industry.

Deal Memo

The first stage of negotiation between a label interested in an act and the attorney who represents an act is covered in a three- to four-page deal memo. If the label, at its weekly meeting of the president and all the departments' vice presidents, agrees to pursue signing an act, then a deal memo is often drafted by the legal department or representing attorneys.

The basic points of the contract are detailed: royalty points for the artist, options, advances, number of recordings, tour advances, and other relevant issues. Deal memos allow each party to determine if and where there is common ground and to establish how close they are to an actual agreement. It is at this point that the label will often investigate your personal life for any negative deal-breakers, like crimes, drugs, or emotional problems. They do not want to invest a million dollars into an unstable artist without being aware of the risks. If they find highly negative information and still want to pursue a deal, the issues will be addressed in the contract in a manner that will protect the financial investment and reputation of the label, and the very best thing you can do is be honest about your issues. Labels look at the big picture of global markets, which most new acts are not aware of. What are their marketing, promotion, branding, and publicity plans and why do they think you're a good fit?

At the same time, artists should be thinking about the label as members of the team of professionals who will make their career successful. It's nice to get a label deal, but it's a waste of time if everyone is not on the same page about your potential for success. As the act, you should be asking does the label really get it, understand me, can they help develop my career, and mentor me through every step of the career development and branding processes? Consider the personalities in the room—do you believe in their people, are they knowledgeable and fun, do you respect their roles as executives, and do they respect your creativity? The real question is, are they the professional contacts in the industry who can help launch your career and sustain it or should you be looking at another label? If you are happy with them, then the next step is to make the deal. And there are many choices, so you need to consider which one is right for you.

The Deal With Deals

Exclusive recording agreements between artists and record labels are often complex, lengthy, and confusing. Traditional major deals offer high exposure and large budget agreements between artists and labels. Each agreement is uniquely tailored to the specific requirements of the label and artists. Terms vary from single, one-time releases to a multiple-album, several-year obligation for both parties. I cannot repeat often enough: *Most recording artists never make any money from their record deals and some labels fail to make any money from a few of their acts.* Thus, it's the few acts who really break through and become superstars who provide the significant profits to cover the others who fail. However, major deals are usually very expensive and most artists never make any royalties. Yes, you read it right—most artists never make any money off their first recording royalties. Even the Beatles didn't make much money from their original deal. According to Courtney and Cassidy, who studied the business of the Beatles,

> *George Martin (Producer and label executive) . . . signed the Beatles to an unusually stingy contract, with only a penny royalty for every two-sided single sold. Of this one English penny, Brian Epstein (manager of the Beatles) and NEMS (Epstein's North*

End Music Store) took 25% off the top, leaving the Beatles with three-quarters of a penny, or three "farthings," to split four ways. So, for every million records sold, the Beatles took home just 7,500 pounds or 1,875 pounds apiece. And this sum was their before-tax income.[18]

Attorneys

Artists should look at a major deal as a shot to become famous, as a chance to see if their artistry and image can build a global fan base. Attorneys are there to represent you, but it is always better to understand each other's positions. The labels will want to do what they need to do to see if the fan base you've already built can be expanded into millions of paying fans on the label's nickel. However, you do not want to become a puppet on a string. Sadly, this happens to many acts as they are basically on the road for years, doing a show every three days just to make money for everyone involved. It is important to get over the brain rush of "I'm going to be rich and famous" and think of it as a business deal.

Recording deals are really more of an opportunity to work together to create magic and profits than merely becoming someone's cash machine. So, what you negotiate in the deal is very important and it's best to make sure you understand what each paragraph actually means. This is the emotion business, but it is built on the profit and loss statements at the labels. Most attorneys representing acts think labels are shafting the acts. And they will shaft you if you're dumb enough to sign a bad deal. Demand opportunities where the act and label can carefully engage together to make both successful. If you don't, there are plenty of attorneys who will be happy to talk to you and stroke your ego as long as they are paid $200–$500 per hour.

Of course, that's the way an attorney representing an act is going to think. So find one who understands both sides of the business and who is respected. There are usually two sides to a coin, and the label's side is putting up a million or so bucks on creatives (who are sometimes considered a little flaky), trying to sell unproven talent to consumers' emotional preferences. It's not as if we're selling hot dogs or cars! It's a difficult business and that's just one of the things that make it fun. At the same time, the act needs to approach the deal as an opportunity of "what can the label can do for me" and the label needs to understand that without the act, there's nothing to sell.

Deal Points

Many of the multipage deals have the same issues listed; sometimes they are just stated differently. However, when an act signs a deal, what are they really committed to accomplish? Here are some issues you'll usually find in all contracts between recording artists and labels. Labels offer contracts reflecting that they put up the front-end money to pay for the expenses of advances, recordings, development, promotion, publicity, and tour support. All expenses are seen as a 100% recoupable, if promotion and publicity are tied to the tour support. In exchange, the label will own 100% of the master recordings and a negotiated percentage of all generated revenues from the act's touring, merchandise, album sales, streaming (which may go 100% to the label, if not initiated), music publishing (usually 75% of the total income—100% of the publishing share and 50% of the writers share), corporate sponsorships, and branding revenues for a predetermined amount of time. Watch out for that crazy lifetime commitment that makes the label money even when the act has passed away. Let's take a look at some of the important contract terms each is agreeing to provide.

- *Exclusive recording agreement* means that the artist may record only for the label and must have permission (rarely granted) to record with another act on another label.
- *Description of services* explains what each party is responsible for, including the artist (you) and the label (company).

You will often read in a contract the words "you will cause" which means "you, the artist" are responsible to complete the master recordings to be recorded and produced, plus hiring of the producer, and all other third parties, including the audio engineer, musicians, and background singers.

- *The term of the agreement* usually requires a commitment from the act for as long as seven years. However, it is usually divided into an initial period of one year, one album (or less), and then six option years with an album or less each year. The artist usually is required to give up his or her right to option out of a deal when he or she signs it (labels demand exclusive rights to determine if the contact will be extended year by year after the initial period). Thus, the label will review the P&L statements weekly to determine if the act is gaining in popularity and revenues or declining. If the act is not making the formula amount of income required by the accountants, then the label will not pick up the option and the deal is over.

- *Commitment album and other expenses* mean that the label is committing to pay only for the first recordings required during the initial period, not the full seven years of recordings, tour support, and so forth. They pay only year to year based on how successfully the act is generating revenue. If the first album and other revenues from touring, merchandise, and so forth hit the formula amounts, then the label will pick up the act's option and then all the second-year expenses are defined as commitment responsibilities of the label.

- *Recording elements*: The label is going to pay all the bills so it really wants to make sure the songs, producer, musicians, and studios are all approved by the A&R department. Dates and times of the sessions must also be approved by A&R before any sessions as well as the songs selected to be recorded. The artists are responsible for delivering the master tapes to A&R for approval. If they are rejected, the label usually pays the bills, but the artist is then required by the contract's terms to create an acceptable album out of their own pocket. The major labels are signatories to recording unions so all musicians must be with the AFM and singers with SAG-AFTRA, and all must be paid master scale. Here what you learned earlier in this book about budgeting really matters. If a musician or singer on the session is not a U.S. citizen, then the artist is responsible for making sure they have a green card. The act is also responsible for submitting a full *recording budget* to the label for approval *before* sessions start, and they also have to provide all documentation, mechanical licenses, and master copies (in various formats) to the label when approved. The royalty artist with the approval of the label hires the producer, who in turn, as we read in Chapter 8, must hire the remainder of the musicians and singers, and all of this has to be accounted for in the recording budget. In addition, if the royalty artist plays an instrument, he or she must be a member of the AFM and if he or she sings then a member (in good standing) with SAG-AFTRA. The label also determines the number of songs on the album and the full length of time (minutes). An example is 12 songs and 50 minutes.

- *Material*: Songs (in most cases) must be new and never released in any media prior to the recording (first-use rights). Oldies and old versions of previous hits must have A&R approval, which is rare. Greatest hits albums will not count a commitment album, and the label will retain the right to release none to as many as it wants during the term of the deal.

- *Advances:* Artists may be paid an advanced amount of money when they sign the deal. On a major label deal, a new royalty artist may receive $50,000 to $150,000 depending on the type of genre and target market. Sometimes the advance may include money for touring to help the act pick up a tour to support single releases to radio, streaming, YouTube, and other media outlets. The act is usually paid half when they sign the deal and the remaining 50% when the album is accepted by the head of A&R. Additional advances are paid on each option period album, which is also tied to a formula that ranges between a low and high figure ($100,000–300,000), if the label decides to pick up the option. The figures increase with each option and advances are also paid on greatest hits albums. All advances are recoupable out of album sales and percentages of the artist's other revenue streams. Advances paid to superstars are often in the millions. Michael Jackson and Madonna both received about $100 million in advances on some of their deals. But then they didn't receive any additional advances until the labels had recouped the full amount out of album, touring, merchandise and other artist revenues.

- *Royalties*: Artist royalties are called "points" on album deals and percentages on all the other act revenues, as outlined in Chapter 7. New royalty artists make 12–15 points (percentage of the suggested retail list price) on the 40/20/40 splits of hard platforms, such as CDs. They also make a percentage of the wholesale list price on Internet downloads from Amazon or iTunes, for example, with the 70/30 splits. Superstars are often paid 20 points or more. Album royalties are still recouped at 100%, and all other percentages from tours, merchandise, corporate sponsorships, and others are used to recoup the debt created to launch the act in the first place and then as profits to the label once the act is making money. The percentage paid by the artist back to the label (as a percentage of their net from merchandise, touring, and so forth) is negotiated. Albums and all merchandise sold outside the United States will usually be paid at only 50% of the artists' points and percentages.

- *Accounting*: Artists usually receive a semiannual or quarterly statement informing them of the financials tied to their deals. It is often sent to both their business manager and personal manager. Royalty artists usually have the right to have an accountant review their label "books" once a year, although they often have to request the audit in advance and in writing.

- *Independent contractors*: The royalty artist signing the recording deal is a not full-time employee. They are hired as a *work-for-hire*, which means the label will retain the copyright of the vocal and music performances in the master recordings.

- *Image/brand*: The artist's stage name, identification, likeness, bio, and photos of the artist are all required to be copyrighted. Also, when required, these same things may be trademarked by the artist and/or label and may be used by the label in advertisements, promotions, marketing, and exploitation of the act and their songs, recordings, merchandise, and live shows. The artist usually grants a global exclusive right for the label to use or sublicenses all logos, likenesses, and stage names for the manufacturing and exploitation of the act and their products.

Reality of Innovation

Consumers' continuous use of illegal downloading and streaming has reduced the labels' economies of scale to a point that makes it difficult to break even financially, even with reduced budgets and limited artist rosters. Thus, most major labels are morphing into entertainment labels (that can be more appropriately referred to as music companies) by signing artists to 360 deals that reduce their risks of doing business in a changing financial landscape. The shift is on the way to destroying the traditional business model of the music industry. It is very difficult for traditional labels to fund new artists (often at $500,000–$2 million) when consumers rarely buy albums and streaming providers pay such low royalties.

As negative as all this may sound, the creative destruction of the business is also leading to many new entrepreneurial business opportunities. Money invested in your recordings, promotion, publicity, and marketing is just the "shot in the arm" many acts need to get discovered by consumers. Without the label's money, most artists remain unknown or limited to local fame. In these digital times, deals (depending on the amount of touring, tickets, merchandise, and endorsements the artist company provides to the label) allow artists a chance to widen their fan base and the label a quicker method to recoup its financial investment and both to share profits generated by artists' success.

360, 270, and 180 Deals

Labels' break-even points have dramatically become more difficult to clear without album sales and increases in streaming. As a result, labels have demanded a percentage of the artist's other revenues, such as music publishing, concert tickets, and merchandise. Many feel this is a rip, but the labels must have a shot at making a profit on their investments; so, let's follow the money. Some of the labels become full-service suppliers of tour merchandise in order to increase revenues and other investments. About the only thing labels can't do is personally manage the act, due to the 2004 Spitzer vs. AIG lawsuit, which determined that would be a conflict of interest. Other recoupable income included in 360 deals can be anything from song copyrights to ticket sales and corporate sponsorships. The first job of the attorney representing the act is to reduce the sources of revenue the labels grab, such as in a 360 deal, which is a percentage of everything. A 270 deal and a 180 deal simply mean the artist was able to maintain more revenues, sharing less with the labels. The artist's negotiating power in all types of deals is based on his or her fame, track record, and revenue-generating potential for the labels. All new artists can expect a 360 deal; known artists may be able to land a 270, and really famous artists might get a 180. At this point in time, very few get a traditional recoupment deal allowing them to retain all revenues from corporate sponsorships, ticket sales, and merchandise.

Traditional Recoupment Deals

Labels use artists' negotiated royalties (per unit sold) to pay the debt of the recording costs, advances, and mechanicals (the all-ins), which are 100% recoupable. In addition, the marketing plan bills (promotion and publicity) are charged to the artist's account at 50% of the total expenses. However, if the label declares the marketing expenses "tour support" (and they usually do), then the marketing bills are also 100% recoupable. The traditional business model has the label paying back the entire artist debt before the artist receives any royalties. Sadly, most artists never break even, much less receive any royalties from their record deal. However, the label puts its money into an act's creativity (on the hope they will make a profit) and then it pays it all off itself. On the average, it takes between 75,000 and 200,000 units sold (depending on the total debt) before the label breaks even, and about 500,000 units or more sold (a gold record) before the artist breaks even. On the front end, who knows where the money will be generated. One thing for sure is that unless you're Taylor Swift, it's probably not going to be through album sales, which, you are aware, are in the toilet. However, you never know what's going to happen. Labels still construct their deals for recording unit sales on the old, traditional recoupable math structure. The deals are usually skewed three to one to the labels' advantage, yet they are the ones putting up the money, while the act just has to show up.

Other Types of Deals

The business of record labels starts with a great recording of a great song tied to a sellable artist. The business of artist management and the corresponding corporate sponsorships, concert tours, merchandise sales, TV roles, and film careers usually begins after the release of a sellable recording and the corresponding branding and fame. Therefore, having some type of label deal is extremely important for the ultimate success of most artists. Actual deals are very specific and often cover many legal points not discussed in the deal memo. Therefore, do not be surprised to find actual contracts to be 75–125 pages in length. It is also very important to read and understand what you are signing, as most contracts can tie up an artist's career for at least seven years.

During negotiations, the label should include all points verbally promised in writing. Budgets should be established for recording expenses, promotion, distribution, and publicity. Financial advances must include tour support and basic living expenses, as you are probably going to work for hire while completing the recordings. A work-for-hire agreement means that the label will own your performances of the master recordings. The deal will probably span seven years, with an album release required in the first year. The following six years will be options based on album sales. Thus, you want to make sure the deal from the label will give you an opportunity to successfully record, distribute, promote, and publicize your recordings and image so that the label will pick up the option for the next album. From the label's perspective, it is really signing you to a one-year minimum deal (with additional option years) to see if you can catch on, become popular, and make it money. After the label has found and signed a sellable talent and recorded them, then the marketing plan, consisting of distribution, promotion, and publicity, is launched.

Development Deals

Development deals are offered to talented vocalists, groups, and bands perceived as having the potential to develop into major recording acts. It's a chance for the artists to professionally develop their talent at the label's expense. The label pays for extra musicians, studio time, audio engineers, tapes, computers and effects equipment, and the producer. If the final master tape is accepted, the act is offered a recording contract. The cost of the recording is recouped through record sales. If it is rejected, the recording is routinely defined as a tax write-off. Some music publishers also offer promising vocalists development deals. Publishers pitch the demos to record labels in exchange for part of the artist's points when the artist is signed.

Independent Deals

Indie labels range from the major personality-driven labels (e.g., Mike Curb/Curb Records, David Geffen/Geffen Records) to artists' labels (Madonna/Maverick Records) to novice recording artists and musicians. It is another opportunity for many niche artists to be recorded and distributed. Amateur recording artists, musicians, and producers often work together (and may or may not share the expenses) to record albums that can be used to promote a band, generate a local buzz, and sell off the bandstand at a live performance. Former major label artists usually retain the names and addresses of their fan club members and then use the lists to sell their self-produced CDs and digital downloads through iTunes, Amazon, CD Baby, or other outlets. Alternative music formats are frequently self-produced on computers and sold at conventions, festivals, church services, and concerts. Regional and genre independent record labels sign recording artists who may have only a niche client base in a specific region of the country or genre of music. Examples include Texas music, Latino music, or Carolina beach music. Sometimes major labels offer successful regional and genre labels affiliate status (part ownership or part of the profits in exchange for worldwide distribution) or buy out the labels to gain their recording artists' contracts.

Partnership Deals

Music companies and labels are more aggressively seeking to structure partnership deals with independent, newly successful acts (with local and regional success). They can use their proven worldwide branding and marketing networks to quickly market, promote, and provide publicity of an act's products and tours through new and traditional media. The artists who have built their own fan base, unit sales, merchandise, and sponsorships have a proven sellable product. In effect, the act has test-marketed their "it" factor and found success with consumers. The music companies and labels (knowing the act is successful) can then very quickly "break" the act into national radio, touring, new media, and traditional markets. Sometimes the deals involve venture capital (outside money matched with the labels), plus the recording masters and a variety of negotiated issues; however, artists may be able to "recapture" their copyrights at the end of the deal.[19]

Master Lease Deals

Once you've become a superstar, why not do it all yourself? It is not hard today to record your own tracks and release a master tape with a major label. Let them do the distribution, promotion, publicity, and sales and give them part of the profits in return. Of course, you'll need to have a high level of fame and platinum unit sales. However, the real advantage is that you as the artist maintain the copyrights to your songs and song performances on the recordings. Depending on their success, you can reissue them later (every ten years, perhaps) or sell them for an average income over several years, compounded by a multiple between 5 and 15.

Direct Distribution Deals

Some acts have benefited from remaining in control of all aspects of the production, recording, marketing, and distribution of their products. Garth Brooks and the Eagles have profited from recording their own albums and marketing them directly to retail stores, like Walmart. They've taken out the intermediary (label), reduced all expenses, and can therefore claim all current and future profits. Of course, you better be a businessperson if you plan this type of deal. You'll need to own or control your copyrights and be free of other commitments to industry entities.

Shark Deals

The recording label shark deal is a rip-off, just like the music publishing shark deal. Recording artists should never pay money for a recording contract or deal unless the artist wants to continue to own his or her master recordings. In addition, if approached, you (the artist) should always confirm that the distribution is through a major label or mega entertainment company (e.g., Sony Red) and that it has a budget for promotion (radio airplay) and publicity (trade and popular press publication). If not, pass on the deal, since all they really want is your money, and their business model will not provide profitable sales figures. Remember, it is the record label that pays the bills, takes the risks, and provides tour support money. A legitimate label will cover the costs of recordings, pressing of the CDs, and digital downloads, as well as the marketing, promotion, and advertising of the artists and recordings.

Vanity Deals

Vanity labels are run by artists who have started their own label and pay all expenses or find an investor to pay for the studio musicians, audio engineer, and producer, as well as the mastering, marketing, promotion, and pressings and/or downloads. Most vanity recordings lose money because they fail to receive radio airplay and cannot generate enough sales to cover the production costs. In addition, most are of poor quality and lack promotion, publicity, and distribution. However, occasionally, a successful vanity label may become an independent label once consumers "discover" them and start to buy their recordings. However, distribution through CD Baby, the Orchard, and iTunes rarely generates enough sales to break even unless the band is aggressively touring and quickly building a sizable fan base and buzz.

Financials

Artists usually want to make a living doing what they love, which is, of course, being an artist. However, another way to think about it is that the labels are in the business of making money, and artists are in the business of getting famous (so they can also make money). Labels provide huge investment financing that they have to repay (to themselves) out of selling your recordings. They may make money (if they can sell enough of them); you may not. In fact, you probably won't. However, you're being given a shot at world fame, and it has not cost you any money. So, make the best deal you can, create a wonderful recording, let the label do its thing (promotion, publicity, and distribution), and then tour and make all the money you can while the fans still love you and your music. Once you have become famous, the power has shifted to you and your manager. Now, do you want to sign another 360 deal or start your own label and stand-alone business? Your job as a new artist is to become so successful that the choice is yours.

Package Fee

In the old days, a certain percentage of vinyl records would break inside the packages when delivered to the retail outlets. The retail outlets or stores couldn't sell broken records, so the labels told the retail outlets to throw the broken records away and replaced them with new inventory. Labels would not pay royalties on broken records and it was too costly to determine how many were broken in every shipment, so they decided to just take an average amount off the number shipped. Artist were "charged" for the broken records (even if none were broken) by taking a percentage off the *suggested retail list price* (SRLP) used as a royalty base to determine actual royalties. The unscrupulous thing about all of this, of course, is that CDs and digital downloads almost never have "breakage" problems anymore, yet the labels still use this practice in contracts. Now, they simply call it a "packaging or detail storages fee." Fees can be as high as 15%–25% of the SRLP, which will reduce the artist royalties on each unit sold, and, in turn, increase the time and number of units that must be sold before the act's debt account at the label is recouped. This is one of the methods labels use to skew their financials in a deal favorable to them. Powerful superstars have been able to "rid" their contracts of packaging fees, yet new acts and independent or small labels that make small profits off of niche markets with niche artists still usually include them.

Gut Judgment

Of course, songs published by nonmajor music companies are often accepted depending on the presumed "hit value" of the song. It's complicated and often a gut feeling about how widely accepted a good a song might be; remember "Silver Bells" and "Tie a Yellow Ribbon Round the Ole Oak Tree" as examples. The same is true for many acts being pitched to labels before the Internet and streaming, including the Beatles, Madonna, U2, and Lady Gaga (L.A. Reid dropped her before her first album). Beyoncé, Elvis, and Kanye West were turned down before someone got the "click" in their gut to say let's give them a shot.[20] Untried songs and acts are in the same gamble, and it's the executives who usually use their guts and experiences to determine which to sign. Yet, it is still a very risky business as it takes money to develop an act and place and record a song with the perfect artists. In other words, a wow song is not necessarily going to sell well if the act (no matter how famous) can't sing it. That's why A&R and producers have to marry a wow song to the right artists with their artistic abilities, brand, and image in mind. Not easy! It's also why producers select the songs to be recorded on an album instead of the letting acts themselves determine what is recorded. The acts often think everything they write is a wow, which of course is rare, unless you are Lennon and McCartney. Because of the creative risk involved, coupled with the significant up-front money to create entertainment products, labels often base their recording budget mechanicals (for use of the songs) on the controlled composition clause, known as the CCC.

Artist Royalties on the Controlled Composition Clause

The CCC is another method labels use to reduce their financial risk because it enables them to reduce the amount of money they pay on first-use rights for songs from music publishers. If the royalty artist is the songwriter, cowriter, owner (publisher of all or part of the song), or first to release the song (first-use rights), then the label saves money by paying only 75% of the statutory rate for the mechanical license, which they need to acquire to be able to record and distribute their recording (℗) of the song (©).

Figure 9.2 Remember the controlled composition clause from Chapter 6 on music publishing. If the act being signed with the label writes their own songs or records others' songs (first-use rights) then the label budgets its music publishing royalties mechanical license payments at 75% of the statutory rate. However, look at what happens to the songwriter, who may have originally written the song and who is expecting his or her 100% share of the songwriting royalties or 4.5¢. However, if the CCC is used the writer now receives only 75% of that or 3.45¢. Sometimes the major artists may change a word or two and claim cowriting shares. That reduces the original songwriter's share to a lousy 1.7¢ per units sold.

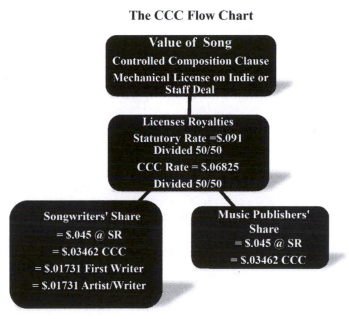

The CCC Flow Chart

Value of Song
Controlled Composition Clause
Mechanical License on Indie or Staff Deal

Licenses Royalties
Statutory Rate =$.091
Divided 50/50
CCC Rate = $.06825
Divided 50/50

Songwriters' Share
= $.045 @ SR
= $.03462 CCC
= $.01731 First Writer
= $.01731 Artist/Writer

Music Publishers' Share
= $.045 @ SR
= $.03462 CCC

That means the writers or publishers are faced with 25% fewer royalties if the copyright holder (the publisher) agrees. It is a hard decision because a major recording artist may sell millions and spark performance royalties from radio airplay. Alternatively, the song not recorded or recorded by a lesser-known artist who will accept the full statutory rate per unit won't generate as much money due to less airplay and fewer album sales. Oh, decisions, decisions. The labels want to pay less for the mechanical royalties. The advantage to the royalty artists on the recording of the song is that they pick up a percentage of the songwriting royalties even if they just change a word or change something minor (e.g., comma or period), which qualifies them as a cowriter. When that happens, the label is required to pay the CCC rate instead of the statutory rate.

But we are not done yet! The labels may require the act to sign over their share of the royalties to help recoup the debt more quickly. If the royalty artists are also the songwriters, then the label will want to publish the song through its own label-associated music publishing company. That way the label pays less under the CCC for the mechanical license and the music publishing royalties share is paid to them. Let's take a quick look at how a recording artist is paid from album sales to give some context to our discussion of deals and contracts.

Artist Royalties (Points) on Recording Sales

Let's say that an artist's royalties are established as 15% of the SRLP (suggested retail list price) of $16.95. The artist would expect to receive about $2.54 for every unit sold. Yet, the breakage or packaging fee reduces that royalty by taking 10%–25% off the SRLP. Let's say the packaging fee is 20%, so the label takes 20% off the SRLP of $16.95, or minus $3.39. This determines the new royalty base of $13.56. Then, the label takes 15% (points determined in the contract) of the new adjusted royalty base of $13.56 to determine the actual artist's royalties, $2.03.

Wait! What about the producer? Most producers are paid their royalties for producing the album out of the artist's royalties. So, let's say the producer receives 2% (usually 1%–3%) of the artist's royalties. That reduces the artist's royalty points from 15% to 13%. The adjusted artist's royalty points are now 13% of the adjusted royalty base of $13.56, which is just $1.76 per unit sold. The artist thought he or she would receive $2.54 per unit sold, but in reality, the artist really receives only $1.76 per unit sold (and that is after recoupment by the label). Also, labels still credit all royalty points to the artist until directed by the artist to pay the producer points out of their royalty points. In other words, the artist and producer will not receive royalties until the label pays all recoupable fees out of unit sales.

Artist Royalties (Percentages) on Generated Revenues

In the 360 deals, royalties are totally negotiable based on the act's previous success in making money for the label and for them. Most 360 deals contain the same stipulations of the traditional deals of recoupment, yet they add in a percentage of all other revenues that may be generated by the act to help recoup the label's investment. Obviously, the label knows from its accountants how much money to invest compared to its potential investments before it offers a deal. The riskier the act or smaller the niche market, the smaller the advance, recording budget, and tour support that will be offered by the label. It won't even consider merchandise until development and smaller acts start to achieve various financial levels of income, and then the amount or percentage of those "other" sources of money is preset and determined in writing. Sadly, most acts and their personal managers and attorneys always negotiate from abstract high figures, acting as if their act will be the next superstar cash machine.

The labels take a different view, knowing the prospective royalty artist needs their money, advice, contacts, promotion, publicity, and marketing to have a shot at launching his or her career. They take a much more conservative view and offer the first deal with low royalty points and percentages. If the first album and other sources of revenue explode, then they build in increases on album points and revenue percentages in favor of the act.

As we leave this chapter, it is time for a heart-to-heart talk about the realities of the music and entertainment business. The public is fickle and constantly maturing into different entertainment desires, needs, and definitions of what is fun and entertaining. They also don't want to pay for an unknown or emerging artist unless the person or the product is breathtakingly hot with a huge buzz. Crazy, I know, but it's a risky business that takes millions in investments to sign and create the products, and only a few are able to deliver the corresponding financial profits. The executives know a few superstars and hit movies will cover all the ones that fail to click with the public. Now that the act is signed and the album completed, or in the case of the movie business, the film is a "wrap," let's drop the other shoe and see what happens.

Notes

1. Goodman, Fred. "Prologue: The Opening Bell." In *Fortune's Fool: Edgar Bronfman Jr., Warner Music, and an Industry in Crisis*, 309. Vol. 1. New York, NY: Simon & Schuster, 2010.
2. Ibid.
3. Ibid.
4. Ibid.
5. Ibid.
6. Ibid.
7. Ibid.
8. Ibid.
9. Ibid.
10. Ibid.
11. Ibid.
12. Keaton, James, and Franklin Paul. "Vivendi to Buy Seagram, Deal Creates World's No. 2 Media Company—June 20, 2000." June 20, 2000. Accessed November 2, 2015. http://cnnfn.cnn.com/2000/06/20/worldbiz/vivendi_deal/.

13. Ibid.
14. "Vivendi Living Together." 2015. Accessed November 3, 2015. http://www.vivendi.com/home/.
15. "Who Owns What—Columbia Journalism Review." February 14, 2013. Accessed November 17, 2015. http://www.cjr.org/resources/?c=disney.
16. Moore, Frances, and Alison Wenham. "Looking for Industry Knowledge? Our Range of Resources and Reports Cover Everything from Analysis on Digital Music, to How Labels Invest in Music." Investing in Music. March 1, 2014. Accessed November 3, 2015. http://www.ifpi.org/investing-in-music.php.
17. Ibid.
18. Courtney, Richard, and George Cassidy. "Revenue Streams." In *Come Together: The Business Wisdom of the Beatles*, 287. Vol. 1. New York, NY: Turner, 2010.
19. Private interview with industry sources (2010).
20. Marshall, Barry. "10 Major Artists Who Were Rejected by Record Labels." The Richest. July 11, 2015. Accessed November 6, 2015. http://www.therichest.com/expensive-lifestyle/10-major-artists-who-were-rejected-by-record-labels/9/.

PROMOTION/ PUBLICITY AND MEDIA   10

Executive Decisions

The purpose of this chapter is to help you understand how the entertainment business is financially riskier than most other types of commerce. Everything is tied to everybody else doing their jobs at a wow level! Creative writers have to write the inspiring songs, scripts, books, and so forth. They have to find the agents and publishers who can license them to the major labels, film studios, book publishing companies, and others. Then, in this chain of events, the executives have to decide, based on their collaborations with the creative teams, when to place the products they select at the perfect time and place, and the best price. They know their company's financial success is determined by these decisions. No pressure there!

Remember we are selling entertainment products as stimulants for consumers' constructed emotions. It's expensive, with high entry-level boundaries, and about as risky as walking on quicksand. Everybody's success, including social and mass media programming, often depends upon gut reactions or the label executives' combination of intuition and experience about what might sell. They frequently use research and analysis into consumer behavior in an effort to better predict how to make a crazy profit. Once a musical artist is chosen and signed, the label teams and executives usually develop a conservative total for all expenses and compare them to what they consider reasonable potential long-term career profits. That's one of the reasons labels sign acts for up to seven years, but in reality they can be dropped anytime before the option year extension. Movie and TV production company executives and computer game software experts use the same process, except there are more decisions to be made and higher front-end financing required before signing these deals. Industry executives know that signing a deal is just the beginning of the trip into a truly risky business, and not anywhere near the end of the line to success. However, they don't see it as a risk. Instead, they usually perceive it as an opportunity to provide a product to the world that might improve the quality of life and at the same time make them rich. Good for them!

Rock-Paper-Scissors

The tough part of this process is answering the simple questions. Are we going to make any money? Believe it or not, we'll know within a few days. The label has spent between $1 million and $1.6 million on the chance that consumers might dig the product (a song off an album in this case). The label executives will be looking daily (as will the personal manager) to determine from the research data and *Billboard* charts if there are radio station airplays, videos on YouTube, streams on Spotify and other services, and any record sales. Last, they check to see how many people are stealing it (hard to believe, but they know) and where those people live. They are simply looking, hoping, and praying for an indication that the album and single being released is getting traction. They will check for Internet hits on social media and glance at the reviews in the trades and popular press. If the executives perceive some positive movement toward consumer acceptance, then they take the next step in the process, which is often pouring more money into the project to get the act touring. If it's a new act then they may "buy on" to a major artist's established tour, depending on if the consumers buying or stealing the products are the same or similar to the superstar's.

Roll the Dice

Artist and product launch promotion and publicity budgets are based on the *all-ins* part of the contract. Remember, the *all-ins* are the 100% recoupable expenditures the label has invested in the act, including recording budget, advances, mechanical licenses, and platform or digital platforms. As an example, if the recording budget is $200,000, the advance $100,000, and the mechanical licenses and platform pressings or digital storage costs $100,000, then the all-ins total $400,000. First, the budget line for promotion (of a new act) usually falls between 100% and 200% of the all-ins. We also need publicity (which many people think is free), budgeted at 50%–100% of the all-ins. Publicity also costs significant money when done well as someone's got to write the publicity copy, book the act on TV shows, plant the news article and gossip in the trades and popular press, and so forth. *Publicity and promotion work together, as one gives consumers a chance to experience it (e.g., song or trailer of a movie) and the other tells consumers why they might enjoy it.* Thus, if we pay $400,000 for the all-ins and add promotion at 200%, then we'll budget another $800,000 for it. Let wrap this up with a 50% of the all-ins for a budget line for publicity (equal to $200,000) and we've got our final budget of $1,400,000. Remember in the chapter on labels that our investment budget amounted to $1.6 million, so this budget is conservative.

Just as the people side of the entertainment product creation is interconnected, the budgeting tells its own story about how the business of the deal and the product might make or lose money for the investors, the label or production company, and the artists involved. It costs money to make recordings of music or films, more money to promote them, and even more money for product distribution or touring. The total cost is usually figured before the first foot of film is shot or the musicians even walk into the recording studio.

Risk vs. Profits

The daily relationship between risk and threats to success drives the executives to buy bottles of Maalox and Tums antacids. Forget the talent. Now the conversation is about what constitutes a financial risk vs. potential profits. What is the competitive advantage the act needs to grab as much of the market as possible? The executives figure it out and help the act enhance both through artist development. Many famous recording artists and movie stars just don't get it. But it is the executives' efforts and concerns about identifying threats, here understood as competitors' new acts and films or shows, and how they might reduce the potential income of their acts and movie launches that actually help make the seemingly ungrateful stars rich and famous.

Note the complexity of the creative processes and think about how these are all interconnected to bring a creative work to the public.

Timing

Timing becomes extremely important, as the goal is to launch the product or act at the best time in the market. It was much easier when the labels made their projections based on album

Figure 10.1 The creative to business process in the entertainment industry is complex. In this figure, the center circle represents the author's creative content. Because of the copyright laws, the creator owns it; however, creative don't always know how to sell their works and they need representation (the inner circle). The agents, publisher or managers decide on what they think is sellable and take those to the major industry corporations (the outer circle) who may then choose to license or by the product and hire production teams to develop it.

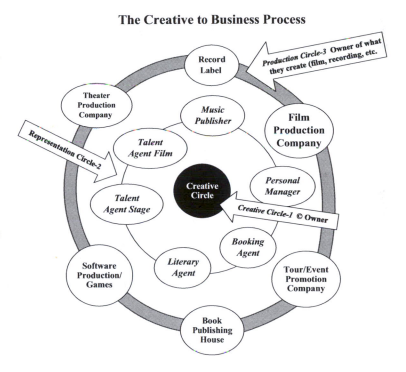

The Creative to Business Process

sales. It wasn't that hard to predict the "ifs" of album sales and the generated royalties from retail and digital downloads and from music publishing royalties, especially when the label had a part of the publishing deal tied to the break-even rates based on artist royalties. The same was true for the movie industry until the impact of the Internet caught up to them and they began to see the same types of financial issues caused by illegal downloads and streaming that have nailed the music labels.

The Cash Machine Poker Game

Since it has become difficult to sell albums, the projections have shifted to the more difficult "ifs" of branding and image-based revenues, such as ticket sales, merchandise, sponsorships, interactive streaming, and other sources of revenue. The innovative technologies of audio and video streaming have not generated enough revenue to pay all the copyright holders, recording artists, and songwriters, who formerly benefited from CD and DVD sales. However, the goal of the labels and movie companies has remained unchanged as they continue to discover and promote products consumers perceive as emotional "lightning in a bottle," with the potential for hundreds of millions to billions in profits. The industry needs it, because less than 5% of products released turn into never-ending cash machines. About 80% of the deals break even or make a small profit for the labels over several years. With movies and TV shows, it is even more difficult to generate profits as they cost millions *more* to make and additional millions to promote and market. Thus, the film business usually sells foreign distribution rights before the first foot of film is shot in order to generate seed money in a project.

The Rejected Rocket

Some of the greatest hit movies, songs, recordings, actors, and scripts were initially, and sometimes repeatedly, rejected before someone saw their potential for success. Ironically, the off-the-wall products industry executives thought were too risky became the biggest historical moneymakers, once they got their shot. The TV series *Cheers* was at the bottom of the ratings for years and then slowly grew into one of the most successful syndicated shows ever. Gene Roddenberry, the producer of the TV series *Star Trek*, gave me a great example when I met him in the middle 1970s. My friend the late Don Cary had known Mr. Roddenberry from his old Hollywood days and suggested I and some others meet him when he was visiting to give a speech.

Mr. Roddenberry played one of his black-and-white pilots for a new science fiction space show that was rejected by network executives. It had a female captain and a Vulcan, Mr. Spock. After Mr. Roddenberry's speech and showing, I got to meet him and I asked him a couple of questions. The pilot was the third one he'd made and pitched, and when the executives saw it, they finally agreed to stand it up with two big ifs. First, they said nobody would believe a woman could be the captain of the Enterprise spaceship, and they also demanded that he drop the guy with the "weird" ears. Every time before he'd pitched a script or pilot to the networks he'd been rejected. He would then just go back to work to earn enough money to write and film a new script. He said, "I will make you a deal. I'll drop the woman captain, but I must keep Spock." They agreed and the rest is a history of an entertainment and merchandise franchise that includes ten movies and seven TV series. It happens all the time, so having the passion and determination to go on sometimes pays off.

Risk Management

At the same time, inside the company, from the CEO to the lowest intern, everyone is usually aware of the business risks involved in launching a new product and in keeping existing ones fresh before the consumers. If we were selling Hondas we'd compare last year's sales figures to the country's GDP (gross domestic product) to predict how many Accords, Civics, and other cars in the line we'd need to make next year for sale in various countries around the world. It is not so easy in the entertainment business. Remember, it is a much riskier business as it's based on consumers' flaky emotional needs to experience and pay for something they connect to, often in an unarticulated way. Labels try to control risk through management processes as they launch the act or product, often budgeting a minimum investment in promotion, publicity, branding, and marketing to determine within a weekend for movies and two weeks for recordings if they've got a potential cash machine or a dud. Labels and entertainment production companies look at the "overnights" and weekly reports to see how or if action on the charts, sales, spins on radio, and streaming is generating any profits. Based on those results, the label executives will decide whether to pour more money into the marketing or dump the project and move on.

Artists failing to achieve profitable unit sales or other revenues are customarily dropped from the label after the first few poor-selling record releases. High-profile movie producers, experienced scriptwriters, and others in the launch of new entertainment products who have the bad luck of failing or only marginally succeeding may be associated in the executives' minds with bad judgment, and if they fail again, gone, baby, gone. The label fulfills its contractual obligations for the commitment albums and then it passes on picking up the artist's option. If the executives put more money into the marketing of the product, then the level of the risk management is also increased. Crazy business.

Figure 10.2 indicates how complex the industry is when you look at it as a business. It has many moving parts and these, as is common in organizations, can change depending on the life cycle of the business and its profitability.

Obviously, when, for instance, major labels have spent millions to make an artist successful, and they do not appreciate an artist who wants to jump to another label after they invested and worked hard to make them famous. It's the same reaction across the entertainment industries. Performance and generated revenues are also tied to employees who fail to sign or generate profitable products or sales. After continued decline in profit, both productive and nonproductive employees are usually dismissed. It is not uncommon for the entire staff to be replaced for not producing profitable acts, movies, products, and recordings. It's the same situation in the film business, except for the professionals who actually make the films, who are represented by unions. The union professionals are the best in the world; thus if a product fails it's usually because of poor executive decisions, timing, a bad script, or poor marketing and risk management of the process. It's not their fault, but if the industry suffers, they don't get any calls and, therefore, no work. If there's a problem with losses

Figure 10.2 The internal operation of major entertainment companies is built on the flow of communication between executives and departments. From the top down (president to interns) everyone has to work together to create, market, and sell a successful product. In addition, other companies, such as media and the Internet, are used to promote, deliver, and generate revenues.

Internal Operations

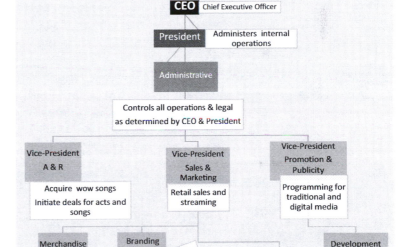

overshadowing the profits, then the employees at the companies and the union members who create the products are all facing unemployment.

Products

As we've discovered, the most difficult decision executives have to make is how much money to spend on the entertainment products they are financing. They base it on the potential profits in this risky business. Many things can go wrong, such as the actors not clicking, poor production, and the lack of great story, producing, or directing. They all have to work together to create the magic that is described as a bolt of lightning in a bottle. That's why they often spend the big bucks on the most talented professionals to create the song, recording, computer game, book, and so forth that catch the essence of the imagination of the popular culture—wham, bam, billions in profits.

Every day, people move in different directions, events happen, political situations change, and what was hot one day is dead the next. In addition, it frequently takes more than a year to produce and market a movie and months to finish and market an album. And all the time, the wants and needs of consumers are changing. As you've seen in previous chapters, the budget to create the actual entertainment products is expensive at the professional levels. Low-budget movies cost $10 million or more, and albums at the master-scale levels in the recording budget will often top $200,000. Now that we've got something to sell, how are we going to help our world of consumers discover it? And more importantly, what's it going to cost in addition to the recording/movie production expenses? Let's once again look at the typical record label departments but now see how they work from the perspective of risk and management.

The Debt—The Cost of Doing Business

Signing, recording, promoting, and marketing acts and their products are expensive. We spent about $200,000 in our example of a recording budget. Labels spend more or less

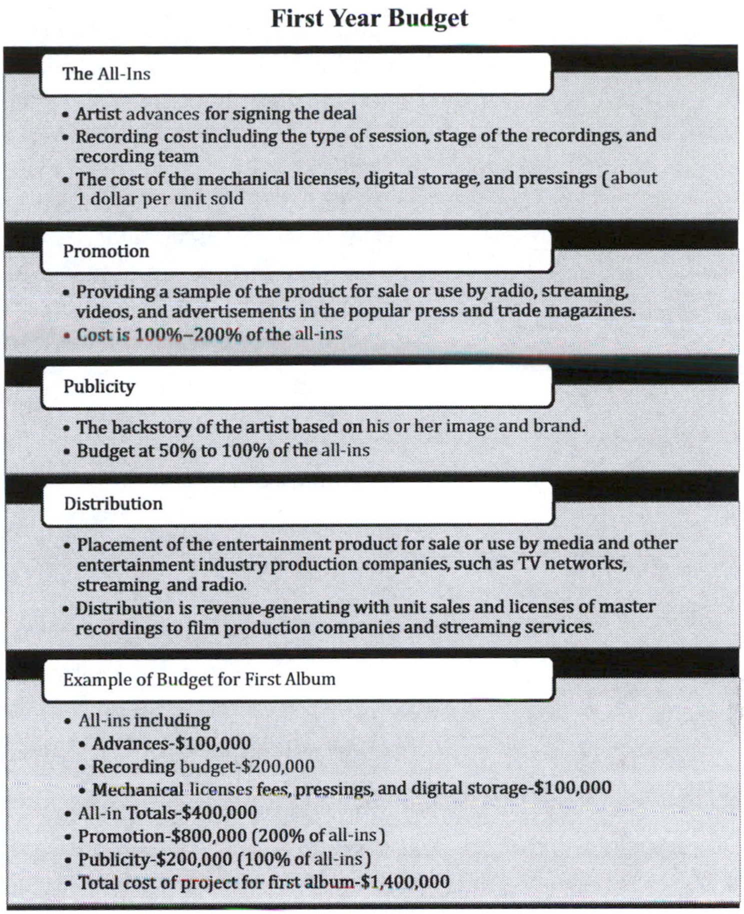

Figure 10.3 The budget for the first year on a typical recording artist is around $1,000,000 to $1,200,0000, with the all-ins totaling $400,000, promotion another $400,000 to $800,000 and the publicity about $200,000. The labels will know with one to weeks after spending the money if the act will take off or fail. More money is poured into the acts that find the traction of a buzz with fans and media.

First Year Budget

The All-Ins

- Artist advances for signing the deal
- Recording cost including the type of session, stage of the recordings, and recording team
- The cost of the mechanical licenses, digital storage, and pressings (about 1 dollar per unit sold

Promotion

- Providing a sample of the product for sale or use by radio, streaming, videos, and advertisements in the popular press and trade magazines.
- Cost is 100%–200% of the all-ins

Publicity

- The backstory of the artist based on his or her image and brand.
- Budget at 50% to 100% of the all-ins

Distribution

- Placement of the entertainment product for sale or use by media and other entertainment industry production companies, such as TV networks, streaming, and radio.
- Distribution is revenue-generating with unit sales and licenses of master recordings to film production companies and streaming services.

Example of Budget for First Album

- All-ins including
  - Advances-$100,000
  - Recording budget-$200,000
  - Mechanical licenses fees, pressings, and digital storage-$100,000
- All-in Totals-$400,000
- Promotion-$800,000 (200% of all-ins)
- Publicity-$200,000 (100% of all-ins)
- Total cost of project for first album-$1,400,000

depending on potential profit margins. Movie companies start in the millions. Once again, let's look at the music industry to determine a typical budget for a project, but this time including the cost of branding, marketing, promotion, publicity, and distribution. Now that we know what the marketing, promotion, and publicity plans are, let's take a glance at what it is going to cost. Basic budgets for launching new acts are based on the all-ins.

The All-Ins

The all-ins (short for all-inclusive) are all the expenses on the artist's careers that are 100% recoupable. Additional budget lines are recouped at 50% (meaning the label pays half and the act's royalties pay half) on promotion (e.g., music videos) and publicity. Typically, all advances, the recording budget, and mechanical licenses fees are charged back to the act's debt account at the label at 100%. Thus, if we gave the act a signing advance of $100,000 and the recording budget totaled $200,000 and we decided to press, digitize, and sell 100,000 albums each with ten songs all under five minutes at the CCC rate ($0.06825 per song times 10 songs, which equals 68.25¢ per album), that equals $68,250 for the mechanical license for 100,000 ten-song albums.

However, the pressing and digital storage servers also cost money (the more we press the less cost per unit). Add the artwork, shrink-wrap, and so forth, and the cost to the label is about $1.00 per unit ($0.6827 for mechanicals on ten songs and about $0.31 for the platform or server expenses). Thus, the all-in account debt in the artist's name is $100,000 (for advances paid to the act for signing the deal), $200,000 for the recording budget (creating the recordings and paying for the producer, musicians, engineer, and singers), $68,250 (for the mechanical licenses on 100,000 albums), and another $31,750 for pressings or storage, for a grand total of $400,000. And we haven't sold anything yet as we know we've got to add the promotion and publicity expenses from the projected marketing plan.

Baseline Promotion and Publicity Expenses

Promotion and publicity are expensive, and the label usually pays 50% of all expenses, as stated on their contracts. Here's an example of the crazy cost. Music videos are now budgeted much lower as they do not push (sell) as much product as a few years ago. Indeed, YouTube posts and mash-ups that receive lots of airtime are free to the label. They love the word *free*! Still most acts want and need a music video or two, costing $20,000 to $100,000, depending on the status of the act and the purpose of the video. Michael Jackson would spend more than a million on some of his videos, probably not realizing that half of the million was attached to his account debt at the label and the other half was coming out of his pocket.

The 50/50 makes most artists reconsider their video budget downwards to the lower figures. In addition, most artists simply do not have that much money at hand, and thus the label runs the entire cost of the video through tour support that makes all promotion and publicity expenses 100% recoupable. Radio station tours and consultants (trying to get recordings added to playlists) also cost significant money. Regional promotional experts' costs start at $5,000 per region, and then additional payment and bonuses may be provided with each station added. National promotion can easily cost half a million to a million for a two-week campaign.

Billboard has a circulation base of about 17,000, including the executives, creative, and broadcasters who run the industry. If we want to place an advertisement in the weekly magazine it's expensive as so few people read it. Yet, the people who read it are the most powerful decision makers in the industry. Who are they? According to DJG Marketing, their average age is 47, their average salary is $212,000 annually, their net worth is more than a million, 68% have a college degree, and 25% have a postgraduate degree.[1] Want to know more about them? Forty-three percent of *Billboard*'s print readers also visit Billboard.com once a week or more, 71% of *Billboard* readers spend 30 minutes (or more) per week across all *Billboard* platforms, 49% of readers have senior/upper management job titles, and 63% of readers are business purchase decision makers.[2] So what will it cost? That depends on several things, including size, such as full-page, two-page spread, six-column, or half-page horizontal, vertical, or square.

The labels have many options in connecting to the target fan base. Advertisements in the popular press, trade magazines, and mass and social media are typically expensive, depending on the circulation and industry importance of the provider, magazine, or website. As an example, according to *Advertising Age* costs vary from media to the number of impressions (people views/number of times consumers hear something).[3] Do you want to advertise nationally or regionally? Once again the costs vary and so do the results.

How much money do we have to spend on promotion and publicity? The costs will be scaled differently by region, publications selected, and size/style of the ad copy. For instance, one online website, Quora.com, says that the average cost of a full-page ad in *Variety* is between $3,000 and $4,000, but if you go on the *Variety* website, which shows you that its brand group is composed of over 400 products, it tells you to call for a quote. You get the idea. If you want to get a ballpark idea of what product placement costs in this trade publication, you can find a starting point in a consumer site, but you have to work with the publication directly for the real costs and decide how much money you have and, more keenly, want to spend on promotion and publicity.

Promotion and Publicity Budgets

Entertainment corporation executives have to control their budgets and wisely choose to spend money on promotion and publicity. The goal is to light a fire or start a buzz based on a few fans "discovering" the act or product and then using social media to spread the word. Spending money in the old days it was easy—get the recording on radio and the act on TV network talk shows. That still happens, yet radio airplay is not as powerful as it used to be and many fans don't listen to radio, watch TV talk shows, or read newspapers. So how can labels connect? Social media, of course! Still, it all costs money and we have only a limited budget and a small time period (usually two to four weeks) to determine if the act will be successful.

The Failure of Promotional Economics

Here's where things get hard to understand. Labels provide broadcast radio with free copies in exchange for promotion, but they almost had to blackmail streaming companies and demand payments. Excuse me, but what the heck is going on here? In addition, streaming services are quickly accelerating with younger listeners, who listen less to radio broadcast stations. So why are the broadcasters getting the music recordings for free and streamers are paying?

Labels claim both have a promotional advantage for launching new acts and products. Yet, the audience they want to reach is moving toward streaming. Network TV is having the same situation with viewers who are quickly leaving for Internet and streaming services. It seems the old free promotion of recordings given to radio may be a thing of the past and certainly needs to be reevaluated as to its value of breaking acts and products. This will of course depend on genre, as country music still seems to have a reasonable following but some of the others may not have much of an audience remaining.

It's harder to steal a digital movie, yet many are doing it every day, using illegal sites and bit.net software. Actors, directors, music producers, and even union musicians and recording artists are also losing royalties tied to unit sales in the signatory agreements with the entertainment production companies. It simply makes the future more difficult for both the creative and business sides of the industry. In the 1950s and 1960s radio airplay was so important to the labels that some got caught illegally paying stations to play certain recordings. That's called payola and is not legal. Clearly the value of airplay has diminished, and as a former CEO of Sony told me once, he's not going to continue to pay high fees to promotion men/women to consult stations into playing certain products. That will also increase the need for street promotion, where people are hired (or volunteer) to spread the word about new acts and songs.

Free Goods

How can labels and distributors motivate retail outlets to stock unheard-of entertainment products, such as newly

401

signed acts and low-budget movies? Labels provide free goods (free units that can be sold) to retail outlets to stimulate sales. Retailers usually want some guarantee from distributors that they will not lose money or be stuck with albums they can't sell. Thus, the labels (through the distributors) usually give a certain amount of new and unknown artist releases to the retail outlet (as much as 25%) when they buy and stock a minimum inventory. Additionally, retail will usually demand 100% return on the unsold units. Thus, the retail store agrees to stock and sell the new release, it can sell or give away free units it did not pay for, and the label will accept returns on all the bought units that didn't sell.

Promotional Copies

Digital copies transmitted to radio stations and networks for broadcasting are not free goods that are given to retail for inventory and are not to be sold. It still costs the label money, as it has to pay for the creation of the copies and songwriter and music publishing royalties per copy provided. Artist royalties are not paid on promotional copies; in fact what often happens is artists are charged a promotion fee of up to 15% off the top of their royalties (when determining recording points times the royalty base), which means the act receives less money on each unit sold to compensate for the units given to media for promotional purposes.

Winning the Game

The risks industry executives undertake are severe, yet the profits when they succeed are amazingly strong. It's a little like hitting the jackpot in Las Vegas, except it costs millions to play the game instead of a couple of bucks. It's the team of experts at the labels, stage, book publishers, and movie companies working together that actually makes winning the game easier. It's the experts working together who make the industry successful. The famous stars get the credit, but it's the executives and all the hard-working employees at the company who make it happen. Now that the product has been produced, how can the company alert consumers it is available for use and sales? In the entertainment industry, the definition of marketing is more than the typical definition of the *four Ps*: product, price, placement, and promotion. Let's take a quick look and make sure to notice the additional costs of promotion and publicity.

Artists and Repertoire Department

As we learned in Chapter 8, A&R, the department of artists and repertoire, is the part of the label's business structure that is engaged in new artist or song acquisitions and in making sure the new acts are consistent with the brand of the label. Because they are working with the creative talent and the legal and financial teams in-house, A&R to some extent is risk central. Choose the wrong thing and you can negatively impact your company's bottom line; choose the right thing, and you can make a lot of money.

Development

Artist development, like A&R, takes on another layer of significance in the risk management and media conversation since they are responsible for the development of the artist's image and brand, which will later be projected in music videos, on TV, in social media, and finally, in the entire business plan surrounding the launch of the song or album. The importance of all the parts of each of these areas is played out every time a new act or a new song is considered for inclusion on the label. If you don't promote the artist, the album, the song, or the live act, then there will be nothing to sell, so marketing and promotion styles are essential to the successful management of the artist. It's also associated with matching the right songs to the perfect recording artist. It is essential that we follow the market share in advertising through such sources as the marketing reports published by the Pew Research Center on consumer behavior or in eMarketerDaily and other similar media analysis websites to understand and learn where ads and publicity photos will be seen. Central to successful promotion of the artist is what we can learn from the Nielsen data on consumer consumption of entertainment products, which will be discussed ahead.

Marketing and Sales Department

The weekly sales figures related to the marketing plan are used to see how well the product placement is doing in generating sales. It may let the label know that more of one kind of advertising is needed and less of another, or, more drastically, that the brand of the artist or the advertising associated with a single song might need to be overhauled. As we learned already, the marketing and sales department needs to pay close attention to psychographic research data trends to scout the common and uncommon media and retail outlets where the product can be successfully placed. It is now time to take a closer look at the idea of branding the act, as that is why the label spends time on artist development, marketing, and publicity.

Branding

Branding is the use of the act's emotional connection with the fans to profit on consumers' preferences, responses to the act's image, and name recognition. Remember why consumers want to be entertained? It's much different than most realize. We've already examined the psychology of entertainment and how we as consumers use our subconscious minds to help ourselves understand (feel) something. As an example, the goal is to have everything you say and do as an artist help consumers define you as a person in their minds. All of us want to be loved deeply; welcome to the Beatles singing "I want to hold your hand," with their unique manner and sound. The brand informs the image and these combine to connect with the consumer.

The public discourse surrounding popular culture continually moves as an emotional pendulum between good and evil, love and hate, anger and kindness, and a few other emotions that many of us are stuck with throughout our lives. Listening to music, being entertained at concerts, and watching the love scene or car chase in the movie let us emotionally "connect" with something inside of us that we feel is enjoyable. Now, though we see Elvis's dancing as harmless, the censorship rules were such in the 1950s and 1960s that network executives were afraid of being shut down for something close to what some might consider *immoral use of the public airwaves*, a concern led by conservative religious and political leaders. Today's performers who show only partially covered body parts gyrate and twist in ways that might make Elvis, were he still with us, blush. Labels and artists sometimes consider over-the-top PR stunts to get the consumers' attention, such as Lady Gaga wearing raw meat on stage and in her music videos. It has to be a wow to get a lot of people to respond emotionally strongly enough to buy the product, want to steal it, or throw a little money at it by buying merchandise and concert tickets, but that's what the professionals in the industry accomplish every day.

If you become the next superstar actor, rock star, or whatever, then it's because of how the fans "feel" after listening to or watching a movie or show. Thus, casting the perfect actor into the movie script with the director who "gets it" is difficult. Then you have to hire a great stage crew and creative technicians, who also have to create the perfect listening and visual environment. Roll the dice, spend the millions, and profit only when the magic happens. The bottom line is simple; fans perceive great entertainment products as having value.

It costs money to accomplish a true brand connection, and the best are often defined with one word. As an example, if we are talking about cars, then when we talk about Toyota the word reliability comes to mind. Say Volvo and the word is safety, Porsche and the words experience and fun come to mind. When it comes to human beings it's even more difficult. But here's the bottom line: a well-defined brand gives you the advantage in the competitive marketplace. If the consumers know you as rebel, lover, sexy, funny, solid, ideal friend, or hero, then the product branding will be matched to the image.

Publicity

Also, as we learned in Chapter 8, there is a difference between a well-orchestrated publicity campaign that is likely to help an artist reach major markets and one that the artist has launched him- or herself. Of course, after the product is a hit, risk can be further mitigated by "shameless self-promotion" that gets the artist before the public and even into new venues, such as singing the National Anthem at the Kentucky Derby. First the act is seen and heard by millions on live TV, and then their names are often dropped into news stories or clips in the sports report on the news, and picked up by racing trade publications. The importance of publicity cannot be overstated in sustaining interest, as long as the publicity remains positive for the artist and for the label!

Radio Tours

To set up the release of a new recording or movie, recording artists often take a "radio tour" and the major cast members of new film releases will hit the talk show circuits. The purpose of the radio tour is to introduce the new artist to the power brokers (program and music directors) at the stations, who make the decisions on which recordings to play.

Labels view the free airplay as promotion and the most important way to break an act or new album release. Radio views the recordings as a method to build an audience who want to listen to "the hits." Radio stations then charge advertisers for the commercials they "air" based on the number of people listening to the music. The location of the station, the type of transmission (AM, FM, or digital), the power or wattage of the signal, and the genre of music or news being broadcast often determine the audience size and the cost of the 30-second or 60-second commercials. Stations would rather play the "established hits" to build an audience unless they can be the first to air a breakthrough song and act ahead of others. The movie stars usually avoid radio and use TV as consumers quickly recognize the "star" and listen to the talk show conversations.

Market Research—Nielsen

There are many ways consumers can access entertainment products. Table 10.1 shows you some of the more common and popular ones.

Nielsen Marketing is one of the top research companies in the industry. Major labels and other entertainment companies pay

Table 10.1 Finding an Audience. We've got to find consumers first so they might buy or use our entertainment products. The average use of media by consumers is over 11 hours per day in the United States. Which media they use depends on their demographics and psychographic lifestyle choices. Executives use this type of information to determine product placement, advertisement, publicity, and promotion.

Media/Device	Total Number of Users	Total Medium Monthly Time in Hours/Minutes per User
Traditional TV	284,817,000	151:33
DVDs/Blu-Ray Devices	182,725,000	5:36
Game Consoles	98,664,000	9:15
Multimedia Devices	53,236,000	4:42
Internet on Computer	192,875,000	30:36
Watching Video on the Internet	138,502,000	12:13
Web Application on a Smartphone	170,303,000	44:32
Watching Video on a Smartphone	128,432,000	1:53
Listening to AM/FM Radio	260,099,000	58:10

Source: Nielsen (2015).[4]

Table 10.2 Personal Device Use Dayshares. A deeper analysis also lets us see the best time to reach an audience with our entertainment products is in the evening for TV, early in the morning for radio, and around noon for computer and smartphones.

Devices	7 a.m.	Noon	8 p.m.	Midnight	5 a.m.
TV	36.3%	35.7%	62%	66.7%	49.4%
Radio	39.8%	33.4%	9.6%	4.8%	27%
TV-Connected Devices	1.5%	3.2%	6.4%	8.3%	3.3%
Computers	6.5%	10.6%	7.3%	6.3%	5%
Smartphones	12.3%	14%	11%	10.1%	10.7%
Tablets	3.5%	3%	3.8%	3.8%	4.6%

Source: Nielsen (2015).[5]

it for research to help develop their promotional, publicity, and marketing plans to launch an act or a new book, computer game, movie, recording, or other platform of entertainment.[6] However, the size of the audience and devices consumers use typically change during the 24 hours in a day. An example is the Nielsen report of May 2015 that tracks an audience ranging from 30,644,555 at 4 a.m. to 144,184,562 at 9 p.m. That means the number of people using different devices also changed hourly, as Table 10.2 appears to show. However, about 78% of the audience disappears between 9 p.m. and 4 a.m., so the data provided ahead reflect only the percentage of the total American audience at a *specific time* of the day or night; yet the percentage is derived from various totals.

Media Promotion

After the master tape is recorded and accepted by the label, a marketing plan that coordinates publicity, promotion, distribution, and sales is activated. To recap, the marketing and sales department coordinates the label's efforts to ensure a successful launch of the product in retail and digital markets. The promotion department coordinates the label's plan for radio airplay, streaming, videos, popular press, trade magazine advertisements and charts, plus website information, and other forms of consumer involvement, such as contest tie-ins and radio station tours. At the same time, the publicity department works on the placement of stories in the trades and popular press (tied to the act's image and brand), schedules TV and radio talk show appearances, and uploads photos and information for an active presence on social networks and stories in approved tour date media.

Here's a short description of the music label's departments' duties on the promotion and publicity side of the business. The same happens for much of the other products in the industry, yet films use expensive trailers on social media and commercial broadcasting, plus cable.

Media Publicity

Publicity boosts promotion by providing the backstory of the artist, which is tied to his or her image and brand through appearances on talk shows, articles in the trades, and popular press, social media, and other outlets. As we read through the different types of media and promotional tours, remember that the artist is also selling him- or herself at the same time by being (acting or actually being) who the consumer thinks the artist is. Thus, stories appear in the media and so do the acts. What do they talk about? Themselves, in a positive, friendly manner, and they are also excited about the products they are selling and tours they are on. The unspoken works are "come see me" and "buy my products and tickets." The people love you for whom they think you are (in their heads) and what they perceive is often what they get. It makes tons of money and always works best when the image, brand, talent, and personality are based on facts, not made up. It's hard to fake honesty on that little TV screen.

Talk Show Circuit

Talk show moderators have been provided with a script of questions the movie star or singer has already seen. The purpose is to blend just enough private stuff (gossip) and then drop into the discussion information about the new movie, book, or whatever that is going to be released on a certain date. Radio receives the digital copies free and labels beg, scream, and twist the arms of the program directors for airplay. Here's another comparable metric for everyone ages 2+ for the average week during the first quarter of 2015.[7] The labels and studios want to make sure they market their products and acts on the correct media that will increase the chances (and thus reduce the risk) of consumers discovering what they've got to sell. The artists usually have agreed to all promotional and marketing plans (availability) when they signed their deals. The radio or TV networks pay the base union scale for their appearance, plus transportation, housing, food, and limos. It is amazing how different radio and TV stations are, as some are AM, others FM, some are high-power stations reaching millions, others are low-power day timers (operating only during daylight hours), and still others are very low-power campus stations. The industry's goal is to have as many people "discover" the new acts and products (recordings) as fast as possible. Therefore, whatever the power of the stations, the ones located in the main population centers of major cities are most important. Second are lower-population areas and last is just getting some airplay anywhere, however it's possible.

Social Media

Why not let consumers promote the acts and entertainment products themselves? Social media allows consumers to do the job for the labels for very little money. Hey! As an example, *Billboard* magazine has a weekly subscription of only 17,000, which is viewed by 115,000 viewers, but it has 15.2 million fans (views monthly) and music industry insiders who read the .com and .biz sites. *Billboard Mobile* has 6.7 million a month who use the website and charts, 12.9 million a month who use the networks for social connections, and it also sponsors 20+ events every year to help target industry insiders with fan-based information.[8] Every successful artist today is texting, tweeting, and doing everything else they can think of to connect with the fan base. It makes the artist/fan connection "seem" more personal and stronger.

Social network sites such as Facebook and many others allow consumers with common interests and preferences to construct their own virtual reality tribes of friends and fans. The owners of the websites then use the membership totals to price and sell advertisement impressions (based on hits) to companies to market products correlated to the memberships' usage (click) patterns. Of course, the music and entertainment industry has many albums, films, TV shows, and so forth they want to sell, and they use the sites to spread the brand and start the buzz. Social networks give established artists and labels a direct way to connect their products to consumers.

Taking a look at the bigger picture, we'll see that social network members are in effect doing the labels' promotional work for free. Love it! It costs the labels very little to place stories and ads where social media users can discover them. If they click on it and like it, then they start to "spread the word" to their social network "friends." Once a fan base is established, labels and artists quickly use the connections to build a larger fan base and buzz (excitement for the act). Some sites have tried to establish legal and free music downloads based on advertising hits (viewers/impressions) per thousand. Advertisers pay an amount of money based on the number of viewers who use the site for music downloads. Sadly, most of the smaller ones are failing. MySpace, Facebook, Twitter, and others have failed to deliver the desired unit sales through streaming and downloads, yet they are providing another method for major labels to use the *enthusiasm of fans* tied to technology and innovation to increase marketing and profits.

Marketing Plan

Labels, movie companies, networks, and others use whatever information they can acquire to create a marketing plan. Before the acts are signed at labels or movies are shot, the executives usually discuss the suggested size of the market and the promotional processes to reach the consumers. If it's a no-brainer, then there's not much of a problem as the executives simply review previous sales figures and analyze the types of consumers who most bought music, downloaded it legally or illegally (remember Sony), or attended concerts. If it's a new act or risky story line for a movie, the entire project will be riskier and jobs may be on the line. It often costs more on the front end to break an unknown artist than a known superstar. However, if the act or movie (*Star Trek* was my earlier example) turns into a cultural phenomenon, then the cash rolls in for many years. Everyone gets a promotion (we wish)!

Having musical and vocal talent does not guarantee a hit recording. Consequently, A&R directors seek the types of acts who not only have talent but also can sell records, merchandise, corporate sponsorship, and live event tickets through a clearly defined image, plus a brand. There is a difference, and record labels make and lose millions of dollars each year gambling on which acts connect with the music-buying public. Recordings are packages consisting of great songs that fit the image and persona of the artist, supported by an excellent vocal and musical performance by the artists and musicians. Successful acts establish the credibility of the A&R person who signed them. Failures can cost them their jobs. Successful signings provide additional record producing opportunities. Album productions enhance an A&R person's status as a record producer within the company and industry. Multiple successful deals and session productions (with the corresponding gold and platinum records) make the A&R producer a powerful, wealthy individual, often famous within the industry.

So, let's pick up where we left off after developing the recording budget in Chapter 8. Now, we're halfway there in the launch of a new act. We've completed and accepted all the bills for the creative side of the recording and development; now the business operations consisting of promotion, publicity, and distribution kick in, working together to hopefully start a fire of emotion within the consumers' minds. Executives in the marketing, promotion, and publicity departments usually determine marketing plans. How are we going to help consumers "discover" the act or product and what's it going to cost?

Prioritizing Media for Product Promotion

Did you know that all radio and television stations are not the same? Some are very powerful and can be heard by millions of people. Have you ever been driving and turning the radio stations and pick up one from Montreal in the middle of the night? Remember that before CD players, radio was all we had for entertainment in our cars. Film companies, record labels, major studios want to reach the largest number of people possible when they are promoting recordings films and other entertainment products. Thus, the power of the station's signal and where it is located become extremely important. As an example, a 1,000-watt radio station, which is a very weak signal, will have more listeners if it is located in downtown Los Angeles. While a very powerful AM station, say 50,000-watts, may be heard by fewer people because it is being broadcast over a sparsely populated area. Production companies rely on the marketing and sales department to identify, using consumer marketing reports, the outlets where they have the best chance to connect the potential consumers to their products. Another example is the different lifestyles of the consumers including race, language, income, education, and psychographics that need to be matched to their entertainment preferences. Trying to sell an English-speaking record to a Spanish or Japanese market is riskier. Accordingly, the amount of money labels, studios and artists are paid depends upon how many consumers discover their entertainment products. Thus, market research experts need to understand the technical underpinnings of how media, including radio stations, the Internet, and others, work so they can make the best decision in matching their entertainment products to consumer lifestyles through different forms of media.

Radio

Radio from the 1920s to 1930s was programmed with talk radio and live musicians and singers in live performances. Recording tape and direct to disc recordings made a difference until vinyl records were released for media play and consumer markets. Radio still receives the latest recordings free and just has to pay for the music publishing and songwriting royalties usually collected in blanket license fees from ASCAP, BMI, and SESAC in the United States. According to the Federal Communication Commission (FCC) there are 15,470 full-power radio stations in the United States, of which 11,380 are commercial AM and FM stations and the remaining 4,090 are FM educational stations, usually tied to PBS and universities.[9] There are also UHF commercials and educational TV stations, FM translators and boosters, and low-power TV stations that bring the total users of the airwaves to over 31,000.[10]

Analog Radio Signals

Analog radio waves are provided in two different forms (AM and FM) that use the broadcast signal wave as a carrier wave to embed the program or music being transmitted from the station tower. Amplitude modulation (AM) modulates the peaks of the carrier wave and frequency modulation (FM) changes in the frequency. Changing the amplitude or shape of the waves (AM) or the number of cycles of the wave (FM) means that if the carrier wave is diverted or blocked then you lose the signal.

Digital Signals

Digital signals offer many advantages in the quality of the sound, picture, and signal. Instead of the music program being carried on a wave and then being played by a receiver, the digital signal transmitted is simply binary codes of zeros and ones. The digits or signals are carried on a bandwidth 1,500 times wider than AM and FM, so the signal is much stronger. The computer chips in our TVs and our radios convert the binary code (numbers) into sounds we hear as the programmed music (refer to Chapter 3 for more on this). Therefore, the quality is excellent. There are radio and digital TV stations that carry several different signals at the same time, called multiplexing, which enables the sound quality to be consistent across the platforms.[11]

Digital Receivers

We receive the digital signals broadcast and transmitted to our TVs, radios, and cell phones from digital radio, TV stations, cable, and satellites that are converted into the recordings, movies, computer games, phone conversations, texts, or whatever by the computer chips in our devices. The key to the advent of small devices is literally size! The smaller the device, the more convenient it is to use any time, any place, anywhere. Love it. However, this explosion has also made the marketing of entertainment products more difficult as the market *now* has many consumers of all races, ages, social economic status, and locations with access to the products. That's a great advance in technology. But there's also a bad side, which is, if we've got a few 100 million or so people who don't give a flip about property laws or copyrights, it makes the executives' jobs of marketing their entertainment products for profit (a return on their investments) much more challenging. It's a double-sided coin as the passive market of consumers who have to buy or use (watch or listen to films, recordings, and so forth on media) the products has switched from the retail (controlled) markets to the lawless Wild West of taking what you want for free.

Media Promotion

Industry executives use media as promotional opportunities to break new acts and products to the consumer market. Remember, the label's job at this stage of the game is to help the public "discover" and "get excited" about their entertainment products and acts. Promotion is the second part of the four Ps of marketing, starting with *product* (which we've already created by spending millions to sign talent and create), and then *promotion* (which is to create and grow a market for the products and acts through airplay, media impressions, videos, advertisements, and other activities). The last two Ps are *price* and *placement* (distribution). The labels, movie production companies, studios, computer game companies, and book publishing companies all use the four Ps, but it costs money, of course, to create the electronic press kits, advertisements, promotional videos, news stories, and so forth, and more money to get them "placed" on the right types of media at the right time in order to best "reach" the targeted consumers.

Programming

Radio stations have varying degrees of importance to labels and movie production companies for promotion depending on the station location, signal power, audience size, and broadcasting hours of operation. Believe it or not, some stations are only 1,000 watt AM and others 50,000 watt FM. The quality of the broadcast signal is poor, yet the signal is strong on AM compared to a higher-quality but weaker signal on FM. Digital signals are mandatory on TV, satellites, and cable, and there are a few radio stations that are starting to use it. However, not all cars and homes have digital radio receivers in them so it will take a few years to complete the change to digital radio.

Day Parts

Radio listeners are divided into average listeners per hour (quarter of an hour by Nielsen Research) and ages such as 18 or older or 55 or older. As an example, in May 2015 nationally on all radio there were 35,600,000 Americans listening to radio stations at 7 a.m., compared to only 5,663,300 listening at 11 p.m.[12] The number of people listening to radio on all devices is the highest during what the industry likes to call "drive time" or when people are in their cars driving to work. Thus, the cost of advertising at that time in the morning (6 a.m. to 9 a.m.) is much more expensive than 9 p.m. to midnight as the amount of people listening is only about 16% of the drive time audience. These 2015 results are presented in the Nielsen Research pie chart (Figure 10.4), which provides an average of all radio formats' popularity in the United States.[13] Though we don't think radio matters that much anymore, the Nielsen data suggest otherwise, so here is some information on the major radio markets. Remember, there are just three major labels active in the United States today, so imagine the competition for market share as you read the descriptions of these broadcast companies.

iHeart Media (Clear Channel)

Clear Channel now is under the iHeart Media banner, with more than 245 million listeners each month in all the nation's

Figure 10.4 Different people listen to different types of radio formats depending on their age, demographics, and psychographics. Of people 18–34 years old, most enjoy Adult Contemporary at 10% of the audience, Classic Rock at 8%, and 20% enjoy Pop/Contemporary. Those figures change as people get older, as News, Talk, and Information is usually the most popular with much older listeners tuning in on AM radio stations. *Source*: FCC (2016).[14]

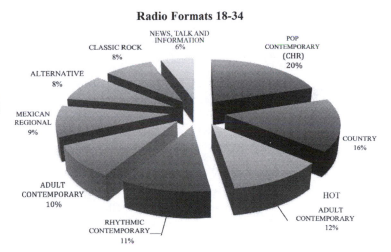

Radio Formats 18-34

NEWS, TALK AND INFORMATION 6%
CLASSIC ROCK 8%
POP CONTEMPORARY (CHR) 20%
ALTERNATIVE 8%
MEXICAN REGIONAL 9%
COUNTRY 16%
ADULT CONTEMPORARY 10%
HOT ADULT CONTEMPORARY 12%
RHYTHMIC CONTEMPORARY 11%

major radio markets through its 858 stations and digital iHeart Media services. It has another 450 million listeners monthly worldwide.[15] Its programming is available on more than 150 AM and FM stations, HD digital radio channels, and the Internet through iPods and cell phones, and is used by navigation systems in cars and global positioning systems (GPS), such as TomTom and Garmin. The company's operations include radio broadcasting, syndication, and independent media representation and outdoor billboards. iHeart Radio has a free digital music service of more than 2,000 live radio stations that allow listeners to create a custom station based on their favorite artists and songs.[16]

Cumulus Media, Inc.

Cumulus Media has about 454 radio stations in the top 90 U.S. markets that reach an audience of 245 million weekly. Westwood One (the radio production side of the business) syndicates radio programs to over 8,200 stations, with agreements with the National Football League, the NCAA, and the Grammys, Country Music, and Billboard Award Shows.[17]

Cox Media Group

Cox Media Group is a broadcasting, publishing, direct marketing, and digital media combination of companies that includes 14 TV and 59 radio stations. It also owns seven daily newspapers and other types of publications and digital services. It owns Valpak digital marketing, a leading advertising company. It has about 52 million weekly customers, including 31 million TV viewers, 14 million radio listeners, and 3.5 million print and online readers.[18]

NPR (National Public Radio)

NPR is an internationally acclaimed producer and distributor of noncommercial news, talk, and entertainment programming. A publicly supported, not-for-profit membership organization, NPR serves a growing audience of 26 million

Americans each week in partnership with more than 925 independently operated, noncommercial public radio and TV stations. It has a budget of $188.7 million and about 800 employees.[19]

Target Marketing

Once again, the executives in promotion have to target the station and their audience to the preferences of the listeners. As an example, who watches more TV broadcasts, blacks, whites, Hispanics, or Asian Americans? Each has different preferences based on ethnicity, cultural norms, divergence, and economic factors. The answer may surprise us, as according to Nielsen (2015) blacks watch broadcast TV an average of 51:23 minutes per week, compared to Hispanics at 29:13 per week, and 18:44 for Asian Americans.[20] Radio stations' broadcast formats are based on the ethnicity and divergence of their potential audience. Programming is usually tagged as Pop Contemporary Hit Radio (CHR), Country, Hot Adult Contemporary (AC), Rhythmic Contemporary Hit Radio, Adult Contemporary (AC), Urban Contemporary, Mexican Regional, Alternative, Classic Rock, Sports, and News, Talk, and Information.[21]

The variance in radio is also tied to gender and age. Accordingly, the number-one format for persons six years old and over in 2014 was News, Talk, and Information. But also in 2014, the number-one format for persons aged 18–34 placed News, Talk, and Information on the bottom of the list and Pop Contemporary Hit Radio (CHR) on top. Many sources provide "charts" on how well entertainment products such as recording artists, TV shows, computer games, and movies are selling based on units sold. *Forbes Magazine* has a different chart as it lists the top entertainer or celebrities every year based on revenues but also website hits and the mention of someone's name on TV, radio, and social media. What you'll learn is that when major personalities hit it big, we really do mean millions of dollars.

SoundExchange

Most of the royalties being collected from the media are for the "use of a song" paid directly to the copyright owner(s) and songwriter(s) of the songs (©) by ASCAP, BMI, and SESAC.

However, if the recordings of the songs are "transmitted digitally" over the Internet, such as on Pandora, webcasters, cable TV music channels, or Sirius XM satellite, then *digital performance royalty licenses fees* are collected and paid to the *featured* (royalty artists) and *nonfeatured performing artists* (e.g., musicians and background singers), and the *label* (the owner of the copyright of the recording ℗).[22] This is a common practice in most of the world; however, the NAB (the National Association of Broadcasters) has a very strong lobby in Washington, DC, that has successfully represented broadcasters for an approved exception to this licensing requirement. Sound-Exchange collects royalties and fees ($3 billion since it was commissioned by Congress) and pays them to its 110,000 members, who are the labels, musicians, singers, and royalty artists.[23]

AirPlay Direct

AirPlay Direct is a company that provides secure delivery of broadcast-quality recordings and artists' press kits digitally to radio programmers and film and television music supervisors (who seek songs and recordings in their visual productions). It is free to advertise, although the company has paid membership fees for additional uploads and advertisement banners. Established labels, independents, and even wannabes can use its services to upload new recordings. Broadcast and Internet streaming services around the world may download recordings to play in rotation. The top 50 singles and albums downloaded are charted by the number of station/streamer downloads. The music programmers who are impressed with the music and audience reactions may provide "radio creds," which are then added to the "charts" if you've got enough plays and streams. There are several of these types of services, and it's an inexpensive way to try to launch a career and to see if your recordings and talent have any market value. Analytics are provided that show station locations and plays. Contact information for the station is provided, and for the few program directors who like to be in search of the next big thing, this is a very valuable service. If the label finds its music is getting a positive response in Germany, then obviously it will start booking a club tour. There are also opportunities to provide telephone interviews and Skype content that may be fed directly through the stations to listeners.

Television

About the same time radio stations were hitting the airwaves, a 21-year-old guy named Philo Farnsworth invented the first television picture. However, David Sarnoff of RCA later became known as the "father of TV" as he marketed the first sets, and many just assumed he also invented it. The first really bad distorted images were broadcast at the 1939 World's Fair in New York, and the first major league baseball game was telecast shortly thereafter, with almost nobody watching as most consumers could not afford the TV receiver (called a set) in their homes.[24] It took almost ten years for the networks to build an audience before World War II was over, the economy had improved, and programming was available. Most of the shows were live and shot head-on with one camera and very cheap staging and lighting. As the TV audience grew, the live touring vaudeville market started to dry up.

When the center of entertainment in the 1890s was the player piano, the main source of profit in the music business was sheet music and piano rolls. Sheet music was sold next to the cash register at the local five and dime and was delivered by the same distributor as books. When the railroad systems were completed in the United States, actors, comedians, and musicians hit the "rail" to stage small shows in almost every little town on the way. That was the development of the vaudeville circuit, and it was about the only form of entertainment other than books, pianos, and a few radio programs. As radio started to increase in market share many of the vaudeville performances jumped off the trains and started to perform live on the radio. When TV really started to take off in the 1950s, the vaudeville stars moved from radio to TV. Radio had to come up with something to program, so they started accepting major labels' free recordings, which led to the first stages of multiple sources of entertainment programming available to consumers.

Networks

Chains of radio and TV stations owned or associated with one umbrella company are known as networks. As the technology improved, new stations (both radio and TV) had to be built where the people lived and the market existed, depending on the signal strength and population of the area. The business model is still the same today: acquire programming, place it on the airwaves, and then sell advertisements to pay the bills and gain a few profits.

The Federal Communications Commission

Nobody owns the air over the United States so the government had to issues licenses to radio and TV stations to "serve in the public interest." The stations were provided with a broadcast frequency just as radio stations are permitted to broadcast on the AM and FM spectrum. The purpose, of course, was to control the number of stations and programming and to also assure a clear signal so that one station's signal did not interfere with others'. Consumers soon loved the free programming and national and local news, and starting spending many hours daily in front of the black-and-white TV screens. The FCC's job has exploded with the Internet and the entrepreneurial uses and new business opportunities that have been created. Congress will soon have to make decisions about the FCC's reviewing powers of the Internet in the United States and all of the corresponding legal issues of access, taxes, and authorship of programming, bandwidth, and, of course, copyright violations.

Movie Production Companies

Movie companies have their own labels: the Walt Disney Company owns Walt Disney Records and Hollywood Records. But there's also a bigger picture to the story and that's what we hear when we watch a movie or TV show. Music beds enhance the emotion of the shot and the presentation of the line. Then there's the opening and closing credits, and let's not forget the meet cute plot point, where the iconic lovers meet in some crazy, unusual way that builds the foundation of the story. *Music supervisors* select the songs, recordings, and music that best fit the director's and the producer's visions of the film. Having a song in a movie can break a new artist into an upper level of success practically overnight. Just ask Hayley Williams of the rock band Paramore, and you'll get my point. Their 2008 song "Decode" in the hit movie *Twilight* helped bring a new fan base to the group and Paramore's fans to the movie. It also got them on the soundtrack album with promotion and publicity that helped promote both the movie and them. According to Williams (2008),

> *I didn't get into the "Harry Potter" series, even though I love the movies, "Twilight" really caught my attention and held it. I'm really excited to see the book adapted to film and excited that our band gets to be a part of the phenomenon. . . . I chose the title "Decode" because the song is about the building tension, awkwardness, anger and confusion between Bella and Edward. Bella is the only mind Edward can't read, and I feel like that's a big part of the first book and one of the obstacles for them to overcome. It's one added tension that makes the story even better.*[25]

All forms of media, broadcast, print, transmissions, cell phones, and Wi-Fi, are powerful business and promotional systems of entertainment and communication. Everyone uses them: politicians, actors, recording artists, advertisers, news sources, manufacturers of products of entertainment, and even national and local personalities.

Entertainment Media and Politics

Remember how political the Gutenberg printing press was in the 1500s? With the advent of this invention, anyone could print a book about almost anything or anybody. The same thing happened with the invention of the radio and TV in the United States. In the late 1940s, Senator Joseph McCarthy (1908–1957) of the U.S. Congress held hearings of the House Un-American Activities Committee, as it was called, concerning the impact of possible communist control of the media. World War II was over, but there was a concern that the Soviet Union wanted to take over the world and might be using broadcast media to get its message against American values out to the people. Sadly, in America, people in the entertainment industry were getting fired and blacklisted based on their political beliefs, and Congress's effort to protect the people from the threat of communism caused many talented creative people to lose their jobs and careers. It has taken years for some people's reputations to be restored.[26]

The Quicksand

Digital transmissions have become an important part of the marketing plan for the labels, but not because of promotion. Labels receive payment for every stream on audio and video, as they are the copyright holders of the recordings, but the stream is to only one person instead of being broadcast by a radio station to thousands. Songwriters and music publishers still receive public performance royalties collected by ASCAP, BMI, or SESAC. This is the new *quicksand* facing the industry when it comes to management of risk and potential profits. The risk is, quite bluntly, that anything digitized and transmitted will probably be stolen, in which case the labels receive zip on their investments, or it will be streamed and the label will receive approximately 6/10ths of a cent per stream. The difference in gross revenues between a person buying an album and a stream has resulted in billions of dollars of revenue losses. The labels could recoup a lot of their business expenses out of album sales, but the volume of such sales no longer exists. Successful recording artists, once the losses had been recouped, started making money on every album sold. That business model is gone.

Distribution—Retail Platforms

Labels still make money from the sale of the recordings they place into retail, and while labels determine the suggested retail list price, the cost of creating, promoting, and marketing recordings seldom determines the selling price of an album. Consumer demand and profit levels are more important. Wholesale and retail markups between retail outlets and distribution depend on the following:

- The type of distribution (retail, department stores/racks, or one-stops)
- The type of platform on which the music is recorded (CDs, DVDs)
- The sales propensity of the artist (a hot superstar, a high-ticket item, a known artist with current radio airplay, or title albums with recurrent radio airplay)

Retailers often ignore the SRLS (suggested retail list price) in favor of their own marketing strategies, offering discounts on some items and price increases on hot-selling or hard-to-find recordings. However, the SRLS is used to determine the wholesale price. Retail outlets use the range (difference) between wholesale (what they paid for the product) and retail (the price they quote to the public) to generate profits. Accordingly, brick-and-mortar retail outlets that sell CDs for $16.95 more than likely paid about 60% or approximately

$10.17 for that CD. The distributors are paid about 20% ($3.39, in this case), and the labels recoup the remaining 40% or around $6.78 per album.

Traditional retail outlets have suffered financially as consumers have more choices for their entertainment dollars. In addition, consumers know they can buy the same album for several dollars less at another store or online (which is instant and more convenient), and that's if they plan on paying for it at all. Traditional retail includes the following.

One-Stops

One-stops are the brick-and-mortar stores and the mom-and-pop places that are now mostly out of business or bankrupt. They simply cannot compete with the lower economies of scale of digital downloads.

Mass Merchandisers

Best Buy still places the CDs and DVDs in the front or middle of the store and prices them low to draw in customers, and then encourages them to buy a larger-ticket item, such as a computer, TV, or home appliance.

Rack Jobbers

Walmart, Kmart, and Target place CDs in racks in their stores, rent the floor space, and pay a small commission on each unit sold. Unlike the one-stops, with their large inventory and slow turnover (and therefore, higher prices), rack jobbers stock a smaller inventory of mostly hits that will sell quickly at a lower price. The racks keep getting smaller as most of these companies believe they will stop selling CDs within the next three to five years.

Record Clubs

BMG Direct and *Columbia House* have been two of the best-kept secrets in the business as they have often been used to unload overstock items for a huge profit. The CDs are sold at higher prices (after the discounted membership units, which are used as a tax write-off by the labels), and the customer pays the distribution cost (shipping and handling). The artists usually only receive 50% of their royalties, which is an industry standard. Columbia House and BMG Direct, which are owned by Sony Music, have quickly moved into digital downloads.

PI Advertising

"Per-inquiry advertising" is those late-night attempts to sell the overstocked albums through television. Older consumers who do not shop in record stores or online still buy old recordings that seem to stimulate fond memories of their youth. Television and cable systems play the advertisements for free when they have unsold air or broadcast time available. They are then paid a small percentage of the money consumers send in when they buy the products.

Steve Jobs

Many in the industry see the late Steve Jobs and his devices at Apple as "the man who saved the music industry by killing the album." When the labels signed up to iTunes they got a wonderful 70/30% split in the price paid, but consumers were also allowed to download "singles" from the albums, which almost wiped out album sales. Before, the labels were collecting about $6.78 on a retail album sold at a brick-and-mortar store and almost $7.00 online. But consumers started buying the singles they preferred off the album instead, meaning the label made about 70¢ per single instead of the $7.00. Talk about a reduction in financial gross income! What made it worse was the illegal snatching of unreleased products (CDs and DVDs) from the pressing plants by employees and then the uploading of them to bit.net sites. Now, the labels and movie company losses only increased. Meanwhile, Jobs and Apple made billions on the devices they sold and consumers got used to getting their entertainment products for zip.

Changing Attitudes

However, the products consumers often want for entertainment are still available through streaming, as the safe harbor laws allow the ISPs protection while creating an method for consumers to acquire entertainment products legally instead of stealing them. Technological innovation developed to the point where streaming is technically of high quality, but at the cost of billions to the copyright holders of entertainment products. My hope is if consumers develop a "habit" for streamed content, there is at least a chance that labels and artists will make some money. But uncontrolled and unregulated downloading of artists' copyrighted material will make it hard for them to preserve what is legally their property.

The best thing that could happen at this point is for consumers to get used to paying for a streaming service just as they have with cable TV. In many ways that is what is great about Netflix as we have a menu of what's available and can pick and choose what we want to watch, when we want to see or listen to it. We pay for that service (probably not enough) but it provides a way for consumers to pay pennies instead of dollars to be entertained. Let's look again at Honda Motor Company: we walked onto the lot and decided which car we're going to drive away. Guess what—they won't let us do that unless we buy it or lease it. Maybe leasing is the solution, which is similar to the streaming model, yet we'd need to pay enough money to make it feasible.

Three Blind Mice

What's really going on is some companies, such as ISPs, are turning a blind eye or facilitating copyright violations on a massive scale to the point where the industry executives had to blackmail them to comply (by paying millions up front) or be shut down. Still, billions of dollars have been ripped off and the major entertainment companies have lost their equity in copyrights that are protected by law. The safe harbor rules have moved the responsibility of digital copyright infringement from the businesses that make the products available to be downloaded or viewed to the consumers, who either don't care about the legalities of their use of the content or are unaware of any potential legal issues with their activities. All future sales and profits from owning the copyrighted entertainment products are close to worthless as the market is gone. The money spent in developing the entertainment content is lost when the consumers can acquire it for free. Here's an example: let's say Honda Motor Company spends billions to design, produce, and market new cars. Then consumers can go to any of the dealerships and take any car they want for free. How long do you think Honda would stay in the car business?

The Sharing Economy

One of the real psychological trends in the world today is younger people's belief that ownership of entertainment and physical property (such as cars) is actually a burden. If they can get it for free, who needs to own it? In some ways, this is really a cool ideal now that we have the technology to share things quickly by renting a bike, or jumping into someone's Uber taxi. Still, it's not right to steal others' property. But in this business it has clearly happened, so how are we going to make money now?

Labels Become Music Companies

To survive, record labels had to morph their old business models into marketing and branding companies that marketed artists to consumers instead of albums for profits. That's why we have the 360 deals as the labels still have to sign talent, record albums (for promotional reasons), and promote the peanut butter out of it in order to *break* an act. However, since the revenues from album sales are in the toilet, they have had no other choice than to fund and then receive a percentage of all the other artist revenue streams. Superstars at the time of the transition hated it, but when their deals expired, what they got offered were 360 deals. New artists and wannabes think 360s are a rip until they try it on their own and realize that in most cases only label executives can accomplish the task. We'll probably still call them *labels*; however, they are now much more than the old definition as they now include funding and profiting from concert ticket sales, merchandise, branding, and corporate sponsorships. The transition to the new types of deals has not been easy, but the economies of scale have changed the royalties (which stay the same on album and single recording sales) to a percentage of all artist revenues from their entertainment businesses. Other than that, about the only other change is that the packaging fees that usually come off the top of the digital sales have been dropped.

Websites

We've probably clicked on a label's website to check recording schedules or news, merchandise, careers, and label stuff. The bottom line is label artists usually have two sites, the label's and their own. Since 360-deal labels want to make sure their investments are controlled and protected, merchandise and other fan base connections are provided by the label. It's all about building the *brand* and giving fans a feeling they are a part of it. Executive names, pictures, and bios may be included. Latest news and publicity buttons show the headlines and a "read more" button provides a click away to the full story. Always written to feature the acts' and the label's image and brand in a relational manner, the *publicity hit* is based on the fans' emotional excitement and belief that they know the act and want to help them become more successful. The goal is to feed the fans' enthusiasm and let their emotional connection to the label and act build a buzz, sell stuff, and make a profit. Featured videos are available as well as links to the act's official website and online store. Press kits, videos, pictures, blog links, and bios of the artists are supplied at the act's site, which may or may not be operated by the major label. The online store link connects the fan directly to a label-provided merchandise store or third parties, such as Amazon.com. Let's just say they want to make it easy for the fans to purchase act- and label-related products. Professional websites cost $10,000 to $250,000 to set up and operate.

Distribution—Digital Platforms

Digital downloads are sold by providers such as iTunes, Amazon, and many others, including phone services, such as AT&T. Downloads are priced close to wholesale (60% of SRLP). As in our previous example, a $16.95 SRLP album is downloadable for about $9.95. This has forced the SRLP of $16.95 to consumers down to $12.95 to $10.95 in many big-box retail stores, but it has also reduced the size of the inventory. In addition, digital downloads bypass the cost of brick-and-mortar retail outlets and of shipping and pressing, and (most importantly) the split is 70/30 (label/provider), as opposed to the 40/20/40 of retail outlets. Thus, the label receives *slightly more money* on digital downloads than on traditional sales. In the case of a digital download, the label receives about $6.96 (70%) on a $9.95 album instead of the $6.78 on the traditional brick-and-mortar unit with the outdated SRLP of $16.95 at a one-stop or $12.95 at a rack jobber.

Applications (Apps)

The future may be direct sales of recordings, tickets, and merchandise by recording artists through their own app that provides a direct link to fans. Free or low-cost apps simply take out the intermediary and reduce the cost and time consumers need to acquire products. Apps also provide a direct marketing connection to the fan base. However, potential fans need to know who you are and they need to have discovered your music before they will buy the app (even if it's free). Thus, cell phone apps are great for known acts but not very useful for new acts until a marketing, promotion, and publicity program has been successful. Once an act and their recordings have been discovered by consumers, labels may need to increase the quality, speed, and value of the "discovery." The process is tied to the consumers' actual emotional connection(s) to the songs, recordings, and images of the acts. The label or music company is in the background, just trying to profit off its investments. Cell phone applications are a great way to provide a more personalized pathway to the act that will hopefully enhance the emotional connection and increase the buying of an act's products.

Digital Subscription Services

Labels are starting to make significant money from digital subscription services, such as Rhapsody, Spotify, Pandora, Amazon Prime, MySpace, and Yahoo. The key terms to remember are *interactive* and *passive* when referring to customers and *tethered* or *untethered* when talking about labels.[27] Some of the digital music providers allow interactive streaming, where consumers can choose what they want to listen to or download (e.g., Pandora for listening and iTunes for 30-second clips and buying). "Passive" means that the listener has to listen to whatever is offered, as with traditional radio stations streaming their signal over the Internet. "Tethered" and "untethered" are label terms that define whether the label is giving up certain rights, allowing the purchaser to make unlimited copies (untethered), allowing the recording to be downloaded and used for a limited time, or allowing consumers to make a limited number of copies on various devices they own (tethered).[28] According to Passman,

> To record companies everything that isn't a CD is an electronic transmission, which is the industry term for digital delivery (whether by download, streaming, or otherwise) . . . A digital download, which is also called DPD (standing for digital phonorecord delivery) is a transmission to the consumer (via Internet, satellite, cell phone) . . . tether . . . is a fancy word for "leash," meaning the record company doesn't give you complete control over the download.[29]

Digital Streaming Services

Why buy music or even steal it when we can stream it instantly for a few pennies on our computer, cell phone, or mobile device any time we want it? I keep asking that question over and over, but that is really about what monetizing entertainment has become: How can the creative writers, actors, musicians, and singers, and the business executives, employees, and stockholders, make money in today's psychological environment? Consumer behavior trends (their use of cell phones and other devices) appear to indicate a major cognitive shift in lifestyles and ethics. The key players in the game have been Rhapsody, Spotify, Pandora, iTunes's Beats and Apple Music, Amazon's Prime Music, Tidal, Google Play Music, and of course many others. YouTube Music is a new entry available on both iPhone and Android devices. It provides programming around the old concept of top 40 radio, called the Daily 40 singles. It is presented in both the audio and video formats, and just as with top-40 radio programming, its purpose is to provide multiple streams of the same top 40 singles in an effort to increase and maintain its consumer base.[30]

Spotify

Streaming music sites merge the idea of acquiring music over the Internet (Rhapsody and Spotify) with social networking, tied to the labels' desire to make money and break new artists. The *streaming sites* (e.g., Spotify) sync everything on to all our devices. Amazon.com offers its Prime Music and digital store, with its corresponding phone app and Kindle reader. Spotify is a European model that offers music for free with advertisements, or for a low monthly fee without advertisements, and has millions of songs we can listen to almost instantly. It has about 20 million paying subscribers ($10 per month) and 55–60 million free users who tolerate the advertisements. However, the service is quickly growing its audience base as consumers change their listening habits around streaming. The common denominator is the music. It is the glue that connects the consumers to the acts and the labels intend to profit from that connection, along with the investors of the website provider. Spotify also has a "dashboard" function that provides artists' managers and labels with analytics (algorithmic curating) about the location, demographic, and minor psychographic lifestyle of their listeners. Very cool for decisions that need to be made about touring dates and locations, ticket sales, promotion, and merchandise offers.

Streaming Revenues vs. Sales

Recording artists make $1.50+ on album sales and about 8¢–12¢ or more on the sale of a one-song recording. Labels gross about $7 on an album purchase and about 70¢ on a single sale, out of which they have to pay mechanicals, artist royalties, and other administrative fees. How much do they make from one stream? Hold on to your hat, as this will give us a clue about the economics of streaming for the artist and label. It depends on if the Internet service provider or ISP—Spotify or Pandora—is paying subscriptions or drawing down advertising-based revenues. As an example, if the customer buys a single CD or digital download, the label will owe the royalty artist close to 8¢–12¢. But the label will be paid only 0.12¢ from Spotify on an advertising revenue stream (free to the consumer) and 0.59¢ on a subscription stream. Both streams are paid less than one cent and the artist may be paid zero depending on the deal with the label. Pandora pays SoundExchange and then it pays the label far less (0.067¢); however, in this case, the artist also receives 0.067¢. To counter the loss of revenue, the major labels receive 70% of all of Spotify's generated revenues. In addition, they own 15%–18% of the company, thus getting another bite out of the profits. It would take about 178 streams on Pandora to equal the money paid to the artist and over 1,000 to equal the revenues the label would have grossed if the consumer had purchased the single instead of streaming it.

The recording acts often complain that they are not getting their fair share, yet they signed their contracts. The labels pay what's in the deal or they'll have to spend too much money on attorneys. So, the labels have to be paid by the streaming companies to license the rights to the master recordings to provide programming on their interactive streaming sites. The acts may not receive any money, as the royalties may not be defined in the deal as album royalties.

Labels who are the copyright holders of the recordings have had to strap on their "bad guy" caps and threaten to shut down some of the lowest-paying sites, such as YouTube. That places the labels or copyright holders of the recordings in the position to (shall we say) "blackmail" the violators, such as YouTube, when some kid mixes a recording owned by a major label with some stupid video. That's why every so often we'll see a message posted on YouTube that states "content no longer available." The labels are just defending their property rights and, of course, they should be paid. They made their point with Spotify, which now pays 70% of its net revenues (gross income, minus business expenses) to the label, and in addition, labels also own about 18% of the company. That gives the major labels some control of the business and additional profits on top of the licensing fees. Sadly, this has created a series of interrelated problems.

First, the digital streaming royalties being paid to the labels does not cover the expenses (cost of doing business), and, therefore, the artist royalties (recouped to pay the debt) are absent. So, the old business model of album sales is broken and not coming back.

Second, many artists are now signed to 360 deals, yet they may or may not be receiving all the money due to them as the old contracts may not pay them a high enough percentage from the "other" resources. Remember, it's all about risk at this point in the game. Any newer acts have a very hard time securing reasonable percentage deals until their level of success is established.

Third, due to the advanced technology, consumers may acquire programming through low subscription services or just steal the stuff using bit.net. This situation has been about a twenty-year problem that has gutted the economic value of the production companies and label that owned the copyrights. So many of the labels and movie industry's valuable copyrights are now worthless as consumers already have them or can find them free online and pay zip for them. The same thing is starting to happen as consumers make the move to Netflix and Amazon Prime.

Fourth, most of the Internet streaming companies claim to be "safe harbors" as long as they take down copyrighted recordings, which they often fail to accomplish. This puts the labels in the position of demanding money for the master recordings they own and they can ask any amount of money they want, which puts them in the position of almost "blackmailing" the safe harbor streaming services.

Fifth, the artists may or may not be paid in exchange for the lost revenues from the increases in streaming and decline in album sales due to the types of contracts they are committed to or might sign. The royalty artists who may have millions of streams may end up with a whopping check of about $5,000, as the SoundExchange rates do not cover the royalties from album sales.

Streaming Payments

Labels usually demand all the streaming royalties up front before they approve any streaming deals. That means they are receiving hundreds of millions of dollars before they know what is going to be actually streamed. Artists want to know what share of the money paid is theirs (if any) even before the labels may know. Crazy business. Yet it is the label's millions that got the act's careers launched and it is a little bit of mystery about who's getting paid and when. Honestly the labels have taken a lot of heat from acts that feel that the difference between the losses of album sales royalties and streaming royalties is a rip-off. Labels are making their first efforts this spring (2016) to address the situation with Sony Music Group and Warner Music Group, both suggestions based on packaging fee reductions and a percentage of the sales price of the streaming stock, if and when the labels decide to sell it.

YouTube

YouTube to the industry is a double-edged sword. On one side, the site allows consumers to post copyrighted songs, videos, and so forth without permission, payments, or even licenses. Sure, they are operating as a safe harbor, but they have been very slow in abiding by the law. Thus, lots of videos are posted and only later taken down. One view or listen to a copy (audio) is a violation. On the other side, YouTube provides viewers the industry perceives as a potential market, and, thus, they use it for promotion. At the same time, this situation places the labels in the position of having to threaten YouTube for money or shut it down. This is sort of a sleazy way of doing business as the labels would rather be paid for their copyrights, but since that hasn't happened yet, they have been forced to "do their best" in this unregulated vacuum.

What would happen if the legal shoe drops on YouTube? According to a January 2015 article on *Pollstar*, industry insider Irving Azoff is getting ready for that to happen: "mogul Irving Azoff formed performing rights organization Global Rights Services and ended the new year rattling his formidable sword at the online video service over 20,000 songs, for which he figures YouTube owes $1 billion."[31] Google, which paid $1.65 million in 2006 in a stock trade with owner/founders Chad Hurley and Steve Chen, now owns YouTube. Sequoia Capital is the third player in the game as they invested in Google and now have about 30% of YouTube.[32] Just before its own deal, YouTube signed deals with Warner Music Group, Sony Entertainment, Universal Music Group, and CBS Corporation to avoid a threatened lawsuit for copyright infringements. Not a surprise—either pay us money or . . . we'll sue you for copyright violations. Guess who's in control in this case? We'd think it is the copyright holders as the labels were able to get at least some money for the illegal use of their recordings. However, laws have been broken and the same appears to be the case in the Authors Guild lawsuit against Google, except two different judges have ruled for Google and against the copyright owners.

YouTube—Safe Harbor Status

It is estimated that about half of the digital data between the U.S. and European countries is protected under the Safe Harbor Act. However, in early 2015, the European Court of Justice rejected the law and the EU Court of Appeals reconfirmed that decision on October 6 of the same year.[33] Congress has not yet responded to European court actions. Let's take a closer look at what the Safe Harbor Act is really about and why it's so controversial.

Congress was faced with a balancing act in 1998 when they passed the Digital Millennium Copyright Act into Title 17 of the U.S. Copyright Codes. To ensure that the copyright laws would not restrict future innovations in technology, they added the Online Copyright Infringement Limited Liability Act or what we call a safe harbor. It limits the liability of providers such as Google, which owns YouTube, that have billions of copyright infractions. With the videos it could be movie clips, photos, or an amateur video mixed with a copyrighted song or recording. To operate under the safe harbor status, companies (e.g., YouTube) are required to make services available for copyright holders to *notify them* if any of their copyright properties are being used illegally.

The catch is that the ISP is then not responsible for copyrighted materials being transmitted on their channels. If the people who use the services and who violate the law don't have any money, they are not worth chasing down for a lawsuit. Under safe harbor, the copyright owner has the burden of notifying the ISP instead of the other way around. It's the copyright owner's responsibility to find the infractions, the name and contact information of who should be alerted at the ISP, and then file a complaint. If any of the required information on the forms is missing, the complaint is ignored.

European High Court

It now seems the safe harbor worm may be turning as the European High Court ruled against the safe harbor laws in early 2015 and it was upheld October 6, 2015. That means the fight has really moved into the court systems in Europe as the rulings may hamper ISPs' business model to operate legally, unless they become responsible for streaming any copyrighted products. If that happens in the United States, many streaming sites, such as YouTube, would have to seriously restrict their current programming and possibly shut down.

The Black Box

The "black box" account is the place where labels place unclaimed royalties and revenues that may be paid in the future. Some of the streaming revenues may end up in the black box depending on the individual contracts between the company and their roster of acts and possible changes by Congress in the copyright laws.

Gold, Platinum, and Diamond Awards

In the halls of many of the executive offices in major labels are the old vinyl gold records awards hanging on the walls. Success was visually hanging right in front of me as I'd walk down the halls to meetings. In his home, Jimmy Bowen had a large room plastered full of the gold and platinum records he'd been awarded from producing many famous artists and their best-known hits. He'd produced everyone from Frank Sinatra and Dean Martin to Hank Williams, Jr., and George Strait. His career (1957 to today) has been working as a performer, composer, producer, and label executive.[34] A great combination if you've got the talent. Seated in his guest room, full of the awards, I realized how imposing this must be for a new person. Record awards are presented by the RIAA (Recording Industry Association of America), which is the trade organization that represents labels. A *gold award* is for the pressing and selling of 500,000 units, *platinum* is for a million, and *diamond* for 10 million.

Billboard Charts

Billboard magazine reports airplay differently as it looks at the news of the industry and the *chart placement of the recordings* being played, streamed, and sold nationally and internationally. Instead of charting and ranking "who is listening and buying" it examines the "popularity and sales" of the actual recording in various forms of media and retail sales. The industry executives use the charts to determine market play and label strategies. Consumers use the charts to see what's available and becoming popular. *Billboard*'s weekly charts include the following:

> The Hot 100 *ranks the* top 100 *single recordings released each week in the United States based on radio station airplay, sales, and streaming. The songs listed may have a green arrow on the left of the chart indicating an upward movement, a horizontal gray arrow pointing to the side indicating no chart movement or a red downward facing arrow resulting from a lower placement then the previous week. Clicking on the small arrow next to the chart placement number reveals additional icons and information. A small microphone icon indicates the single as having the biggest gain in radio station airplay. A small icon of a CD disc indicates a gain in sales performances and a streaming icon declares*

a gain in streaming plays. A small dollar sign icon represents the increase or biggest gain in digital sales for the single. The single recording's chart placement is also displayed for last week, its peak chart position, and the number of weeks it's been on the Hot 100 Chart.[35]

The Billboard 200 ranks the top 200 selling albums in the United States based on sales and streaming. It's rare to have albums played on broadcast radio. Thus, this chart for the ranking of the top 200 bestselling album does not include airplay. The same icons are used on this chart except the small microphone icon indicates the gains in airplay. Many of the charted albums and singles also have a direction button on the right side of the chart listing that links to streaming.[36]

Other charts include R&B (Rhythm and Blues)/Hip-Hop, Pop, Dance/Electronic, Country, Rock, Latin, and Christian/Gospel, plus many others based on genres and world or territory indicators. The *Hot 100* and *Billboard 200* charts may list any genre of music from the individual charts once they hit the sales, airplay, and streaming to place onto the national charts.

Buzz Angle

A new research database labels and industry insiders use to determine marketing plans and promotional and publicity decisions is BuzzAngle.com. Nielsen and *Billboard* provide weekly charts and news, but Buzz Angle provides the charts daily. In addition, Buzz Angle software provides industry insiders data and charts based on the researchers' selections, based on song, album, sales, streaming, airplay, and social media metrics in a unique "snap-shot based on real time analysis." In addition, it provides retail data on physical and digital sales, including 225 record stores, direct-to-fans sales, venue sales, on-demand streamers, and radio station airplay. This is great information for labels, managers, acts, and concert promoters making plans about acts and products. Buzz Angle's CEO, Jim Lidestri, states,

Record Day (2014) was the first time we were able to say, "You really should be looking at things on a daily basis because that's how events happen." We showed what happened with a Super Bowl performance around Super Bowl Sunday. What was the impact for Katy Perry and Bruno Mars (appearing on the Super Bowl Halftime Show) on Super Bowl Day itself? With their performance what happened when AC/DC played Coachella? Here's what happened that day and the day before and the days after. Anybody who does a tour, and plays a city, that's one day. That's not a week. That's a day. If someone appears on any one of the late night shows, it's a day. You want to find out what happened on the day. If you run a Twitter/Facebook campaign, you run it on a day, and you want to see what happened. So yes, I am glad that you picked that up. It was the first time that we could say, "Hey, there's value in the daily data." A lot of value.[37]

There are other industry sources used by executives to determine how well an artist and product are selling (current and past), and then we can use the information to set up and predict future success. *Celebrity Access* is a little over $50 a month, yet it provides touring information and contacts for managers, acts, venues, transportation, and lighting and sound technicians. *Billboard* and others have both printed and online research that provides similar information. *Pollstar* is the trade magazine for touring, and we'll take a glimpse at it in Chapter 12, on touring.

Variety

Variety is the weekly trade magazine that reports box office revenues and business news about the film, movie, music, theater, and other areas of the entertainment industry. News articles on the business of production, broadcasting, transmitting, and streaming networks are detailed. Film projects, production schedules, casting calls, and updates and stories about Hollywood actors, directors, producers, and other personalities are provided. On the website, movie clips and interviews of new movie releases are provided. Digital entertainment news is detailed for production, features, business, and computer game products and business events. Industry executives often purchase advertisements and publicity to support the Oscars, Emmys, and other types of award shows and nominations, hoping the buzz in the magazine and website will gather votes. Winning an Oscar, Emmy, or Golden Globe Award usually increases notoriety, professional acknowledgment, and revenues. This in turn means the creative people, writers, directors, and executives will receive more money for their entrepreneurial job skills, knowledge, and studio salaried positions. Job listings are provided, although most of the job opportunities on this side of the industry are gained through word of mouth, track records of success, fame, working relationships with industry insiders, and executives' experience. On the lighter side, if you have a few extra million, *Variety* also provides information and advertisements about famous actors' homes and businesses that are for sale. Sure, that will be the day—right.

In this chapter, we have looked at the various ways artists and entertainment business executives can creatively and strategically promote and publicize their products. Within the advertising channels, from the print trade magazines to the streaming services, there is a good deal of variety, and it is consumer driven. Promotion and publicity campaigns are dependent on the ability of the A&R and Marketing and Sales departments to be on the same page when promoting an act or a product. They have to be well informed and nimble enough to rework their efforts in light of consumer responses to placement and sales. When people talk about the "costs of doing business," this is what they mean- will the time and effort we put in as an entertainment company to our personality and product really pay off when it comes to show openings, concerts and merchandise sales? If you are afraid to roll the dice, then you need to spend more time with the internal administration of the company to learn how the laws and media can help you plan for success. You may also want to think about the kind of representation you or your act or product will need to reach the market you want. Read on to see how you might just be willing to take that plunge and go for your dream, with the right help from the right people.

Notes

1. "Billboard 2015 Media Kit." Billboard.com. November 1, 2014. Accessed November 19, 2015. https://www.billboard.com/files/media/bb_2015mediakit_050115.pdf.
2. Ibid.
3. Ibid.
4. Ibid.
5. Ibid.
6. "Insights." The Total Audience Report: Q2 2015. May 1, 2015. Accessed November 11, 2015. http://www.nielsen.com/us/en/insights/reports/2015/the-total-audience-report-q2–2015.html.
7. "Insights." The Total Audience Report: Q1 2015. June 23, 2015. Accessed November 12, 2015. http://www.nielsen.com/us/en/insights/reports/2015/the-total-audience-report-q1–2015.html.
8. Ibid.
9. "FCC News, Broadcast Station Totals as of September 30, 2015." FCC.gov. October 9, 2015. Accessed November 5, 2015. http://transition.fcc.gov/Daily_Releases/Daily_Business/2015/db1009/DOC-335798A1.pdf.
10. Ibid.
11. Woodford, Chris. "Radio." And Digital Radio. 2007. Accessed November 13, 2015. http://www.explainthatstuff.com/radio.html.
12. "Insights." The Total Audience Report: Q2 2015. May 1, 2015.
13. Loynes, A. *2014 Nielsen Music U.S. Report.* 2015. Accessed November 13, 2015. http://www.nielsen.com/content/dam/corporate/us/en/publicfactsheets/Soundscan/nielsen-2014-year-end-music-report-us.pdf
14. "IHeartMedia, Inc." 2015. Accessed November 24, 2015. http://www.iheartmedia.com/Pages/Home.aspx.
15. "Why Radio." Rab.com. Accessed February 21, 2016. http://www.rab.com/whyradio/Full_Fact_Sheet_v4.pdf.
16. Ibid.
17. "Cumulus Media." 2015. Accessed November 24, 2015. http://www.cumulus.com.
18. "About Cox Media Group." 2015. Accessed November 24, 2015. http://www.coxmediagroup.com/about/.
19. "NPR Fact Sheet." NPR.org. June 1, 2015. Accessed November 24, 2015. http://www.npr.org/about/press/NPR_Fact_Sheet.pdf.

20. "Insights." May 1, 2015. Accessed November 11, 2015. http://www.nielsen.com/us/en/insights/reports/2015/the-total-audience-report-q2-2015.html.

21. Ibid.

22. "SoundExchange." 2014. Accessed November 24, 2015. http://www.soundexchange.com/about/.

23. Ibid.

24. "TV History." 2013. Accessed November 14, 2015. http://www.emmytvlegends.org/resources/tv-history.

25. Montgomery, James. "'Twilight' Exclusive: Paramore to Contribute Two New Songs to Film's Soundtrack." MTV.com/news. September 19, 2008. Accessed November 17, 2015. http://www.mtv.com/news/1595214/twilight-exclusive-paramore-to-contribute-two-new-songs-to-films-soundtrack/.

26. Stephens, Mitchell. "History of Television." Accessed November 14, 2015. https://www.nyu.edu/classes/stephens/History%20of%20Television%20page.htm.

27. Passman, Donald S. "Advanced Royalty Computations." In *All You Need to Know About the Music Business*, 163. 7th ed., New York, NY: Free Press, 2006.

28. Ibid.

29. Ibid.

30. Olivarez-Giles, Nathan. "YouTube Music Opens After a Year in Beta." *Wall Street Journal-Digital*, November 12, 2015, Digital ed., Digital sec. Accessed November 18, 2015. http://blogs.wsj.com/digits/2015/11/12/youtube-music-opens-after-a-year-in-beta/.

31. Speer, Deborah. "2014 in Review." Pollstarpro.com. January 9, 2015. Accessed November 25, 2015. http://www.pollstarpro.com/NewsContent.aspx?cat=&com=1&ArticleID=815796.

32. "Google Buys YouTube for $1.65 Billion." Msnbc.com. October 10, 2006. Accessed November 25, 2015. http://www.nbcnews.com/id/15196982/ns/business-us_business/t/google-buys-youtube-billion/#.VlY374Qp-JU.

33. Risen, Tom. "How Ending the Safe Harbor Law Threatens U.S. Businesses." Usnews.com. October 5, 2015. Accessed November 25, 2015. http://www.usnews.com/news/articles/2015/10/05/rejecting-safe-harbor-law-threatens-us-and-eu-businesses.

34. Unterberger, Richie. "Jimmy Bowen." Allmusic.com. 2015. Accessed November 26, 2015. http://www.allmusic.com/artist/jimmy-bowen-mn0000354600/credits.

35. "Music: Top 100 Songs Billboard Hot 100 Chart." Billboard.com. November 21, 2015. Accessed November 13, 2015. http://www.billboard.com/charts/hot-100.

36. "Top 200 Albums Billboard." Billboard.com. November 21, 2015. Accessed November 13, 2015. http://www.billboard.com/charts/billboard-200.

37. LeBlanc, Larry. "BuzzAngle Music CEO Lidestri Takes on Billboard, Nielsen SoundScan With Innovative Music Charts [INTERVIEW]." 'hypebot' 2015. Accessed November 17, 2015. http://www.hypebot.com/hypebot/2015/09/-celebrity-access-interviews-jim-lidestri.html.

REPRESENTATION 11

Artist managers bring representation, administrative supervision, and surrogate control to an artist's complex image and long-term career. A developed *strategic management plan* links the business and creative systems of the artist's career together. Career goals are established based on the perceived commercialization or "branding" of the artist's image and talents. Managers act as gatekeepers, selecting only the recording deals, concert appearances, and staff members that can best enhance the artist's long-term goals. Managers supervise everything connected with the business of the act. *Booking agents* are hired to establish contacts with promoters, who create concert appearances. *Business managers* are added to the team to control the revenues and take care of the accounting, taxes, and investments. *Road or tour managers* hire security, roadies, and everyone else needed to control all aspects of the tour. *Attorneys* are hired for legal opinions and representation concerning legal issues. *Publicists* and *marketing experts* provide artist interviews to retail stores and the mass media to promote the act in addition to the work being done by the label.

The Local Market

If you just want to sing and skip the business of being a popular culture icon, then find employment as a musician, a soloist in choral group, an opera singer, or a BGV (background vocalist), or all the above at the same time. The world is waiting to be entertained, so go for it, and get paid in the process. It is helpful, even in a local market, to know sight-reading and have quick memorization skills, a positive personality, knowledge of the copyright law, music publishing, accounting, marketing, publicity, business management, promotion, and a basic understanding of the music business. It's even better if you can sing in tune, love contributing to the group, and have fun helping others. It important to understand that when you pick up your guitar, sing in a choir, write your book of poems, or shoot a local video that you are a vital part of the entertainment and music business. The next great songwriters, authors, producers, directors, singers, film production teams, and industry executives will come from your generation. So, get on with it and may the best be given their chance to entertain more than the local fans.

Passion for Success

Major label executives often describe artists' passionate drive for success as a prerequisite to offering any deals. They also need proof of a growing buzz caused by strong local performance success, airplay, Internet and social media presence, and a wow showcase. Then, of course, they will investigate your work ethic, personality, and any closet issues that may increase their risks of a reasonable return on their financial investments. If they are going to dump a million dollars into your career, they want to make sure that you'll be able to handle the hard work and mental stress of success. At some point they know, and so should you, that you'll need professional management and legal advice. The bottom line is simple: once the label is satisfied, then they need the act to have personal and business managers, plus top-notch booking agents and in some cases established and highly connected talent agents. These are the professional managers who will help the artist make the important career decisions often required to earn the spotlight of success and a ton of money in the bank. It's also a safety net for the label's investment to know their new acts are working with top industry representation to correctly direct and build the artists' careers.

The Importance of a Buzz

Now you know that it often takes millions for labels to be in the entertainment business, and tens of millions for the stars to be in the movie business. What does it cost the act to work on the road, touring? And where does their money come from? The most important question that needs to be answered is why would a promoter (or a nightclub owner in a small market) want to hire your act to perform? Why would a director and movie producer want to finance your book into a movie script and then into a major film? The answer is simple: they want to make a profit. Artists tend to think it's about them, their creativity, and so forth, but it's really about exploiting an act's talents and fame or reputation in exchange for profit. That how important the buzz is, as it's easier to get hired for a tour or by talent buyers for nightclubs if you've been heard of. It is also a lot easier for film companies to secure money from producers if they have a "bankable" star signed to be in the movie project. How are you going to become known by the people who could employ you and your band?

Fame

Fame is built on the act's creativity and the label's risky investments to profit from the brand—building recordings, promotion, publicity, and marketing that stimulate revenues from many sources. It's the same for the film, book publishing, computer game, and theater businesses. Let's start at the beginning as a creative artist who wants to build a career as a touring musician/vocalist, and possibly if you catch on, you can become a known personality and, who knows, a superstar. Let there be light! When the talent is ready (rehearsed) then it's time to play some gigs. Many performances are free or darn close to it when new acts crank it up. However, once you start to build a following, then you'll be able to sell yourself to bigger venues based on your successful track record. It takes time, sometimes years, as it took the Beatles several years of local gigs and then several months in Hamburg,

Germany, playing many long shows in some of the city's sleaziest clubs to develop their unique sound, performance, and show.

Jumping Off the Cliff

One of the hardest things for anyone is to face rejection. If you're an artist of any type, such as a writer, actor, recording musician, or performing artist, it's even tougher. It's easy to live on the adrenaline rush and think that someday you'll be discovered, and then become rich and famous. It doesn't usually work that way at all. Years ago one of my favorite places for lunch was the Green Hills Grill (which is now closed). I remember sitting next to famous artists and their families lots of times. Little did I know that funny little girl over there at the next table would grow up to become Miley Cyrus. Or that I'd get into a discussion with actress Florence Henderson, who played the mother on the TV show *The Brady Bunch*, or Sam Phillips (at another place and time), who discovered Elvis. There are, of course, many stories of famous people I've met over the years, yet they all have one thing in common. They all had to start somewhere, and you'll have to do the same no matter what you want to do in the industry. So go for it!

Networking

First, you need to get noticed by someone in the industry who can connect you with others in the industry, which will lead you to meeting others. Networking is vital to getting your shot at a career in the business. Of course, don't ask for a picture or an autograph as that is usually seen as a thing fans do and not what serious people who want careers want. Instead, think about what you can contribute to the industry that can make it more successful and more profitable. When it comes to being a recording artist as a musician or singer, you've got to perfect your talents and live performance show before anyone will consider signing or hiring you for gigs. So do it to the point where you're sure you'll be putting on your best show, performance, or whatever. Then find a place that will let you play for free or maybe get a cut of the door (ticket sales). The club owner or talent buyer who places you into a club or performance hall will determine your success by one thing, how much money they made or lost. How much money was generated from ticket sales, booze, food, merchandise, and anything else they can sell? The fans may love, hate, or not even pay much attention to your show, but it's the amount of money generated that counts.

You're probably there for a different reason, to entertain the people in the seats and on the dance floor. The hope is the people who attended will start talking about how much fun they had and how much they enjoyed the band (you). If that happens and the money is flowing, the word starts to spread and the next thing you know, you'll be getting offers to play for more money and you'll be on your way. Isn't the buzz a wonderful thing? However, it'll cost money to put the show on the road and to tour small clubs. Let's look at an example.

Building Your Career Team

In this example, a band played a local gig and made $90 after the expenses for travel, food, CDs, and T-shirt manufacturing were calculated. However, it's also got a $30 monthly payment for equipment, which is not unusual. The hope is that by performing some free gigs and then small clubs for a little money, you'll be able to draw a crowd, build a fan base, and create a buzz. If the club owners start making money you'll be asked back repeatedly. If the momentum grows and the buzz increases then you'll be able to seek representation and start a small club tour. But you're the one who will have to find the gigs and pay for the equipment and all the bills generated by the band. The goal is to create a buzz that grows to the point executives in the industry start hearing about you or your act. Just as industry career professionals are often hired by word of mouth, that's how bands also get noticed. As soon as the industry executives perceive your act as something that's creating a market, they will check it out to determine its potential value. With experience and successful building of the fan base, this is the time when the act may want to write a pitch memo to label executives and invite them to a showcase, as we talked about in previous chapters. If—and it's a big *if*—the industry executives start taking you (the act) seriously and they decide to analyze your potential financial success vs. their investment costs, you may also receive a showcase request, as we talked about in the previous chapters. When that happens you need to start building a team of experts who can help you make the right decisions.

Table 11.1 Band Business Payments. It costs money to get out on the road, so plan ahead and make a budget that will help you determine how much money you need to make to pay the bills.

Per Show Expenses	Cost per Item	Number Sold	Sales Price	Income	Minus Production Costs	Total Income
CDs	$2	4	$10	$40	(–) $8	$32
Merchandise, T-shirts	$6	2	$15	$30	(–) $12	$28
Performance payment	$25 gas and food	1	$100 split of door	$100	(–) $25	$75
Total				$170	(–) $45	**$90**

Equipment	Cost
Audio board and mics	$1,000
Lighting	In-house
Staging	In-house
Video, computer, and software	$2,000
Total	$3,000
Monthly payment on equipment	**(–) $30 per month**

The Next Big Thing

There are many professional industry executives also looking for the next big thing. Consider using an attorney, talent agent, booking agent, and, when you're confident that you're good enough, a more highly successful personal manager. They all represent different types of artists who work in various areas of the business. Want to be a film actor? Then find a *talent agent* who is also connected (sanctioned) by SAG-AFTRA. If you want to be a stage actor then you need representation by an Equity-sanctioned *talent agent*. Want to set up concert tours as a performing artist? Then you need a *booking agent* licensed by the AFM. Want someone to take care of the business and money? Then you need a *business manager* with industry contacts and hopefully an accounting degree. Let's take a look at the creative business chain of events that usually happen in different areas of the entertainment business.

Chain of Representation

As a good example of creative representation let's take a look at the path for a book author to find a literary agent, who then attempts to place the manuscript with a book publishing company.

Major Representation's Power

If you have been offered a major label record deal, then quickly find an *entertainment attorney* with a stellar reputation within the industry. Similar to the way you would vet an entertainment attorney, there are unions for actors who perform the same functions. The key thing is that you find someone who is trustworthy, ethical, and professionally qualified to be part of your creative team. Check with industry sources, such as *Billboard*, *Pollstar*, Celebrity Access, and local union offices, to see if they have them listed in their directories (SAG-AFTRA for singers, actors, etc., AFM for musicians, and Equity for stage work).

Figure 11.1 There is usually a contract required between the author of the manuscript and the literary agents before the agent may pitch the material to a book publishing company. If the agent lands a deal with the book publishing company, then another contract is required between the author (the copyright holder) and the book publishing company that plans to print and sell the book. If the book is a huge success another contract may be required between the author, his agent, the book publishing company, and the film production company (depending on what was sold in the original contact).

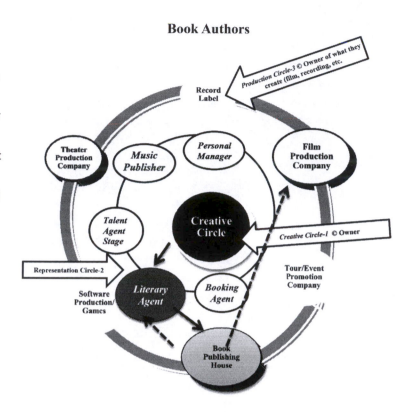

Since representation is based on contracts as well as types of functions, in what follows, we will look at the types of representatives there are for you in the industry and what some of the particular legal and contractual concerns are should you choose a career in the entertainment or music industry.

Personal Managers

While most talent agencies are responsible for the worldwide employment of creative authors, artists, and production talent, personal managers are usually responsible for the career development of recording and performing artists. Personal managers guide and provide advice to newer artists about their career decisions. Ask yourself a question: *what can a*

Figure 11.2 The songwriters in the creative circle pitch their songs to music publishers, who (if they like it and think they can get it placed) offer the writer a single-song contract. Then the music publisher's song pluggers may pitch the song to A&R at record labels to see if they have any artists who may want to record it. Connected song pluggers may also pitch the song to music supervisors for placement in new movies and theater productions. The labels would need to get a mechanical license if one of their acts records the song (probably from the Harry Fox Agency). If a film production company uses the song in its movie, it is required to pay for a sync license (acquired directly or through HFA), and if the film production company wants to use the recording of the song, it will need to acquire a master license from the label.

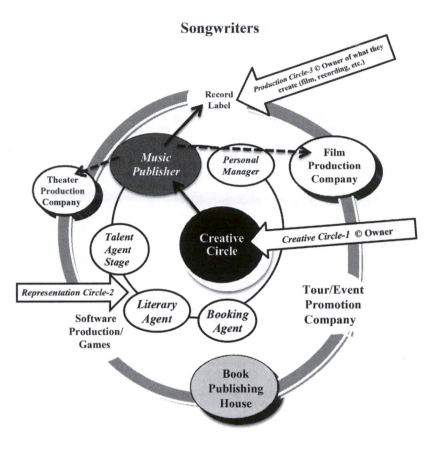

manager do for me that I can't do myself? The answers should be very clear, and now you know what you need to look for in a personal manager. Thus, the first step of a prospective artist is to find the perfect manager!

Advice and Consent

In general, the role of the personal manager is to counsel, assist, and supervise the development of the artist as a professional act. Obviously, to be able to successfully accomplish this mission, the manager needs experience, contacts,

Figure 11.3 A musician, singer, or band in the creative circle may seek representation through a personal manager or union representative. Or, the artist may act as his or her own personal manager when getting started by pitching his or her "show" to club owners or booking agents. Another contract is usually required between the artist, personal manager, and booking agent (depending on each of their job descriptions), with the manager receiving 15%–20% of the gross revenues from each gig and the booking agent claiming another 10%. In turn, the booking agent's job is to then find artists work where they will be paid for their shows. The more successful the artist is at each show increases the price of the ticket and the corresponding artist's payment. Once very successful, the booking agent may pitch the show to a talent buyer or concert event promoter so that the act can make lots of money from the live show and additional bucks from the merchandise sales and corporate endorsements.

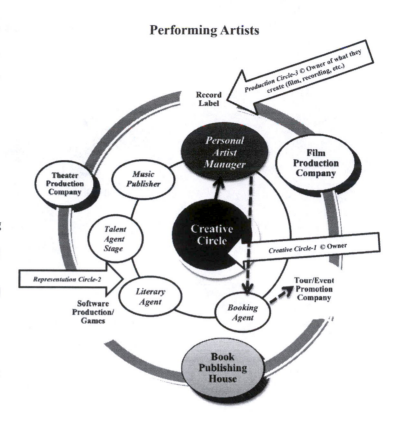

Performing Artists

wisdom, specific industry knowledge (not often found in books), and good old common sense. They need to understand and love business! They should value the power of entertainment and monetizing consumers' emotional connections to entertainment.

However, personal managers are not talent agents and in some states they are banned (as a conflict of interest) from seeking employment for the acts they represent. Their job is to approve work "pitched" to them by booking agents (for concert tours), publicists (for articles and radio, personal, and TV appearances), record labels (for approval of recording

Table 11.2 Industry Representation. Here you will see a brief description of the various types of creative representation found in the entertainment and music industry. A note of caution is to check on their success and track record before signing any deals. Beware of the big talkers, namedroppers, and others who may or may not be able to successfully represent you and your creative products to the industry executives.

Representation	Industry	Job Description	Who They Represent	How They Are Paid
Talent Agent	Film and visual media; tour music industry	Advice; find employment	Actors; recording artist personalities; writers; film production	Percentage
Booking Agent (AFM)	Music for American Federation of Musicians	Advice; tours, musical performances	Musicians for tours and performances	Percentage 10%
Personal Manager	Music industry; tours, image, promotion, publicity, song selection (advice)	Advice	Major recording artists; touring musicians	Percentage 15%–20%
Performance Union Agent (SAG-AFTRA Equity)	American Federation of Television Radio Artists for video; SAG for actors in film; Equity in stage	Advice; find employment	Film, video, and stage actors	Percentage 10%
Business Manager	Music, film, and stage	Control money and financial investments, plus taxes	All	Hourly, salary, or retainer
Attorney	Music, film, and stage	Negotiate, write contracts, advise artist and managers	All	Hourly, salary, or retainer

issues and promotional ideas, etc.), and other business offers. Managers are *not* song pluggers, record producers, or record label executives, so their main job is to build the artist's career through advice, experience, contacts, wisdom, and solid career-based decisions. Artist managers rarely give a "yes" to anything unless it is something that is in the best interest of the artist. They often seek out only those industry insiders who can benefit the artist's image and career.

It is the artist's responsibility to pick the right manager, someone who will make a great impression, be well respected by record label executives and talent agents, and be able to work and communicate well with the key industry figures. Don't hire a friend or relative because their lack of experience and poor decisions will affect your career and income forever! As an artist, you *have* to be able to say "no" to friends and family. If you can't, the manger will say "no" loudly for you. You need professional representation if you want to have credibility within the industry, and friends and family are not the right people to represent you to industry executives.

The Gatekeeper

Because we live in the digital world, opportunities for artists to succeed without a manager have increased. Still, someone has to do the work, take the time to market the act and their products, and build a plan based on a strategic business management model. Select a manager who can help you as an artist create, manage, and accomplish it. A manager is the gatekeeper, keeping all the people, hangers-on, fans, and negative family, friends, and others away by saying, "No, no, no." Time really *is* money in this business, and so are the artist's name and image, which if connected to bad products or merchandise, bad news stories, rumors, politicians, and so forth can be destroyed beyond repair. It is the manager's responsibility to fend off all the negative stuff and approve only people, products, and connections that are in the artist's best interest. Publicity hounds that can attempt to soften a bad reputation and maybe turn it around are expensive. Also, if you are signing a recording deal, the label will make sure you select a manager it approves of and supports. It wants to make sure its investment in you as an artist is going to have a better chance of success by encouraging you to select an established, respected manager.

Trust

The artist usually hires the manager, but signs with the management company. How weird is that? I hire you but then I have to sign a contract with the person or company I hired. I am the boss, but you're going to tell me what to do. Strange— so make sure they really know their stuff and are worth the time, effort, and money. The artist management relationship is built on *trust* in this emotionally high-strung industry. The artist has to trust the manager to make the right decisions, give the very best advice, and build the artist's career, sometimes over the manager's own. The manager has to trust the artist to follow the advice given, stay healthy physically and mentally, work, work, work, and always exploit every opportunity to make use of shameless self-promotion. Before signing a management deal, which may tie you up (the artist) for as long as seven years, establish three goals and have them added to the deal.

First-Year Goal

Set three goals with anyone you're considering as your personal manager. The *first-year goal* is extremely important because it is the lynchpin to a successful career. Your first-year goal is either a major record deal (360) or a planned structure analysis (full financial business plan) of how the manager will launch your career if you're not aiming at becoming a known national or global personality. A full business plan is wise in all cases, yet with a major record deal most of what will be happening and the corresponding financial investments are taken care of by the major label. However, having a business plan at the start of the career will be very helpful to the act and manager. It will help you determine if your career is on track, and if the advances and recording commitments and financial investments are reasonable to fund the launch of your career and to sustain it.

If you're not going for the major label deal, then how are you going to make money? That's the reason for the business plan and the corresponding details on how you're going to be "discovered" (and how much to budget) for promotion, publicity, and marketing of yourself (as an act) to promoters and club owners. At some point you may have to consider paying a booking agent (not connected with the AFM, if you're not in the union yet) to find free or buy-on tours to get you started.

Third-Year Goal

Your second goal actually happens in the third year of your contract with your personal manager. If the manager helped you acquire a major label deal in the first year, you've finished your first album and probably had a couple of singles released. You've also experienced two to four weeks of national promotion with radio (tour and airplay), publicity with local and national TV talk shows, stories planted in the press and trade magazines, social media and website presence, and other types of promotional opportunities. You'll probably also think you're in debt to the labels for more than a million dollars but by now, you know that's their debt, not yours. Still, they will recoup 100% of the all-ins from album sales and a negotiated percentage of all your other revenues (including concert ticket sales, merchandise, and corporation sponsorships) to recoup or pay off their investment and make a profit (they hope). The label, you, and your manager will know by the end of the first year if you're catching on with the public, and the label will have its accountants analyze the books (money made or lost) to decide if it is going to pick up your option.

Options

Labels always look forward instead of backwards when it comes to picking up an artist contract option. Once the initial year is complete (or 60 days before the end of it) the label must determine if it wants to pick up the act for another year (option). After the initial year the deals are optioned by the label as another period of time (usually one year) and number of recordings (usually one).

If the label decides to pick up the next option, then the artist receives another year and another album of which the label will fund all the bills that are approved, and the act receives another advance. Why would a label pick up the option? It quickly looks at the first year to determine the potential growth for the next period.

While we don't generally think of people as lines on a bar chart, let's compare and ask ourselves which artist's option a label should pick (see Table 11.3).

Clearly, Artist A's manager will want a significant increase in royalty points and higher percentages on other revenues as his artist has made the label almost $2 million. Artist B's manager is hoping, begging, and pleading for another shot, but his act has lost the label almost $2 million. Which act will the label pick up? Again the labels are interested in how much money an act will make in the future as they consider the past as gone, gone, gone. Which artist is decreasing in sales and which artists are starting to really catch on with the public? Look at the skew lines and you'll see the label will probably pick up Artist B's option and release Artist A, as his sales are quickly declining.

If the label has picked up the first and second options, then you should be rocking and not a slave to the system. However, if you're making money, the label is making more and it will want more and more out of you to increase its profits. When you're hot you're hot, and that's the time to turn the

Table 11.3 Picking up the Option. Which artist's option should the label pick up? Artist A has made $1.75 million for the label and Artist B has lost $1.75 million of the label's money. Here's a hint: consider the present, project the future, and forget the past.

Artists	First Album	Second Album	Third Album	Totals
Artist A	1,000,000	500,000	250,000	1,750,000
Artist B	−1,000,000	−500,000	−250,000	(−) 1,750,000

fans' enthusiasm into cash if you're really successful. You'll think you've been on the road forever, selling everything with your name and logo on it. Great, but you'll need to also take care of the stress caused by the constant work of being a known personality, and you'll also have to pay for all the people in the band and on your management team.

At the end of the third year, your second goal is to have $3 to $5 million in the bank after taxes. Careers are usually not forever and you'll be working harder than you ever imagined if you're one of the chosen few. So put most of the money away. In all honesty, that's not a lot of money for yourself as you'll have to generate much more money to pay all the bills and all the people working for you and fend off all the relatives and do-gooders who assume you're filthy rich. Just tell them you don't control the money, your manager does, and he'll always tell them no—buzz off.

If you're not going for the big time and want to just make a great living as an entertainer, then you'll need to follow the goals established in your business plan. People get married, have kids, and need to invest in IRAs and 401(k) plans. Therefore, you've got a bright career ahead of you if you're smart. Think about it this way; most of us have to work 40–50+ hours a week to make a decent living. Entertainers have a different situation, although you'll have to find the opportunities, as you're really an entrepreneur entertainer.

Career Musicians

Assume you're a musician in a band playing cover hits. You make $200 a night working four nights a week for three and a half hours. You've made $3,200 a month, working only 14 hours a week. Let's take the month of January off and work only 11 months a year. Now, you've made $35,200 a year working 14 hours a week. Let's say you create your own record album, using Pro Tools, press it, and sell it off the bandstand and on iTunes and CD Baby. It costs you about $2 per album to pay the mechanical license (for the cover tunes), and to press and have the CDs delivered. You also start your own website and sell digital versions, plus merchandise over the Internet. Let's say you sell only 2,000 albums off the stage and Internet combined. But you sell them for $8 each, making a profit of $6 times 2,000 units, for a profit of $12,000. Now, you've made $47,200 and you are still working only around 14 hours a week. Sell some merchandise at the local bars, restaurants, or whatever, and pick up a few gigs, and you'll probably make $60,000 to $70,000 a year and work only about 20 hours a week. However, to do this successfully you'll have to put on your business hat and manage your career yourself. By the way, you're the one with the talent and the smarts to profit from it. Good for you if you make a very good living and have to work only a few hours a week.

Fifth-Year Goal

If you've been in the business as a superstar or major artist and find yourself in the fifth year of your management agreement, plenty of other successful managers will want to represent you. Have you reached your fifth-year goal of $10 million after taxes, free and clear (without debt) in the bank or in secure mutual funds or the stock market at the end of five years? Management contracts are similar to recording deals, as you'll usually be required to give up your side of the yearly options in the deal. The manager makes the decision if he or she wants to continue the deal or not. The one-, three-, and five-year goals written into the deal establish *performance or reversion clauses* that give you an out if the goals are not met. What if your manager helps build your career in five years to the point where you've got only $7 million in the bank or market? The choice is yours.

Early Termination

The artist hires the manager and can also fire the manager, but you'll probably have to continue paying the fired manager until the end of the deal (maybe seven years). That's the reason for the reversion clauses in the deal as suggested in the one-, three-, and five-year goals. They either accomplish it or fail. Sometimes the breakup of an artist and manager is similar to a divorce as lawsuits and counter-lawsuits are not uncommon, with the manager claiming commissions on the artist's future royalties. Unless there has been some kind of criminal conduct by the manager, he or she usually wins, gaining partial royalty payment for a limited time.

Post-Term Management Royalties

At the end of some deals, the manager may write into the deal that he or she will continue to receive a part of his or her commissions based on the idea that the act's future success is due to the management team made during their tenure. Be careful when you see this type of language in the management deal.

> *After the Term, Manager shall be paid, in lieu of the Fee, a sum (the "Post-Term Commission") equal to fifteen (15%) percent of all Gross Income (except for items excluded under the paragraph 7 above) received or accrued by Artist for the first five (5) years after the Term has ended, which Post-Term Commission shall be reduced to ten (10%) percent for the five (5) years thereafter, and then further reduced to five (5%) percent for the next five years. For purposes of clarity, Manager shall not be entitled to any Post-Term Commission (or other compensation of any kind) beyond fifteen (15) years after the termination of this Agreement.*

What does this really mean? If your deal was for seven years, then you've got to pay the former manager for another 15 years at a decreasing percentage of your gross income, even though he's not your manager! That's a total of 22 years. Wow, be careful to say the least. In addition, if your deal ended and then you picked up another manager, you'd be paying both of them, which means a large chunk of every dime you earn would have to be paid to the current and past managers. Don't turn yourself into this type of slave.

Types of Personal Managers

There are several types of personal managers in the business, and their success is generally based on their ability to help artists become rich and famous. Timing is everything, as the act's position in their career (new, established, superstar, etc.) determines the level of managers employed. If you are a new artist without label or music company support, securing a major or "heavy hitter," established manager is unlikely. If you're an established artist with a career buzz and consumer momentum, then there is the possibility of gaining an established management company for representation. Someone in the office working for the kingpin represents the act, which is fine as the power of being represented by an industry-famous manager (or his or her company) opens many doors. Superstars are often represented by only a handful of the most highly established industry-famous (often unknown by the general public), powerful personal managers who have earned respect based on their track record (the successes of major artists they have helped in the past).

Timing

When does an artist really need a manager? The answer, of course, is only when the personal manager can do things for the act that they *cannot do for themselves.* The other side of the question is when does a manager need a specific act? The answer is when the act is ready (creatively, mentally) to dedicate their time and talents to growing their career as large as possible. Possibly they still need to do more, such as create an industry buzz through marketing and media appearances, build a significant fan base, and try to get the best possible label deal. They can't do it by themselves, but the right manager might be able to help them get it done. Therefore, the artist and manager relationship is a partnership, similar to the emotional commitment of a marriage, based on trust, for the purpose of achieving business goals (making money) that each could not have attained without the other.

Power of Attorney

Artist managers bring order to the daily chaos caused by success in the music, film, and entertainment industry. Personal managers legally represent artists through a *power-of-attorney agreement*. It provides personal managers with the legal privilege and obligation of controlling the business and career decisions of the artist. What is really taking place is the personal manager recommends someone to the artist to hire for their management team. Booking agents, tour or road managers, business managers, and approval of SAG-AFTRA talent agents are examples. One of the mistakes beginning acts make is signing a manager who is also their attorney, agent (nonunion), business manager, and personal manager. Putting all those tasks on one person is not very wise if you're serious about a successful career; if they happen to turn out to be a bad guy, you may still be signed to them for all these duties for years. That means you're really stuck and can't hire others until you gain a release from the bad dude. Also, make sure your power-of-attorney agreement is *limited* to specific financial matters tied to your career business duties, not personal or investment funds.

Figure 11.4 Artists often find a point in their careers when they need some additional industry-based help and direction. Personal managers usually sign an act to the management company, yet it is the artist who may hire and fire the personal manager. The contract usually has a "power-of-attorney" clause in it that gives the personal manager the right to conduct the business decisions for the act. The personal managers in most states are allowed to provide only "advice and guidance," so they hire a business manager to take care of the financial revenues and expenses, a booking agent to find gigs for the act by contracting event promoters and club owners, and a road or tour manager to manage the musicians and business on the road.

The Artist Management and Touring Revenue Stream

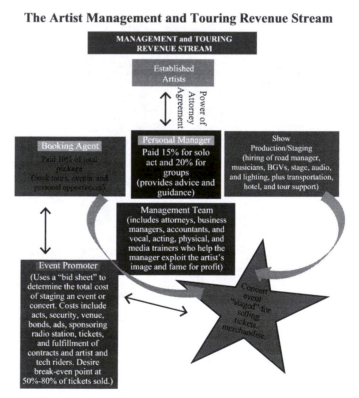

Thus, a limited power-of-attorney agreement permits the manager to lawfully represent (to the very best of his or her ability) the artist(s) in either *specific* or *all business decisions* (depending on how much the artist wants to remain involved in daily decisions). Other duties include record label agreements, music publishing, music producer selections, mass media promotion and publicity through a hired or label publicist, concert appearances supervised by the road manager and arranged by a talent agent through a concert promoter, merchandising, corporate sponsorship, endorsements, charity, and community service. However, guess who is responsible to pay their salaries?

Commission Base

Managers are typically paid 15%–20% of an artist's gross income, including record deals (after recoupment), concert tours, merchandise, and corporate sponsorships. The standard is 15% for single artist representation and 20% for groups or bands. However, the rates may be more or less depending on what is negotiated based on precedent, reputation, and the type of genre (rock managers are often paid 5%–10% more, Christian acts are sometimes paid less, and hip-hop and rap acts' managers are paid whatever they can get). Elvis's manager, Col. Parker, made 50% of everything, which most people in the industry thought was crazy), and sadly when Elvis passed away, he had only a couple of million in his accounts.

Gross vs. Net

Management payment rates are based on the *gross income* of all or part of whatever the act earns. This too is negotiable, but the more powerful the manager, the less likely it is that they are going to negotiate down the percentage of what they are paid or which types of incomes they are to be paid on. In addition, most managers will not agree to work for net, which is the amount of money an act is paid *after all expenses* (touring, recording, agent fees, taxes, etc.) are deducted. Also, the manager's typical business expenses (in the name of the artist) are paid out of the artist's gross income, not by the manager. Business office expenses, travel, payments to business managers, road managers, and booking agents, who are hired by the manager in the name of the artist (remember the power of attorney), are paid by the artist, not the manager.

Business Managers

Business managers are the financial experts who provide the accounting, tax planning, and investment advice for the acts. In good management situations all the money earned by the act goes first to the business manager, who then pays everyone. By putting the business manager in charge of the money, it is easier to control the accounting, taxes, and long-term investments for the act. Business managers should be paid a flat or hourly rate for the amount of work they accomplish and for the success of their financial investments. Some are paid a small percentage of the gross income (5%), but this is rarely in the artist's best interest. It is desirable for a business manager to have a degree in finance, plus a successful track record in investments and money market funds and experience with industry professionals.

Tour or Road Managers

Road managers are responsible for all the daily business decisions dealing with concerts and personal appearances on tour. As a result, road managers handle the local press as well as questions and concerns from promoters, local radio stations, and record retail promotions. In addition, they make hotel and restaurant reservations and handle any issues of concern among the recording artist(s), band members, roadies, and tour support. Road managers report directly to the artist and personal manager, and they are the local communication link between the artist, the promoter, and the manager of the venue. Great road managers often use the road and concert tours for networking, in hopes of starting their own artist management company later.

Booking Agents

Booking agents commissioned by the AFM represent musicians employed as recording artists, session musicians, and concert performers. Equity union–licensed agents represent actors and performers pursuing employment in any type of unionized live theater. SAG-AFTRA approves of agents who represent their members seeking employment as vocalists in recording sessions and as newscasters and TV personalities. The Screen Actors Guild (SAG) represents actors and performers in any type of film and visual media (as opposed to video, which is covered by SAG-AFTRA).

Union-franchised booking agents are ordinarily limited to 10% commission rates, although some may charge more. The unions, of course, bring the power of collective bargaining and legally supported contracts to encourage businesses that hire union musicians, vocalists, actors, and others to fulfill their financial and contractual obligations as originally stated in the deal. The agent holding the license commissioned by the AFM or SAG-AFTRA may employ subagents. They are paid 25%–50% of the 10% gross income the acts receive when "booked" and paid to perform. Film agents franchised by SAG-AFTRA represent acts to be hired by producers, film companies, and, of course, directors (who are the real power behind visual media productions). Royalty artists, vocalists, musicians, and bands are represented by AFM (musicians) and SAG-AFTRA (vocalists) to bar owners, restaurants, venues, and concert promoters, who create, promote, and sometimes profit from concert events and tours.

Attorneys

Managers hire attorneys to structure deals for artist tours, merchandise sales (through fulfillment companies), corporate sponsorships, label deals, appearances, and any other type of business opportunity. Accordingly, some attorneys become very powerful and can act as personal managers after they have established themselves in the business. Attorneys charge hourly rates ($250–$1,000 per hour) or are placed on a retainer, which is a monthly minimum fee based on a negotiated number of hours of work per month or part of the artist's gross income. It is preferable to pay an attorney's hourly rate (as opposed to a retainer fee) to guarantee the productivity of the attorney.

Artist Management Promotion

In the past, radio station airplay was the most important type of exposure for new artists and their recordings. In addition, radio acted as a natural filter to expose only the potentially successful artists and recordings. However, large corporations, such as Clear Channel (which owns close to 10% of all the radio stations in the United States), have made it very difficult to break new acts on radio.

Also, Generation Y (the current 18–24-year-old consumers) has changed the rules of the game by not reading newspapers, watching TV, or listening to radio. Instead, they download their music and listen to what they want when they want it through their computers, iPods, devices, and cell phones. Artist managers and promoters are in the same situation as labels, struggling to find ways to make consumers pay for recordings or at least steal them, attempting to find better ways to exploit their acts and concert events through social networks and consumer technology.

Artist Management Publicity

Artist managers, record labels, professional musicians, and event specialists hire publicists from public relations firms as part-time or full-time employees to plant stories and articles featuring their clients (the artist) in the professional trades and social and consumer mass media. The publicist assimilates images that have been created, molded, or enhanced by the artist's manager and label representatives into stories, news articles, and press releases. Branding is built out of the image, which is the foundation of the emotional connection between an act and consumers. Publicists alert the media to cover the events (which are often created by the management and label representatives) as news or entertainment stories. The stories are then released to the print and mass media with the hope of stimulating a positive correlation between the artist's name and public perceptions. Increase in the public's awareness of an artist and acknowledgment of an artist's creative talents often turn into profits.

The Cost of Doing Business

The big surprise for many bands is the cost of everything when launching your business as a live performance show.

A *public relations campaign*, which is usually provided by a hired publicist, is the process or act of providing material

Figure 11.5 It's complex to be in the band business in the entertainment industry. Look at just a few of the line items that will cost you money. They include your attorney, marketing, representation, equipment, and the cost of the products and services you want to sell. What's it going to cost to actually go out on the road?

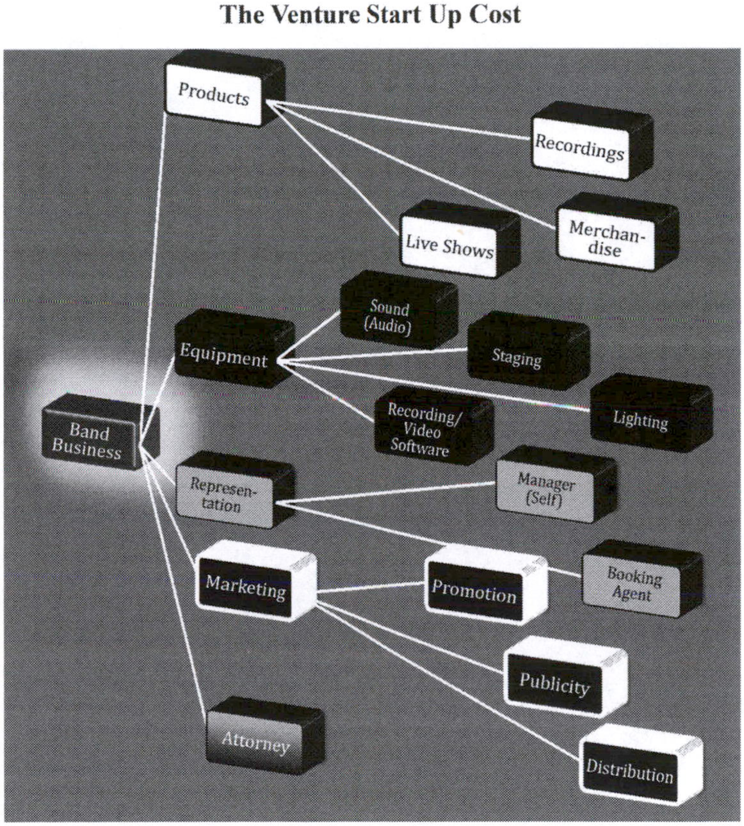

The Venture Start Up Cost

(stories, press releases, etc.) to the social and mass media in an attempt to provide a favorable public image for an artist. The process occurs whenever there is contact with the public or whenever an act has some counter- or bad publicity. Positive aspects of the act are provided to spin the truth or to let consumers choose for themselves what they want to believe about an artist. Bad publicity or negative news can hinder record and ticket sales, so management will often provide the artist's best side to enhance the artist's image with the public.

Courtney Love

Let's compare the artist's and the labels and managers' perspectives on the industry to see where the disconnect between perception and realities might start. Courtney Love's comments are typical, and as you can imagine these types of comments drive label executives and managers a little crazy. According to Kurt Cobain's widow,

> *This story is about a bidding-war band that gets a huge deal with a 20 percent royalty rate and a million-dollar advance . . . What happens to that million dollars? They spend half a million to record their album . . . They pay $100,000 to their manager for 20 percent commission. They pay $25,000 each to their lawyer and business manager. That leaves $350,000 for the four band members to split. After $170,000 in taxes, there's $180,000 left. That comes out to $45,000 per person. That's $45,000 to live on for a year until the record gets released. The record is a big hit and sells a million copies . . . So, this band releases two singles and makes two videos. The two videos cost a million dollars to make and 50 percent of the video production costs are recouped out of the band's royalties. The band gets $200,000 in tour support, which is 100 percent recoupable. The record company spends $300,000 on independent radio promotion. All of those independent promotion costs are charged to the band . . . If all of the million records are sold at full price with no discounts or record clubs, the band earns $2 million in royalties, since their 20 percent royalty works out to $2 a record. Two million dollars in royalties minus $2 million in recoupable expenses equals . . . zero![1]*

Ms. Love is a smart woman who is frustrated by the typical business processes within the music industry. Let's look at both sides of the industry from an artist's perspective and the industry label's and managers' perspective. Honestly, this kind of stuff happens all the time and it's frustrating to everyone. What would it be like to be her manager, trying to explain this stuff to her?

Ms. Love was ahead of industry trends, as she wanted a closer, direct, honest connection with her fan base. Now labels (in most cases) won't even consider giving a new act a shot without an established, strong, growing, strong fan base that can be monetized. Being on social media or having a "presence" on iTunes, YouTube, and all the other sites is meaningless until the buzz generated is real and reflected in serious numbers (millions). Where was she correct in her statement and where does it appear that she may not have been informed or may have misread (which is easy to do) her contracts? The issues in her statement are very common as what attorneys often say is not what the signing acts appear to actually hear.

Recording Budget Mistakes

According to Ms. Love,

> *What happens to those million dollars? They spend half a million to record their album.*

Spending that much money on an album is crazy. In today's world anything over $200,000 is questionable. Most major acts and bands do not use the "band members" to record the album. Remember the Wrecking Crew? They use professional AFM studio musicians who will quickly record the tracks quicker, better, and of a higher professional musical quality. That could save at least $200,000–$300,000 in production expenses and the final album usually sounds better.

Manager Commission Mistakes

> *They pay $100,000 to their manager for 20 percent commission.*

It's a mistake to pay the manager on a recording deal even *if* he or she helped get the deal. The label is not providing the money for her to party, but to help her launch her career with touring, promotion, publicity, and so forth. This attorney must have taken commission on the $500,000 remaining after the band blew $200,000–$300,000 on the master recordings. If the guy indeed gave her bad advice, she paid him for it!

Defining the Advance Mistakes

This story is about a bidding-war band that gets a huge deal with a 20 percent royalty rate and a million-dollar advance.

From her first sentence it appears that she was misinformed about the term "advance." Advances are monies that are paid directly to the act for their living expenses, during the recordings and to (sometimes) help them start generating income on the road. Half of that million went (sadly) to the master tape recording, which is to be paid back out of the album sales by the label to the account with her name on it at the label. She never sees it, as most label executives never give that much money to an artist, as they tend to blow it. The label is putting up a million dollars and giving her a great royalty rate of 20 points (percent of the suggest retail list price) once the bills have been recouped. If the label fails to recoup all its money (that it spent on her) how much does she owe them? Zip! On the other hand, if they were crazy enough to give her the million, then Courtney Love should have owned the master tapes as she paid for them out of her money (property rights). Then the label would have to pay her on every album sold and the deal would be more of a master lease, independent record (her own label) distribution deal.

Retainer and Overpayment Mistakes

They pay $25,000 each to their lawyer and business manager.

They started with a million and then paid half a million dollars for the master recording. The manager "claimed" 20% on the remaining half a million, which reduced the amount remaining to $400,000. They've spent $600,000 and have a finished album nobody's ever heard (no promotion or publicity yet budgeted) and a happy manager who made a phone call for $100,000. I am only kidding about the one phone call, but still he or she didn't do much for the commission. Then it appears that each band member (four) paid for their own attorney and business manager $25,000 for a total of $100,000. But Ms. Love claims they have $350,000 left instead of what should have been a remaining $300,000 ($400,000 minus $25,000 times 4 members = -$100,000). It seems they lost $50,000 somewhere or maybe they had one attorney and one business manager and they paid each $25,000. Whatever! It is still a loss somewhere close to $50,000.

This points to two other languages tied to commonsense problems that often happen when there's lots of money in a deal. Attorneys may demand *a retainer*, which means they are being paid to represent the act if they don't have to do any work. As an example, let's say you pay an attorney $25,000 on a retainer and then they work about ten hours over the next year, looking at contacts. Were they really worth $2,500 per hour? My suggestion is always hire attorneys on an hourly basis, which will end up costing thousands less. I like getting paid for doing nothing and if anyone knows where I can get that job, please let me know. However, famous and industry-connected attorneys often demand the front money before they agree to represent the acts. Are they really worth it?

Stardom and Taxes

> *That leaves $350,000 for the four band members to split. After $170,000 in taxes, there's $180,000 left.*

It appears she either included herself as part of the band or was considered part of the band. She may even be really talented, yet the guys in the band should have been just hired musicians working for her corporation. Why didn't her attorney set her up in a legal form of business that would have protected her and her name and estate, and given her control of the money and assets generated? A good example of a better way to do this is Hayley Williams and her group Paramore. Remember that Paramore is Hayley Williams and the band members work for her. The fans don't know it and that's fine, and besides they don't care.

Most of us forget about Uncle Sam until April 15, but when you're an artist or entrepreneur, then you'll often have to pay on a quarterly basis. Paying the taxes up front is the only way to do it and remain somewhat sane. But now we're down to what's left, which is only $180,000 divided into $45,000 each. Wow, how depressing is that!

Failure to Plan Ahead

So, this band releases two singles and makes two videos.

The record label's executives usually make the decision about which of the album cuts will be released, not the band. What does the band know about the market? How is the band going to get the singles on the radio and who going to inform the trade magazines? It's either the band manager working with the label, together making up the marketing play and budget, or it's the band on its own. It would be very rare for a band to be given $1 million and then be allowed to make business decisions about how to spend the money and how to profit. Most entertainers don't know or have the ability to do either.

Music Video Budget Mistakes

The two videos cost a million dollars to make and 50 percent of the video production costs are recouped out of the band's royalties.

Most videos currently are budgeted $20,000 to $100,000, with the label paying half and the artist paying the other half. Courtney and her band do not have $500,000; they've got only $45,000 to live on. So where's the money going to come from? The answer is the label, and it will just increase the artist account debt by $500,000 and pay the other half a million out of its own pockets.

Promotion and Publicity Expenses

The band gets $200,000 in tour support, which is 100 percent recoupable. The record company spends $300,000 on independent radio promotion. All of those independent promotion costs are charged to the band.

If a label ran the independent radio promotion through tour support, then their 50% of payment (usually stated in the original deal) is dropped and the entire amount for promotion and publicity becomes 100% recoupable out of the band's royalties. About half a million for promotion and publicity is typical for labels to spend on a two- to four-week campaign to determine if the act is going to hit or fail. The label executives (and Love's manager) will be looking at detailed album sales and single radio station airplay from *Billboard* charts, Nielsen SoundScan reports, and other sources to determine if anything is happening.

The Biggest Mistake

All of those independent promotion costs are charged to the band. Since the original million-dollar advance is also recoupable, the band owes $2 million to the record company.

This is the biggest mistake almost all recording artists make. Ms. Love claims the band owes $2 million to the record label. Is this correct? No, she and her band owe nothing! It's the label's obligation to sell enough products (recordings) and now in a 360 deal (recordings, merchandise, corporate sponsorships, and concert tickets) to *repay their own debt* they created in the artist name. What's she got to complain about? The label isn't demanding the money back from her bank accounts. They have to pay (her loan as she calls it) back themselves. I do not know of any banks that would do that for you or me, but if you find one, let me know.

Zip

> *If all of the million records are sold at full price with no discounts or record clubs, the band earns $2 million in royalties, since their 20 percent royalty works out to $2 a record. Two million dollars in royalties minus $2 million in recoupable expenses equals . . . zero!*

The record labels structure their deals with recording artists to favor them instead of the act in this highly financially risky industry. It takes millions to create the products, promote, provide publicity and distribution, and make a potential profit. Thus, most of the deals are skewed toward the label on a 1 to 3 or 1 to 4 ratio, meaning the labels will break even and start making a profit before the artists. Why not—it's the label's money up front. Thus, a label may have to sell 150,000–200,000 albums before it breaks even and 500,000+ units before the act starts receiving their royalties. This was the way it was back in the pre–360 deal days, but as we now know, it is even more difficult for labels to make a profit, as album sales are in the toilet.

It's a funny thing, but if you let people take money from you, they often will and it's her job to make sure it didn't happen. She probably trusted her attorney, who may or may not be working in her best interest. Let's look at the rest of the story.

The Label's Perspective

Labels usually want a long-term agreement with artists, who provide the label and themselves with profits. However, if the act signs a deal heavily skewed toward the label, its (the label's) risk of doing business is reduced and that's what Ms. Love may have been encouraged to do without really understanding what was in the deal. It happens all the time, as some new acts do not even read the deals before signing them. Major label deals are more complex than buying a new home and the paperwork can easily be more than 100 pages.

Ms. Love claims they made zero and the label made $11 million gross and $6.6 million in profits. How could this happen? It's in the details of the deal. She claims they spent half a million on pressing CDs, plus the other expenses she already mentioned in her statement. In addition, she claims the label spent $2.2 million on marketing and thus the total spent by the label was $4.4 million while the band made zero.[2] If Ms. Love's management had told her that most artists do not make any serious money from the record deal, but they can make millions touring, selling merchandise, and making endorsements, because the label had made them famous, it might have been a different story. She also claims the label's music publishing company paid $750,000 in music publishing royalties, yet we do not know what type of single-song contract she signed.[3]

Analysis of 360 Deals

Artists about to sign a deal should understand what Ms. Love and so many of the other famous recording artists have had to find out the hard way. It's a foundation that leads us to understanding the purpose of the 360 deal concept. If labels can't make enough money from their album sales to profit, then they've got to find another way. Follow the money and you'll discover that all the merchandise, corporate sponsorships, and touring-generated revenues were 100% the artist's under the old traditional recoupment deals. If an artist were about to hit the big time, they'd use the *fame* they'd received from the label's investment in album sales and profit (from the touring, merch, and other revenues) even if they never made any money from their record deals. I have had more than one conversation with famous major recording stars who tell me they never made a dime from their recording deals.

However, we do not stand still in a constantly moving universe. The breakthroughs in digital technology have provided many opportunities for consumers to listen to their favorite tracks and even watch movies anytime, anyplace, anywhere, and usually for free if they want to tolerate the advertisements or steal the entertainment products. How are these amazing entertainment companies going to make a profit on their investments (which we already know is risky) if they are not paid for the sale or use of their recordings?

The Solution

The answer is simple as labels, movie companies, creative artists, musicians, writers, and others must be paid (reasonable profits and wages) or the industry will simply dry up. It's the development of a new workable business model (that generates revenues while still producing great entertainment products acquired for free) that is really difficult. But everyone is working on it and hopefully things will start to shake out. Otherwise, what we'll have is a world full of old products and new ones that appear to not be at a very good level of quality from all the computer whiz wannabes.

Here again, we see the 360-deal business model allows us to still have major recording artists and listen to their music. However, the playing field has been shifted from album sales to touring and other sources. Of course, the superstars who were making millions and keeping it all are upset when they have to start "sharing" a percentage with their labels. That's why a few of them have already started their own labels (and hired the former executives to run them) and paid out of their own pockets the millions to create albums, promotion, publicity, and other expenses. They have become their own labels just as they used to be their own businesses, putting on shows and selling endorsements and merchandise. However, many have found out that it is financially riskier and more work than they'd ever believed. They sign distribution deals with one of the three major labels and then have to hire experts to make it all happen.

Honesty is the best policy, and full disclosure about the business of entertainment and the risk, costs, and potential rewards should be clearly explained before acts sign any deals. Management must make sure this will happen if there is to be trust between the two, which of course is what the artist/management deal is founded on.

Stage Names

Artists often pick a stage name (other than their actual birth name) to perform or create an image and to support their brand. There are valid reasons to adopt a stage name, including privacy issues and an opportunity to live under your real name and perform under your stage name.[4]

Table 11.4 Stage Names. Most famous artists don't perform using their given names. Enjoy Table 11.4. There are lots of issues, including taxes and privacy issues. Take a look at the major entertainment industry personality and check out their real names. Then try to figure out a workable "stage name" for yourself if you were to become a "public" personality.

Real Name	*Stage Name*	Real Name	*Stage Name*
Stefani Joanne Angelina Germanotta	Lady Gaga	Born Steveland Hardaway Judkins; later renamed Steveland Hardaway Morris	Stevie Wonder
Ray Charles Robinson	Ray Charles	Anthony Dominick Benedetto	Tony Bennett
Tracy Morrow	Ice-T	Katheryn Elizabeth Hudson	Katy Perry
Calvin Cordozar Broadus Jr.	Snoop Dogg	Beyoncé Giselle Knowles	Beyoncé
Selena Quintanilla-Perez	Selena	Christina Claire Ciminella	Wynonna Judd
Marshall Bruce Mathers III	Eminem	Madonna Louise Veronica Ciccone	Madonna
Ernest Evans	Chubby Checker	Richard Starkey	Ringo Starr
Eleanora Fagan Gough	Billie Holliday	Lester Polfus	Les Paul
Curtis Jackson	50 Cent	Wladziu Lee Valentino	Liberace
John Francis Bongiovi, Jr.	Jon Bon Jovi	Carlos Ray	Chuck Norris
Alecia Moore	Pink	Frank Castelluccio	Frankie Valli
John Anthony Gillis	Jack White (White Stripes)	Farrokh Bulsara	Freddie Mercury
Reginald Kenneth Dwight	Elton John	Steven Victor Tallarico	Steven Tyler
Gordon Matthew Thomas Sumner	Sting	Dino Paul Crocetti	Dean Martin
Victoria Caroline Adams [Beckham]	Posh Spice (Spice Girls)	Marvin Lee Aday	Meat Loaf

Source: "Music Stars Real Names." DigitalDreamDoor.com. January 28, 2015. Accessed November 21, 2015. http://www.digitaldreamdoor.com/pages/realname.html.

Social Media and Press Kits

Press kits, discussed in Chapter 9, help establish the authenticity of new and established artists. Publicists assemble press kits to alert the press and mass media of new recordings, acts, and newsworthy events. A press kit usually includes a black-and-white photograph (publicity photo), a one-page bio of demographic information, artistic and creative accomplishments, and copies of previous press releases to the trade and consumer press. Managers work with the label's promotion and publicity executives to provide a common image and brand.

Personal managers also use broadcast radio and television to market their acts. Variety, late-night, and morning talk shows promote artists and their sellable products. Publicists are again hired to help "place" stories about the artists in the trade and consumer press. They also help generate interest in the artists' backstories (about their brand, life, image), which is publicity that can be used by the labels also on a television variety/talk show. Appearances on popular TV shows reinforce the status of the act and inform the public of new recordings being released, tour dates, and other opportunities.

Star Factor Publicity Budget

The amount of money management spends (it's the act's money) varies depending on the status or success of the act and cooperation with the label. The artist manager commonly meets with label representatives to coordinate publicity. Budgets to be spent on publicity often depend on the star factor (name and image equal the strength of the artist based on past units sold, total revenue generated, and sales projections). At first the money is 50% label and 50% artist, yet many new acts can't contribute their 50%, so the label pays 100% and then claims it as a tour advance in order for it to be noted in the artist account at the label as 100% recoupable.

Talent Agents

Talent agents own or work for agencies licensed by the unions to represent creative artists. They may provide some career advice and guidance just as personal managers do in the music business, but their main job is to seek and acquire employment for their acts. However, in today's music world the talent agent's job descriptions are blending with some of the personal manager's. Usually, an artist's manager hires AFM *licensed booking agents* to set up tours for acts. However, talent agents working at the major agencies (CAA, ICM, UTA, and WME) also establish tours and provide some management advice to representative artists. SAG-AFTRA and the AFM are unions, while CAA, ICM, UTA, and WME are talent agencies or businesses licensed by the unions to represent the union members (creative artists) by seeking employment opportunities (casting calls for film, tour for musical acts, speaker engagement for personalities). Talent agents may seek many entertainment-based employment opportunities (e.g., endorsements, acting, commercials) for the creative acts they represent, while AFM booking agents are usually limited to musical tours and other types of musical performances. Thus, SAG-AFTRA and Equity talent agents have the same *title* as talent agents at CAA, ICM, UTA, and WME, but their *duties and responsibilities* are somewhat limited by the unions their clients represent. Some SAG-AFTRA and Equity agents also provide advice and guidance similar to music business personal managers on a limited basis.

Talent Agencies

Talent agencies control entertainment promotion, publicity, and employment opportunities for many of the most famous acts, authors, public personalities, and artists. They frequently represent the superstars of music and films for productions, tours, appearances, advertisements, and any other types of events or work that benefits the iconic images of the artists. They provide worldwide representation and, in doing so, provide talent to world markets. The four largest talent agencies, known in the industry by their acronyms, are identified and discussed ahead by their distinctive market brands. These are CAA, ICM, UTA, and WME.

Creative Artists Agency (CAA)

CAA was formed in 1975 and is noted for its powerful list of actors and entertainment personalities, including famous directors, sports, business, and celebrity motivational speakers. CAA has about 1,200 employees, with annual sales grosses at about $80 million. According to Answers.com and Hoover's Company Profile, CAA

Represents clients working in film, music, television, theater, sports, and literature. The firm's client list reads like a who's who of A-list stars, including such luminaries as Steven Spielberg, Tom Cruise, Brad Pitt, George Clooney, LeBron James, Simon Cowell, David Beckham, Will Ferrell, and Sandra Bullock. It has also worked with heavy hitting commercial clients such as Coca-Cola and toymaker Mattel. Supplemental services include strategic counsel, financing, and consulting. Its Intelligence Group/Youth Intelligence unit tracks and conducts behavior research for consumers from ages 8 to 39.[5]

International Creative Management (ICM)

ICM was formed in 1975 by the merger of Creative Management Associates and the International Famous Agency. According to Hoovers, ICM

Represents film and television actors and directors, as well as artists in theater, music, publishing, and new media. A major "tenpercentery" (along with CAA and William Morris Endeavor), ICM represents such A-list clients as Susan Sarandon, Al Pacino, Robert Duvall, Beyoncé Knowles, Chris Rock, and Jay Leno, as well as emerging performers. It has offices in Los Angeles, New York, and London . . . Private equity firm Traverse Rizvi Management and Merrill Lynch hold controlling stakes in the agency.[6]

United Talent Agency (UTA)

UTA represents major stars and the behind-the-scenes creative artists in the entertainment industry who *make* the entertainment products we love to watch. According to Hoovers (2015),

> *One agency believes that talent should not be divided. United Talent Agency (UTA) has represented big-name movie celebrities such as Harrison Ford, Miley Cyrus, Rachel McAdams, and Johnny Depp, as well as a client list of directors, producers, and screenwriters, including the Coen brothers (No Country For Old Men) and Alan Ball (American Beauty, Six Feet Under). Beyond the celluloid, the company represents literary authors, journalists, musicians, and creators of mobile, online, and gaming content. It also provides branding, licensing, and marketing services. Founded in 1991, UTA is owned by 20 partners through a structure designed to spread ownership among its top agents.[7]*

UTA also represents many of the behind-the-scenes workers in the film and production business. According to unitedtalent.com (2015),

> *United Talent Agency is a premier global talent and literary agency representing many of the world's most acclaimed figures in every new and emerging area of entertainment and media, including motion pictures, television, music, digital, broadcast news, theatre, video games, books, fine art and live entertainment. The agency is also globally recognized in the areas of film finance, film packaging, branding, licensing, endorsements and representation of production talent. UTA additionally provides corporate consulting, venture funding and strategic advisory services to companies ranging from start-ups to Fortune 500 companies. UTA operates the brand strategy agency UTA Brand Studio and owns market leading broadcast news agency Bienstock. UTA founded and co-owns leading integrated marketing firm United Entertainment Group, a DJE Company, which focuses on branded entertainment for major consumer brands.[8]*

William Morris Endeavor Entertainment (WME)

WME takes a different niche, as it represents talent, public personalities (speakers), and authors. It was formed in 1898 to represent vaudeville acts and currently has offices in New York, Beverly Hills, Nashville, London, Miami Beach, and Shanghai. According to Bloombergbusinessweek.com (2015), WME

> *Operates as a talent and literary agency in the United States and the United Kingdom. It provides music, comedy, lectures, and theatrical and non-traditional tours for college bookings, fairs and special events, international bookings, and corporate and private events. The company offers celebrity voices, radio imaging, TV affiliates, NFL voices, commercial voices, Spanish voices, and animation; and voiceover for promos, trailers, and narration. It allows buyers to schedule talent for music and theatre events, comedy performances, and speaking engagements. William Morris Endeavor Entertainment, LLC was formerly known as William Morris Agency, Inc.*[9]

All of the talent agencies have slightly different business models, as CAA represents many superstar movie and recording musicians. The others do the same, except UTA also represents the movie production teams and financing options for investors. Now that you've grasped the big picture of representation, let's take a look at how representation works with promotion companies and entrepreneurial promoters to create events.

At this point, you have whole picture of how you can be an active and important part of the entertainment and music industry. If you have stuck with me this far, you know that having talent is one thing, and being an intelligent and creative entrepreneur is another, though not unrelated, essential aspect of your future career. You have digested the basics of the all-important copyright law, and my discussion of the implications of digital transformations and innovations, such as streaming, deliberately taking a negative cast, has been to alert you to, without preaching to you, of the dangers of not paying attention to the changes all around you. You have also become acquainted with the many professionals and their functions in discovering, promoting, and sustaining artists and entertainment genres. Also, importantly, you have seen how to calculate costs to manage risks from both the artists' and the label's, studio's, and network's perspectives and how to take your show on the road. Let's make some money!

Notes

1. Love, Courtney. "Courtney Love Does the Math." Salon.com RSS. June 14, 2000. Accessed November 22, 2015. http://www.salon.com/2000/06/14/love_7/.
2. Ibid.
3. Ibid.
4. "Music Stars' Real Names." DigitalDreamDoor.com. January 28, 2015. Accessed November 21, 2015. http://www.digitaldreamdoor.com/pages/realname.html.
5. "CAA." Creative Artists Agency, LLC. 2015. Accessed November 30, 2015. http://www.hoovers.com/company-information/cs/company-profile.CREATIVE_ARTISTS_AGENCY_LLC.0b3efe-9675be1acd.html.
6. "International Creative Management, Inc. Company Profile from Hoover's." International Creative Management, Inc. 2015. Accessed November 30, 2015. http://www.hoovers.com/company-information/cs/company-profile.INTERNATIONAL_CREATIVE_MANAGEMENT_INC.56b56bd6a63f52aa.html.
7. "United Talent Agency Company Profile From Hoover's." United Talent Agency, LLC. 2015. Accessed November 30, 2015. http://www.hoovers.com/company-information/cs/company-profile.United_Talent_Agency_LLC.023c3442ec657319.html.
8. "ABOUT US United Talent Agency." UTA. 2015. Accessed November 30, 2015. http://www.unitedtalent.com.
9. "William Morris Endeavor Entertainment, LLC: Private Company Information." Bloomberg.com. November 30, 2015. Accessed November 30, 2015. http://www.bloomberg.com/research/stocks/private/snapshot.asp?privcapId=872395.

THE CONCERT AND EVENT BUSINESS   12

Consumers may be changing their habits to listening to music streaming services instead of buying or even stealing the recordings they love. But nothing can replace the energy, sparks, and higher level of emotional connection between the acts and fans at a live performance. It's still the emotion business based on a creative system of writers, musicians, performing artists, and others. The business systems are based on the promoters' money, productions, publicity, distribution, sales, and marketing of events instead of albums for sale. The established industry may be changing and evolving, yet the consumers' desire to be personally entertained remains a growing profitable business. For new acts, this is the best time in many years to become a part of the industry and it's a chance to connect creatively with your own generation and to build a fan base. For the aged superstars, it's a great way to make millions in a couple of hours of live performance.

Let's take a quick review of how we got here, from the writer who creates the copyrightable work to making money from the corresponding entertainment products and services.

The first finger is in the air, as someone has created something in the form of a legal copyrightable work others may use to create a wow recording, movie, book, computer game, or whatever. Notice that we do not know yet its true value in financial terms, yet because we created it, we may assume it's priceless, when in reality it may not be worth a cup of cold coffee. The next step is to pitch it to the established

Table 12.1 The Creative/Administrative Business Process. The traditional business model for creativity illustrates how a creative "work of art" is formed, owned and authored by the person, persons or company who make it.

The Creative Circle: Writers, authors, sciptwriters, game creators, photographers, and many others create copyrightable works of art	Creators have ownership and authorship, plus six exclusive rights	Note: If anyone wants to use the creative works then they may buy the copyright or license its use in a book, film, recording, computer game, or stage play
The Representative Circle: Businesses that act as "agents" to pitch the creative works to artists and production companies that want to use the works in their entertainment products	Music publishers Literary agents Talent agents Booking agents	Deals include selling of the copyright for a percentage of the revenues generated for licensing opportunities
Production Circle: Companies that license creative works to be turned into entertainment products	Record labels Film production studios Book publishers Theater/stage Computer game manufacturers	Mechanical License: Direct or The Harry Fox Agency Sync License: Direct or The Harry Fox Agency on small projects Print License: Direct for song books, etc. Promotion: Providing a sample Master License (use of a recording) in a visual presentation

entertainment industry to see if any of the executives want to use it in the form of a hit recording, movie, book, and so forth.

Raise the second finger on your right hand up in the air. The production company found a wow story and song, budgeted the product, hired the production teams, and made the entertainment products, including a major motion picture, stage play, and album from the original copyrighted creative work. However, there's a little problem: nobody knows that the wow movie, play, or album even exists. Sticking it up on the Internet isn't a solution as even if the buzz about it starts, consumers will download or watch it free, and we just lost millions of dollars. By the way, if this is a Disney project, there'd also be a kids' computer game for sale, cartoon versions, and the list goes on. It's time for the branding, promotion, publicity, and distribution side of the business to kick in.

Most executives are financially conservative, as they know that committing $1 million to $100 million can determine the future of the company and their careers.

Table 12.2 The Traditional Production Business Process. Once the licenses for the original copyrighted "creative works" such as a book manuscript, song, stage play, etc., are obtained it's time to make *The Sound of Music* as a play and major motion picture, and the sound track as a major album release. Let's the cash machine collections begin!

Production Circle: After entertainment companies have obtained the licensed rights to use the creative work in their form of business, the companies start the production process.	Determine cost of production.	Analyze the total cost of production including talent, union workers, facilities, and wages and benefits.
	Determine potential market and gross income.	Analyze the potential size of the market to determine how many "units" may be sold, or services (live shows) might be performed. Determine the potential value of the products and services. Find the break-even point by subtracting the total cost per unit or show from the potential gross.
Executive Decision: Decide if the risk of spending the money is worth the opportunity of financial gain. If yes, then "green-light" the project.	Hire actors, directors, producers to supervise production. Sign recording artists for musical recordings. Hire skilled workers to create the entertainment products. Note: Many professional workers (at this level) are represented by unions, such as the AFM, SAG-AFTRA, and Equity.	Create the product or show: • Distribution of hard platforms to brick and mortars. • Provide digital copies to servers such as iTunes. • Provide streaming and media with products.

The Magic in the Message

If you're starting to see how creative individuals and companies use innovation and technology to monetize entertainment products, move the third finger on your right hand up in the air. It is joining the other two fingers you raised in the first part of this chapter. Our products have been licensed and are ready for distribution, for sale, and for use in other media. How are the production companies (who spent millions on

Table 12.3 The Traditional Business, Promotion, Publishing and Distribution Process. Executives have to make business decisions cold about creative productions based on financial projections often determined by accountants and lawyers. The entire world is the market and the question is what's it going to cost and how much money can we make. Most executives are financially conservative, as they know that committing $1 million to $100 million can determine the future of the company and their careers.

The second part of the **Production Circle** is giving the public an opportunity to discover the new released entertainment products.	Promotion	Radio play and advertisements. Trade magazines and popular press advertisements. Social media (paid ads placement). Music videos, such as YouTube.
Executive Decision: Deciding on the image of the product or act is important. Keep it simple in order for consumers to understand the product as an emotion they experience. Tie the image to branding to connect with manufactured products.	Publicity	The "backstory" of the act, book, play, computer game, stage play tied to the image— such as: TV and radio interviews. Articles placed in the trades, popular press, traditional and social media.
	Branding	The process of making the lifestyle of the product or act enticing to the consumers by connecting it or them to significant events, lifestyles, and other products and personalities.
	Distribution	Placing the product for sale and use by media such as streaming is increasingly important, as the licenses fees provide new revenue streams to the labels and film production companies.

making, distributing, and marketing the products) going to generate money? And, following the chain of events, how are the original copyright holders (let's say a songwriter) who created it in the first place and the production company that licensed and paid for the creation of the entertainment product (let's say the labels) going to make any money? In the old days it was simple: sell albums if you were a label, and sell screenings at local theaters, DVDs, or licensed uses by the networks if you were a movie studio. Now, as you know, things are much more difficult due to the innovations of technology, the Internet, and violation of copyrights. Now, how are the labels (and soon the film companies) going to make money? The answer is again simple: any way they can! Entertainment products are just forms of communication that consumers find magical. Of course, if they can attain them for free, most will download everything possible. However, try this as an example for one week. Turn off all devices, don't read any books, newspaper, or printed products, and don't watch any TV or listen to any radio stations. No music, social media, live shows, sporting events, Facebook, Spotify, zip, zero. Now, think about how much we value entertainment products!

The Practical Solution—360 Revenue Streams

If the situation remains the same in the future as it has in the past, the investors of highly professional products will find another way to make a profit or it all will dry up and blow away. Maybe Congress will put some teeth back into copyright laws, but until they do, more revenue will have to be generated from streaming, concert tours, and live events. To be successful, the film companies, labels, computer game companies, and others are morphing their business models toward increasing their revenues through branding and a higher level of emotional lifestyle connection with the fans. Thus, on the production side of the music industry, computers are decreasing the need for audio engineers and, in some cases, musicians. Recording studio sessions are being reduced, and after the basic tracks many albums (and films) are finished using computers to generate vocalists, sounds, musical parts, mixes, and tune quality. If the act hits the excitement nerves in the consumer base (through wow products, promotion, and publicity), then larger revenues are generated out of merchandise, corporate ownerships, branding, and ticket sales. That piece of the pie was solely the artist, but in the 360 deals, the labels receive a percentage of the profits (based off gross), in order for them to recoup and profit on their investments. The film industry is headed in the same direction, book publishing companies are struggling, and who knows what may happen to the rest of the gang. We've got three fingers in the air, but now it's more difficult to raise the other two, as it's harder to make money.

Associated Industries

Concerts and special events provide record labels, promoters, artist representation, media, and venues with another way to profit from the fans' excitement about specific acts. It's the emotional frosting on the cake for the consumers and an opportunity for the acts to grow their reputation, fan base, and bank accounts. It's survival for the major labels, bands, and corresponding road crews, promoters, IATSE production crews, and others. However, the concert and event business is really a separate industry with its own set of entrepreneurial executives and rules. Concert promoters risk their own money on the public's interest to attend a staged event (a concert) based on an opportunity to make a lot of money quickly. Booking agents represent the various acts. Their job is to call on the promoters to book the acts into various tours and venues. Once again, it's an industry with a golden thread connecting the many different types of professionally skilled workers, executives, creative artists, and government agencies with the fans. And if the fans dig your show, songs, personality, image, brand, and sound, there is an amazing amount of money to be made. Let's party and in the process have a great time making some serious money.

Statistics

The worldwide event business grosses about $25 billion a year, with a projected annual growth of 4.5%. It employs over 219,000 people and consists of around 65,000 clubs, venues, bands, and other supportive businesses.[1] The U.S. market in 2014 grossed $6.2 billion. According to *Pollstar*, the leading trade magazine for the concert business, that's an increase over the past few years, primarily due to the increase in festivals. However, in January 2015 *Pollstar* projected that the increase will not make up the differences in financial losses paid to the labels and acts due to decreasing CD sales and digital downloads:

> *The conundrum is that while the public's consumption of music has never been greater, the ability of artists to monetize that demand has been in a steep decline. Digital sales and streaming revenues have not come close to replacing the reduced income from recording and publishing. Eventually there will be a new economic model that works for everyone, but until then the artist community will have to rely on the growing revenues from live performances to make a living. And that's a good thing for the concert business.[2]*

Capital-Intensive

The concert and event industry is a serious money business based on just a few hours that usually determine the profits and losses. The fans buy the tickets and show up or they don't. Still, everyone has to be paid, so the promoters perceive touring as an opportunity and not necessarily a risky business. It's still similar to gambling in Las Vegas, just with tougher odds if you don't know what you're doing. Nightclubs are often a cash business for newer acts, while (at the same time) millions of dollars, euros, and yen are transferred digitally over secure transmission lines from the promoters' accounts (being held by the booking agents) into the act's account as soon the superstars hit the stage. Therefore, the amount of capital required on the front end to actually create an event depends on the size of the venue, the number of ticket buyers, the fame or drawing power of the act, their image and brand, the time of year, location, size of the market and fan base, and other possible things (e.g., weather) that can go right or wrong. As you may suspect, touring exemplifies all the elements of monetizing entertainment. Peel back the layers of the show and you will recognize everything we have talked about so far. The level of your success on the concert stage is a summation of your successes and failures in your career as a wow.

Research

Promotion companies and individual promoters do their research before they sign a deal to employ an artist for an event. As an example, they often use the same data and information we've talked about in this book, such as the *Billboard* charts, SoundScan retail and digital download reports, *Pollstar* box office gross revenue reports, *Variety*, Buzz Angle, and Nielsen. These companies (for a fee) can provide the city, zip code location of the number of streams, downloads, radio plays, CD sales, graphic, and other psychographic information. This type of research provides the promoters with the names of the cities in which their artists are trending and the ones they should avoid. Based on sales data and social media hits, a promoter can estimate the best possible cities and venues to create events that are correlated to the most popular acts available to book for the potential consumers in those cities. *Billboard* and *Pollstar* also provide information on the number of seats available, compared to the number of seats sold plus the concert gross revenues. *Pollstar* has a free service that allows the promoter to analyze all the venues in a city on a daily basis to determine the market competition. It also shows open dates in each city and venue when promoters want to set up a tour.

Festivals

Festivals are huge in Europe, and local bands and older known bands hit nightclubs and performance halls as part of the younger culture's nightly activities. Festivals in the

United States are popular, offering every type of musical performance, from the Monterey Jazz Festival to the Coachella Valley Music and Arts Festival. In 2014, all festivals like these had earnings of a total $183 million.[3] They use their expertise, money, and social and business connections to stage concerts (events) that help generate the positive cash flow opportunities needed by many performing artists and labels (through increased unit sales due to the live appearance of the act). Success is changing the industry, as stated by *Pollstar*:

> [There are] more than 1,500 confirmed events in 70 countries. That would not have been possible just a few years ago but now everyone seems to be trying to get out in front of the crowd . . . The global festival market is approaching saturation as everyone fights to book the same limited pool of artists over an expanding array of events. While the festival business has been generally strong, it wouldn't be surprising to see a bloody market correction that weeds out the weaker festivals with marginal sites. With so many festivals eager to write big checks for headliners, the very manner in which many tours are booked has radically changed. Instead of festival dates being squeezed into tour routing, many artist tours are routed first between festival dates and then filled in with solo headline dates.[4]

The AFM and SAG-AFTRA Contracts

The purpose of creating an event is simply to make money, and artists deserve to be paid for their work. How much the act makes depends on their cost of putting on the show compared to how much they are paid by the promoter to perform. In the previous chapter, we accomplished a simple analysis of what it costs for a new act to actually start performing. Local acts may need only a couple of hundred dollars to perform and be happy. Major superstars cost millions of dollars to launch a multicity concert tour. Where does the money to pay all the bills come from? It's the promoters who risk the money, paying all bills and expenses connected to the event. The American Federation of Musicians (AFM) and SAG-AFTRA use booking agents they license to represent their artists and contacts provided by the union headquarters. Thus, if you're a union member, you'll be paid scale and be supported if a promoter tries to skip out on the payments.

How much money do the promoters pay the acts for their performances? It depends on the fame of the act and the act's ability to draw a large enough audience, buying tickets priced high enough to cover all the bills and provide a profit to the promoter. Another benefit of creating events is the revenues generated by corporate sponsorships, endorsements, and merchandise sales that sometimes boost bank accounts more than ticket sales. The percentages of the merchandise and sponsorship revenue going to the band, venue, or house are negotiated in the contracts between the booking agent, promoter, and general manager of the venue. And because of the 360 deals, a percentage of the gross is also going from the artist's touring income to the label. While the 360 deal is the model labels have identified as the way to survive, the deal ends up taking more money from the artist's bottom line. This is an essential cost of doing business as, in the past, all the money from merchandise and touring went to the artist, paid by the promoter. Now, a part of that money has to be returned to the label to cover the loss from album sales.

Tour Date Approvals

Everything is negotiated by the promoter and the venue managers for the place to hold the event with the approval of the headline artist. The artist's manager must approve bookings (scheduling) of tour dates presented by the booking agent. A deposit or guarantee of 50%–100% of the agreed-upon price for the act is generally required from the promoter to complete the deal. The money is placed into an interest-bearing account. During the concert, the road manager, tour accountant, and promoter tabulate the concert receipts in the back office. The down payment is subtracted from the money owed to the band. Full payment for the band's/artist's performance is required at closing. Successful promoters are known by industry insiders and are often hired for entire national tours, regions, or clusters of cities to reduce the band's risk of working with amateur promoters.

Once promoters determine which acts to book (employ), they match the artists (who are on tour and available) with the venues that have open dates. At the same time, booking agents who represent the artists call regional promoters, corporations, and venture capitalists (who may act as promoters) to generate additional concert dates for their artists. It's a huge gamble as it's usually a couple of hours' event that can be highly profitable or a money drain, depending on crazy things, such as an ice storm, an artist saying something negative that may turn off some fans, sound, lighting, and venue problems, world events, and so forth. It's easy to see why the concert promotion business is considered a very risky business, given the amount of money that will change hands within a few minutes of the first note being played on stage.

Venues

While many promoters rented venues, the business has shifted in the last few years towards larger promotion companies owning their own properties. Many of the larger venues in major cities are the arenas, concert halls, and football stadiums owned and operated by the local government. Professional industry-related experts usually manage the venues, controlling budgets, events, schedules, and personnel. Most of them never make any money for the city, yet the football games, concerts, and sporting events usually draw large audiences of consumers who spend lots of money for hotels, food, merchandise, and other opportunities. The city receives the rental fees for the venues and then taxes from the money generated by the event and consumer spending.

Profit Margins

Promoters subtract their total projected expenses from their projected gross ticket sales revenues to determine their break-even points and profit margins. Knowing their profit margins helps the promoter become familiar with how many tickets they can sell at various prices. That total becomes the projected gross ticket revenues (income). Then using a bid sheet, promoters determine their projected expenses (total cost of creating the event), subtract the projected expenses (debt) from the projected gross income (revenue), and determine profitability. Profit margins are based on how many seats must be sold to break even or make a profit. Many concerts require 50%–80% of all tickets to be sold to break even financially. Promoters hope the profits will be much greater than 20%, depending on the act booked, promotion, and publicity. However, other types of margins are common depending on what is negotiated on the front end and, of course, the type of deal and the number of tickets sold.

Sponsoring Radio Stations

Labels want their artists touring to promote their latest recordings. Promoters and labels often work together to select a local radio station to sponsor the concert. The promoter receives a break on advertisements and the station is allowed to claim it is sponsoring the concert.

Risk Management

What if it rains or snows? What if there is another concert by another promoter the same week as yours? What if a band member gets sick? What if the artist makes a negative public statement or is caught by the press doing something illegal? The concert may still have to be played and the attendance may suffer, which means the promoter gambled and lost money.

Types of Deals

Concert promotion is a very risky business if you do not understand how the system works, as there are often five types of deals that can be structured. The deals inherently have risks for the artist and the promoter, just as the concert itself does. However, according to Mark Volman, of the super group The Turtles from the 1970s (who is still actively touring), four of the following five types of deals are the basic foundation for all the various scenarios that may be negotiated between the act's booking agent (approved by the personal manager) and the promoter of the event.

Straight Guarantee

The act receives a guaranteed amount of money regardless of the success or failure of the promoter to generate an audience. The straight guarantee deal provides the promoter with an opportunity to lock in the act at a fixed cost and therefore increase his or her potential profit based on the success of the event.

Guarantee Plus a Percentage of the Net (Gate)

The act receives the guarantee plus a certain percentage of the net after the break-even point (all expenses have been paid). The percentage of the net becomes a negotiation point for the band to receive additional income on a very successful event and increases the negotiation leverage for the promoter to land or sign a more successful act for the event.

Guarantee Versus a Percentage of the Net

The guarantee versus a percentage of the net deal (after all bills have been paid) provides a base of security versus financial risk as the act receives either the fixed guarantee or a percentage of the net, whichever is greater. This type of deal is a mixture of the straight guarantee deal and the guarantee plus a percentage of the net deal and provides a little more financial security to the act. If the event is poorly attended or loses money for the promoter, the act is still paid the guarantee. If the event is very successful then the act is paid an amount that is better than just the guarantee. It also puts additional pressure on the promoter to control the bid sheet expenses and negotiate a correct percentage of the net (profit).

Guarantee Plus a Bonus

The act receives a guaranteed amount of money plus a negotiated amount or bonus based on the number of tickets sold. The bonus is a variable amount that increases as the number of seats sold hit various levels. This type of deal provides more security to the promoter and act as both receive payments based on the success of the event. The percentage of the bonus is determined between the promoter and booking agent based on the fame of the act (which means it may be easy for the promoter to sell more seats) versus giving the act a higher "bonus," which may reduce the net money paid to the promoter. However, it also reduces the financial risk of putting on the concert or event in the first place due to the act's ability to quickly draw a large audience.

Self-Promotion Deal

When you're getting started as an act, most club owners, booking agents, and promoters will not sign you to perform at a show. Nobody knows who you are. There's no buzz, as was noted in Chapter 11, and the people doing the hiring are not interested in taking those types of financial risks. Self-promotion deals happen at the beginning of careers and sometimes, if you become successful enough, at the end of professional careers. When a new act is unknown, they the act may have to bring a number of fans before the club owners approve their free performance. The owner gives the band a shot in exchange for bringing customers into the club, who will then hopefully buy booze and food. Successful club acts may have to "buy on" to be the opening act for a major artist on a tour—for example, 25 cents per seat for an arena to a dollar a seat for small venues. The purpose is to get the act in front of a new audience and to start building a fan base. If the deal is done fairly, the act should be allowed to sell merchandise to cover the buy-on expenses. Once the act becomes very famous, some will also self-promote. You have to put up all the money for the show, event, or tour, yet if successful, you'll also make all the profits. A few acts have done this by successfully starting virtual corporations to run the business during the tour.

Virtual Corporations

Virtual corporations are sole-purpose companies set up to run the tour, ticketing, and merchandising. Promoters run the businesses out of their hotel rooms, using multiple phone lines, cell phones, fax machines, and email. These corporations often gross millions of dollars during the tour, and when the tour is finished, the companies are closed and "out of business." The Rolling Stones, as an example, have been rumored to rent the entire upper floor of a hotel in Toronto, Canada, running their touring concert business while performing in the United States. All business is conducted on the Internet and when the tour is over, the business is shut down. And the exchange rates between the U.S. dollar and the Canadian dollar provide monies for extra travel and profits.

Merchandising

Merchandising is an integral part of the concert industry. Companies such as EMI, Brockum, FMI, Nice Man, Winterland, and Missing Link create a huge revenue stream by providing and selling merchandise for their artist(s). In the newer 360 deals, the labels control and outsource the acts' merchandise in exchange for profits. The artist doesn't have to do anything except cash the checks. The artists receive 15%–35% of the gross income from merchandise, while the providing company pays for all the merchandise, shipment, sales personnel, booths, tables, and advertisements. Artists approve the design of the merchandise and receive quarterly checks for their percentage of each piece of stock sold. In addition, the concert venue usually receives part of the total merchandise profits for providing space, tables, and advertisement opportunities to the merchandising company. T-shirts, hats, records, CDs, and other types of merchandise endorsed by the artists are considered merchandise. Soft drinks, food, and candy are common concessions sold by the venue, with all profits remaining with the venue. Concession sales locations and commissions are negotiated by the booking agent and approved by the artist's manager before the final contracts are signed.

Artist's Riders

A rider is an additional set of instructions for the promoter regarding specific artist requirements. Riders usually include the size of the stage, lighting, and security and may even include the type of food, drinks, or other necessities the artist wants supplied in the dressing room. Most rehearsals are closed to the public, and in some cases, security will sweep the entire building before the artist's arrival. Security is very important, and the larger the audience, the harder it is to control the fans, who can get a little out of control. Fans do have the opportunity to buy backstage tickets to rehearsals and that gives a small group a close-up view of not only how the act prepares for the show but also how parts of the road and tour support work. Events like these keep interest in acts very much alive as they continue to tour the world, even including Sir Paul McCartney.

The Green Room and Backstage Passes

The green room is called that because it is traditionally painted green. The artists use the green room to host special guests after the shows. People invited will need a backstage pass that will be marked when the access to the backstage is permitted. Having a green room reception is a way for major artists to thank their fans, including members of their fan clubs, and to thank them for their support. The magicians Penn and Teller use the lobby of the theater they own in Las Vegas as their green room. They greet and thank every fan as he or she leaves the venue.

Tech Riders

The production rider is usually included with the artist rider. It covers the specific technological processes and steps that promoters are required to follow for putting on the show.

IATSE crews use the tech or production riders for setting up the stage, lighting, sound, and other show processes. Items include the size and height of the stage, where all the instruments are to be placed, the number of super troopers (spotlights), and even when and where the lights are to be used during the show.

IATSE

The International Alliance of Theatrical Stage Employees, Moving Picture Technicians, Artists, and Allied Crafts is better known by its shorter original name, the International Alliance of Theatrical and Stage Employees, or IATSE, as it's called in the business. It represents about 122,000 employees, such as audio engineers, lighting technicians, set designers and stage builders, and camera operators, and provides the stagehands for live shows.[5] A union house is a venue or studio that has signed an agreement with the performance unions (IATSE, SAG-AFTRA, AFM, etc.) to allow only union members to be employed during the creation of various entertainment productions. Most Broadway venues, sports complexes, and movies made in Hollywood are considered union jobs made or performed in union houses or production houses. Exceptions include the production facilities in right-to-work states that often allow for nonunion productions, yet once the entertainment products are made they can't be sold to a signatory label or movie company unless the differences in wages are paid. According to its website, IATSE's role is to organize and fairly represent creative workers in the entertainment industry: "The goal is to apply the rules of labor legislation in order for the workers to be recognized as a unit and to compel the employer to negotiate a binding contract governing their terms and conditions of employment."[6]

Road Crews

The roadies are members of the touring acts who drive the trucks to each venue, load and unload and often setup, operate, and break down the musical instruments, lighting, and sound equipment. The amount and type of work roadies can accomplish depend on: (a) the IATSE union agreements with each venue, (b) the contractual agreements made with the promoter and artist's manager, and (c) specific items listed in the artist's rider. Roadies or crew members often double as security and bus drivers, and provide other services to the entire act and crew as managed by the road manager (also called a tour manager). They work very long hours at a time and are often hired on their reputation for hard work, respect for the acts, and passion for putting on the best show possible. The trick is that every venue is different, which means they really need to be able to solve strange lighting and acoustical problems before every show.

Live Nation/Ticketmaster

Entrepreneur Gordon M. Gunn III collaborated with two others who had backgrounds in computer programming and in entertainment marketing, respectively, to create Ticketmaster in the mid-1970s.[7] It merged with Live Nation in 2010 and now controls almost all tickets sold at over 3,000 U.S. venues, including those of more than 50 professional sports teams, and is the promoter of 22 of the top 25 global touring acts.[8] The company is divided into four divisions or business models, including concerts, ticketing, sponsorship and advertising, and Artist Nation, a new stage for breaking acts. Its website has truly become a one-stop site for locating tickets, news, and events information. It had more than 1 billion hits in 2014, selling a gross value of $23 billion in tickets.[9] It produced 22,801 events worldwide in 2014, entertaining 58,578,000 fans and grossing $1.3 billion in cash.[10] It grew its business by offering many of the venues front money (in some cases millions of dollars) for the exclusive right to sell tickets to all the events scheduled at the venues.

Scalping

In some states, it is legal to buy a large number of tickets and then resell them at a higher price to consumers before the show, concert, or sporting event. In other states, you'll end up in jail! Ticket scalping costs consumers, as fewer tickets are available at the suggested retail list price (as when tickets sell out quickly), leaving the only tickets available from scalpers at a higher price. The process also makes it appear the event is sold out, which is good publicity for the acts and creates a buzz, and yet if you want to go, you'll have to find a ticket reseller. However, if the event is not in demand, then the promoters benefit as the scalpers may get stuck with the tickets, yet the promoters get paid for empty seats.

AEG Live

AEG Live is just one of the divisions of AEG, a subsidiary of the Anschutz Company, which is really of combination of over 100 major venues in the world. Top locations include the Staples Center in Los Angeles, the O2 World Centers in Hamburg and Berlin, Germany, the Colosseum at Caesar's Palace in Las Vegas, and many top performance halls and theaters in the United States. It also owns three hockey teams in Europe and the LA Kings in the National Hockey League, and it manages the NBA LA Lakers organization. AEG Live is the live entertainment division providing touring, festival, exhibition, broadcasting, and merchandise for tour support and owns, operates, and books 35 significant venues. It produces the largest festival in the United States (the Coachella Music and Arts Festival), plus a touring roster including Taylor Swift, Bon Jovi, Pink, and Paul McCartney.[11]

IEBA—New Talent Showcase

Talent buyers and promoters use the International Entertainment Buyers Association (IEBA), National Association of Campus Activities (NACA), and other showcase opportunities to discover, arrange, and set up concert tours. The organization holds annual events bringing buyers of talent (who may also be promoters), managers, and acts together. The

IEBA is the top nonprofit organization in the nation providing networking and information gathering conferences and showcases for buyers to view acts and sign them up for tours. Hubert Lang and "Hap" Peebles formed the organization in 1970 as they felt "talent buyers need a voice" in the industry. It has grown to over 1,000 members, with a focus on networking information and live music.[12]

National Association for College Activities (NACA)—College Tours

NACA is a nonprofit organization that sponsors national and regional conferences for college representatives to view and book entertainment tours. According to naca.org,

> *NACA hosts a National Convention each February. This . . . event provides networking opportunities with other professionals, students and associate members. . . . Delegates from 950 member schools attend NACA's National Convention, which is the nation's largest campus activities marketplace. During the Convention, delegates can view and book live performances ranging from music and comedy to lecture and interactive programs.*[13]

The Classical "Serious" Music Market

The serious nonprofit music markets of symphony orchestras, opera, Broadway, and dance companies contribute billions annually to the U.S. economy. Many opportunities are available for classically trained, consummate musicians and vocalists. Private and public organizations, churches, universities, public and private grants, donations, businesses, and local volunteer organizations often finance orchestras and various types of choral groups. Many serious music organizations are dependent on these political, social, and music-supporting groups to provide financial support for local musicians. Local symphonies, operas, and so forth are marketed through the mass media by using their public service spots on radio and TV to announce performances and provide stories and photographs to the print media. The Boston Pops has a long history of showcasing popular acts and superstars in their summer and holiday concerts, including Cyndi Lauper. Major symphony orchestras enhance their bottom lines with international touring, which is both extremely expensive and wildly popular and profitable.

Pollstar

Billboard magazine, the most popular weekly entertainment industry publication, covers all aspects of the music industry. Concert promoters, it may surprise you to learn, use *Pollstar* magazine to make concert business decisions. Promoters use the agency rosters to select and locate the artists they want to book. Venues use the trades to publicize their auditoriums for acts, managers, and promoters. Domestic and international news stories detail the artists' lifestyles and their successful enterprises. To make business decisions, *Pollstar* provides a summary of radio charts, stations' play (by formats), a listing of recent concert revenues (by the average gross), the number of shows per week and per act, a summary of album sales (by artist, title, and label), and a listing of touring acts' scheduled concerts.

Award Shows

There's nothing better than hearing your name called as they open the envelope for the Grammy, Oscar, Tony, Emmy, ACM, Country Music, Golden Globe, or any other type of award. It really means you've hit the jackpot of success in your niche of the industry, and it also means *money*! Winning a prestigious award increases your opportunity of going on tour and usually increases profits from sales of your albums, movies, books, tour tickets, and merchandise. Movie stars, no matter what level of talent they possess, can command more money for their next film after winning a major award. Voting for some of the awards is by members only. Others are by consumers, who phone or email their votes into a central location. The best awards that include songs, music, or some form of music production are the following.

The Grammys

The Grammys are presented by the National Academy of Recording Arts and Sciences (NARAS). The Grammys are the ultimate award for excellence in the recording arts. The founding nonprofit organization is located in Los Angeles, with chapter offices in Atlanta, Chicago, Florida (Miami Beach–area office), LA, Memphis, Nashville, New York, the Pacific Northwest (Seattle-area office), Philadelphia, San Francisco, Texas (Austin-area office), and Washington, DC. Voting members include singers, musicians, producers, songwriters, engineers, managers, and business professionals who work at the highest level in the industry. Associate and student members are connected with the business and creative industry but not at the same level as voting members.[14]

The Emmys

The Emmys are the ultimate award in the television industry, presented by the National Academy of Television Arts and Sciences (NATAS) and its sister organization the Academy of Television Arts & Sciences (ATAS). Emmys and Global Media Awards are presented for excellence in news, documentary, sports, daytime programming, creative and digital entertainment, public and community service, technology, engineering, business, and primetime programs.[15]

The Oscars

The Academy of Motion Picture Arts and Sciences, specializing in the movie and film industry, presents the Oscars. They're the definitive award for filmmakers, producers, actors, scriptwriters, editors, songwriters, musicians, technical operators of cameras, lights, audio, and effects, and, of course, directors. Awards are also provided by the Academy to prestigious contributors in the creative arts and business through the Governors, Sci-Tech, and Student Academy Awards, and the Nicholl Fellowship in Screenwriting.[16]

Bid Sheets

Some agencies, such as CAA, require the promoter to fill out a bid sheet online before they submit their request to the agency. According to the CAA and others, items and cost factors on a bid sheet often include the information shown in Table 12.4.

If you've reviewed the first parts of the bid sheet, raise your fourth finger up in the air. Once you get all five up there, you've learned your stuff and you can "high-five" those around you.

The actual bid sheet helps the promoter project the total cost of the putting on the show (talent, venue, advertisement, insurance, bond, IATSE crew, riders expenses, etc.), and then the information gathered helps the promoter and booking agent project a realistic deal for both based on projected expenses and revenue. Everything is negotiable, yet instead of trying to score a huge profit by shafting the other guy, wise booking agents and promoters will attempt to establish long-term relationships that will provide future profitable events and opportunities.

Picking the Headliner

The projected ticket sales part of the bid sheet is the responsibility of the promoter. Smart promoters research the act using BDS for radio airplay, SoundScan for actual unit sales in their specific markets, BigChampagne for legal and P2P downloads (a sign of how popular the act might be), *Pollstar* for tour information, and other mass media and new media sources. Then they scale the house (ticket prices for various sections of venue seating) based on their perception of how hot the act is (emotional connection of the fans to the artist).

Table 12.4 Bid Sheet General Information. Where the act performs needs to "fit" the image and perceived "status" of the act. Also important are security, sound and light quality, dates, the reputation of the venue management, and other performances in (by the same act) nearby venues, that are or may be booked in the near future.

Name of the Artists (you want to hire)	
Agent (who represents the act)	
Date of Show	1.
Beginning & Ending Time of Performance	2.
Public or Private Show	3. Private [] Public [] Other []
Billing	4. Headliner [] Other []
Note: The General Information is the raw information required by the booking agent to determine the basic information of dates, times, and acts the promoter is seeking to hire.	
Promoter (company buying services of talent)	
Promoter/Company	Website
	Address
	Phone Number
	Email Address
Signatory (person signing contract):	Print Name:
Note: Anyone can "claim" to be a promoter. Thus, the promoter part of the Bid Sheet provides information to the booking agent and manager to help them determine the financial solvency, history, key people, and general information about the company before they even consider an offer.	
Venue (place where the concert is to be staged)	
Name of Venue	Website
	Address
	Phone Number
	Email Address
Contact Name (manager of venue & events)	Print Name:

Table 12.5 Production Company Bid Sheet. Production companies are often hired to "put on the show" even if the venue is an IATSE crew for load-in and load-out, sound, and lighting. The production also needs to fit the image and perceived status of the act. Security, sound and light quality, and production company reputation, personnel, and management are important. Managers prefer to work with productions companies that have provided excellent show productions in the past with their acts.

Production Company (production of show)	
Name of Production Company	Website Address Phone Number Email Address
Contact Name (manager of production)	Print Name:

Scaling the House

Scaling the house is the process of determining the ticket prices for various seats (in a variety of locations) in the arena or venue. Festivals often bypass the process by simply charging one ticket price for the entire festival, which may include many acts over a period of days or weeks. When there is assigned seating, fans will pay more money the closer they are to the acts or stage. Once the promoters have estimated the total projected cost of putting on the show by using the bid sheet, then they have to determine how much money they can make by estimating how much consumers will pay for tickets. The hotter the act, the higher the ticket prices!

A simple formula is to divide the total cost of the show by 50% of the seats. The idea is that by selling half of the venue seating the promoter should financially break even. Those ticket prices become the average ticket price for seats in the middle of the venue. Then the promoter will sell the back of the house in either one or two levels of cheaper tickets, depending on the size of the venue and the total number of seats available. As an example, the promoter may reduce seat ticket prices behind the middle of the average seats by

Table 12.6 Bid Sheet Projected Expenses (Cost of Show). What does it really cost to stage an event or concert? The bid sheet (Table 12.6) divides the cost into four different steps for staging an event: (a) preshow expenses, (b) show expenses (for live show), (c) production expenses, and (d) postshow expenses. The promoter will also be required to hire union workers for electricity and city officials (fire marshal and police), have a city and state business license, insurance (in case of fire or other problems), a personal bond (in case of lawsuits), and budget room for city, state, and federal taxes.

Preshow Expenses	Show Expenses	Production Expenses	Postshow Expenses
Advertising $_____	Major Artist $_____	Piano Rental $_____	Stage Manager $_____
Sponsoring Station (Ads & Comp Tickets) $_____	Opening Act $_____	Piano Tuner $_____	Rider $_____
IATSE Crew Load In $_____	Security [] Police [] T-shirt [] Private $_____	Catering $_____	IATSE Crew Load Out $_____
Ticketmaster $_____	Ticket Takers (show) $_____	Chair Rental $_____	Clean-Up $_____
Ticket Printing $_____	Ushers (show) $_____	Electrician & Power $_____	Credit Card Fees $_____
Ticket Sellers $_____	Box Office $_____	Riggers $_____	Housing (Acts) $_____
Promoter Business Legal- Administration $_____	Dressing Room $_____	Forklift Operator $_____	Limo: Other: $_____
Damage Deposit $_____	Medical (on standby) $_____	Equipment Rental $_____	Transportation $_____
Insurance $_____	IATSE Crew Operation of Super Troopers & Audio $_____	Barricade Rental $_____	Personal Bond $_____
Licenses & Permits (individual costs per license) ASCAP/BMI/SESAC $_____ City $_____ State $_____	Fireman (as required by codes) $_____	Rental of Super Troopers & Lighting $_____	Taxes: Federal: State: Local: $_____
Total $_____	Total $_____	Total $_____	Total $_____

Total Projected Expenses $_____

Projected Expenses Preshow $_____ Show $_____ Production $_____ Postshow $_____ Total Cost $_____

Table 12.7 Projected Revenue from Ticket Sales. Promoters take the cost of doing business (found in the third step of their process) and then try to figure out how much money they can generate with ticket sales. Different types of tickets are offered at different prices. Often the closer to the stage, the more expensive, and the back of the house seats (way up high) are much less expensive. Yet, it all depends on the drawing power of the main artist, which is determined as an emotional value to consumers.

Type of Ticket # 1	Type of Ticket # 2	Type of Ticket # 3	Type of Ticket # 4
% of Venue _____	% of Venue _____	% of Venue _____	% of Venue _____
Advanced [] Premium [] Group [] Day of Show [] Student Discount [] Comp (Free) [] Other []	Advanced [] Premium [] Group [] Day of Show [] Student Discount [] Comp (Free) [] Other []	Advanced [] Premium [] Group [] Day of Show [] Student Discount [] Comp (Free) [] Other []	Advanced [] Premium [] Group [] Day of Show [] Student Discount [] Comp (Free) [] Other []
Number of Seats _____	Number of Seats _____	Number of Seats _____	Number of Seats _____
Ticket Price $_____	Ticket Price $_____	Ticket Price $_____	Ticket Price $_____
100% gross $_____	100% gross $_____	100% gross $_____	100% gross $_____
80% gross $_____	80% gross $_____	80% gross $_____	80% gross $_____
50% gross $_____	50% gross $_____	50% gross $_____	50% gross $_____

Projected Revenue From Ticket Sales

Revenue Tickets 100% Sold	Revenue Tickets 80% Sold	Revenue Tickets 50% Sold
Ticket # 1 $_____	Ticket # 1 $_____	Ticket # 1 $_____
Ticket # 2 $_____	Ticket # 2 $_____	Ticket # 2 $_____
Ticket # 3 $_____	Ticket # 3 $_____	Ticket # 3 $_____
Ticket # 4 $_____	Ticket # 4 $_____	Ticket # 4 $_____
Gross $_____	Gross $_____	Gross $_____
Projected Gross Ticket Sales @ 100%	Total Gross (100%)$	
Projected Gross Ticket Sales @ 80%	Total Gross (80%)$	
Projected Gross Ticket Sales @ 50%	Total Gross (50%)$	

25% and then reduce the really distant or "bad" seats by 35%–50% off the average ticket price. To determine the "best seat" ticket prices, promoters simply add the average, poor, and bad ticket prices together to determine a premium ticket price. Sometimes super premium ticket prices are offered on first and second row seats or standing room areas next to the stage. Of course, this is a simple method, and in reality, actual ticket pricing may be more complex depending on the act, booking agent, venue, sponsorship contractual obligations, and other variables. Also notice that we've predicted 50% of the profits going to city, state, and national taxes. However, a few hours of work and a one-night score of $67,000+. My kind of business! If you perceive how this works, put your fifth finger up in the air and think about your future in the industry!

Additional Potential Revenue

Promoters usually negotiate additional revenue sources (money) from food commissions (hot dogs, soft drinks, etc.), the rights and processes for fulfillment companies to sell merchandise, and the act's requirement to hang branding advertisements from corporate sponsorship deals (posters and signs behind the performing act or in the venues). Promoters negotiate between the act, venue, and themselves for some portion of the additional revenue. Often the food revenue goes to the venue and most or all of the merchandise revenue goes to the act. However, a powerful promoter can negotiate a better deal, if he or she is establishing a tour instead of single date, and yet it still depends on the power of the act, booking agent, and manager.

Profit and Loss Statement

The profit and loss part of the bid sheet shows the promoter the projected profits (or losses) based on the total revenues (from ticket sales and all other sources) minus the total expenses (cost of talent and putting on the event) based on the number of ticket buyers tied to the prices consumers paid

Table 12.8 The 5,000 Seat Arena. Let's assume you've done the bid sheet and your total cost for putting on an event is $100,000. How much money are you projected to make in the two hours of the show? I like making about $67,700 for a couple of hours of work and that's after taxes! Let's just say this is my type of business. Notice that the promoter in this case is assuming that a band paid $50,000 or so could draw 5,000 fans, who would pay between $16 and $120 per ticket.

Type of Tickets	Seats	Percentage of Total Seats	Percent Above or Below Average	Ticket Price	Revenue Generated
Average Ticket Price	2,500	50%	------------	$40.00	$100,000
"Poor" Seat Ticket	1,000	20%	(–) 25%	$30.00	$30,000
"Bad" Seat Ticket	500	10%	(–) 40%	$16.00	$8,000
Premium Seat Ticket	900	18%	(+) 216%	$86.00	$77,400
Super Premium (first-row tickets)	100	2%	(+) 300%	$120.00	$12,000
Gross Totals	5,000	100%	------------	------------	$227,400
Cost of Show (bid sheet)	--------	------------	------------	------------	(–)$100,000
Net Totals	--------	------------	------------	------------	$127,400
Federal, State, & Local Taxes, About 50%	--------	------------	------------	------------	(–)$67,700
Estimated Profits After Taxes	--------	------------	------------	------------	$67,700 Profit!

at 100% sellout, 80% of venue, and 50%. This is an opportunity as seen through the eyes and gut reaction of the promoters. They have to make an important decision at this point to commit to the event or to stop before it goes any farther. The decision is usually based on the numbers and the experience of the promoter. Great promoters will know how much room they have to play the game, and if the numbers fail to be there, they will either negotiate with the booking agent for a lower cost for the band or pass.

Table 12.9 Bid Sheet Projected Other Revenues (Food Commissions and Merchandise). Many promoters and promotion companies, such as AEG Live and Live Nation, may use the 100%, 80%, and 50% ratios to determine the amounts of additional revenue that may be generated with food and merchandise. As an example, if the amount of ticket sales is only 50% then fewer people show up to buy dogs and t-shirts, so they may apply a higher percentage rate on each unit purchased. The 80% rate may be a lower percentage per unit sold (going to the promoter), but possibly more total revenues generated as more people attend.

Bid Sheet Projected Revenue: Food Commissions and Merchandise

Food Commissions	Split Percentages	Totals	Soft Drinks []
	Price (– Cost) _____		Hard Drinks []
[] Venue receives 100%			Popcorn []
			Candy [],
[] Promoter receives 100%	% Promoter ____ = ____ per Tic	$ ____	Hot Dogs []
[] Artist receives 100%	% Artist ____ = ____ per Tic	$ ____	Pizza []
[] Split	% Venue ____ = ____ per Tic	$ ____	Other []
Merchandise/Other	Split Percentages	Totals	T-shirts []
	Price (– Cost) _____		Posters []
[] Venue receives 100%			Cloths []
[] Promoter receives 100%	% Promoter ____ = ____ per Tic	$ ____	CDs []
[] Artist receives 100%	% Artist ____ = ____ per Tic	$ ____	Hats []
[] Splits	% Venue ____ = ____ per Tic	$ ____	Other []

Projected Revenue From Other Sources (if any)

Food Commissions	$ ____	Added Gross at 100% _____	
(if any)		Added Gross at 80% _____	
		Added Gross at 50% _____	
Merchandise	$ ____	Added Gross at 100% _____	
(if any)		Added Gross at 80% _____	
		Added Gross at 50% _____	
Projected Gross From Food Commissions & Merchandise @ 100% of tickets sold	Total Gross (100%) $		
Projected Gross From Food Commissions & Merchandise @ 80% of Tickets Sold	Total Gross (80%) $		
Projected Gross From Food Commissions & Merchandise @ 50% of Tickets Sold	Total Gross (50%) $		

Table 12.10 Event and Concert Projected P&L Statement. If you've been following along with the concert promotion process, you'll probably understand how stressful it can be waiting for the people to show up and buy stuff or face the music with a red face and learn from the experience of the money lost and move on.

Profit & Loss Projections		
100% Sold Out Revenue Tickets $_____ Other $_____	Total $_____ (minus) Expenses -$_____	$
80% Sold Out Revenue Tickets $_____ Other $_____	Total $_____ (minus) Expenses -$_____	$
50% Sold Out Revenue Tickets $_____ Other $_____	Total $_____ (minus) Expenses -$_____	$

Table 12.10 shows you the variables in the idea of a successful tour—it is not a given.

If you've been following along with the concert promotion process, you'll probably understand how stressful it can be, waiting for the people to show up and buy stuff or face the music with a red face and learn from the experience of the money lost and move on.

Your Right Hand High in the Air

If you've read the book and confidently feel you better understand the industry, then make sure you've got your hand and fingers reaching toward the sky. If you've got a sincere passion to be a positive, contributing part of this exciting business, take your hand (that's reaching up in the air) move it behind your head and pat yourself on the back.

Welcome to the time of your life! Blessings, L.

Notes

1. "Concert & Event Promotion in the US: Market Research Report." Concert & Event Promotion in the US Market Research. November 1, 2015. Accessed December 1, 2015. http://www.ibisworld.com/industry/default.aspx?indid=1960.
2. "2014 Year End Special Features." *Pollstar*. January 10, 2015. Accessed December 1, 2015. http://www.pollstar.com/news_article.aspx?ID=815827.
3. McIntyre, Hugh. "America's Top Five Music Festivals Sold $183 Million in Tickets in 2014." *Forbes*. March 25, 2015. Accessed November 25, 2015. http://www.forbes.com/sites/hughmcintyre/2015/03/21/americas-top-five-music-festivals-sold-183-million-in-tickets-in-2014/.
4. "2014 Year End Special Features."
5. "IATSE Labor Union, Representing the Technicians, Artisans and Craftpersons in the Entertainment Industry." About the IATSE. 2016. Accessed December 1, 2015. http://iatse.net/about-iatse.
6. Ibid.
7. "Ticketmaster." Ticketmaster.com. 2015. Accessed December 2, 2015. http://www.ticketmaster.com/about/our-history.html.
8. Rapino, Michael. "Live Nation Entertainment Reports Fourth Quarter and Full Year 2014 Results." Livenation.com. February 22, 2014. Accessed December 2, 2015. http://s1.q4cdn.com/788591527/files/doc_financials/2014/4Q/LYV-Q4-FY-2014-Earnings-Press-Release-FINAL_v001_p6t8oq.pdf.
9. Ibid.
10. Ibid.
11. "Company Overview." Aegworldwide.com. 2015. Accessed December 2, 2015. http://www.aegworldwide.com/about/company overview/companyoverview.
12. "IEBA 2015—International Entertainment Buyers Association—Nashville, TN." 2015. Accessed December 2, 2015. http://ieba.org.
13. "NACA Events." Your Best Campus Tradition. 2015. Accessed December 2, 2015. https://www.naca.org/AWARDS/Pages/YBCT.aspx.
14. "Grammy Pro, Join Us." GRAMMYPro.com. 2015. Accessed December 2, 2015. https://www.grammypro.com/join/about.

15. "National Academy of Television Arts & Sciences." Television Academy. 2015. Accessed December 2, 2015. http://www.emmys.com/content/national-academy-television-arts-sciences.
16. "Oscars.org Academy of Motion Picture Arts and Sciences." 2015. Accessed December 2, 2015. http://www.oscars.org/.

APPENDIX A

Contracts

Appendix A provides three sample contracts for songwriters, music publishers, recording acts and labels, and performing live artists and personal managers. The contracts are provided by entertainment lawyer and Belmont University associate professor Rush Hicks, Esq. These documents may not be copied and used in any form except for educational purposes and reviewing.[1]

Single-Song Music Publishing Contract

This Agreement is made as of this day of _____, year, and by _____, and between _____, whose address is _____, ("Writer") and ("Publisher") whose address is _____.

Assignment of Copyright

1.1 *With respect to the following musical composition(s) ("Original Works")*

Title _____   Writer(s) _____   Percentages _____

WRITER does hereby agree to and does assign and transfer to PUBLISHER all rights of whatsoever nature known, or which may hereafter come into existence, including, but not limited to the exclusive rights set forth in Title 17 U.S.C. section 106 (1977) in and to said Original Works invented, written, and conceived, arranged, composed, created, or originated by WRITER and the right to secure copyrights thereon throughout the entire world in the name of PUBLISHER and all renewal and extension copyrights thereof and all right, title, and interest both legal and equitable in and to the same, it being understood and agreed that said Original Works and copyrights thereof and each and every right in said Original Works, whether now known or hereafter to become known covering the use or any manner or type of use of said Original Works, are and shall be the sole and exclusive property of PUBLISHER.

1.2 *PUBLISHER* shall have the right to grant licenses for the reproduction, printing, recording, arrangement, or performance of any Original Works without restriction.

1.3 *PUBLISHER* shall have the right, in its sole discretion, to retitle, translate, arrange, and otherwise edit, review, and adapt said Original Works created hereunder. In the event said Original Work is an instrumental composition, PUBLISHER shall have the right to have lyrics written for said Original Work by a writer or writers designated by PUBLISHER, said lyrics shall require only the approval of PUBLISHER.

1.4 *WRITER* agrees that WRITER shall deliver to PUBLISHER Original Works created hereunder in such a form as is deemed suitable by PUBLISHER. Unless otherwise specified by PUBLISHER, any form, as established by prior course of dealings between PUBLISHER and WRITER, or usual and customary industry practices shall be deemed suitable.

Grant of Rights

At PUBLISHER's request, WRITER agrees to execute an Assignment of Copyright between WRITER and PUBLISHER for every Original Work created hereunder. WRITER agrees to execute any other documents as requested by PUBLISHER to effectuate such transfer of ownership. Upon WRITER's failure to do so, PUBLISHER shall have the right pursuant to the terms and conditions hereof, to execute such separate agreements on behalf of WRITER. Such separate agreements shall supplement and not supersede this Agreement. In the event of any conflict between the provisions of such separate agreement and this Agreement, the provisions of this Agreement shall govern. The failure of either party to execute such separate agreement shall not affect the rights of either party.

Power of Attorney

WRITER hereby irrevocably constitutes, authorizes, empowers, and appoints PUBLISHER, and PUBLISHER's successors and assigns, WRITER's true and lawful attorney in WRITER's name, place, and stead to execute and deliver any and all documents, which PUBLISHER may from time to time deem necessary to effectuate the intent and purpose of this Agreement, this power being coupled with an interest and irrevocable for any cause or in any event. Such power of attorney shall include, without limitation, such documents as PUBLISHER deems necessary to secure to PUBLISHER and PUBLISHER's successors and assigns the worldwide copyrights for the entire term of copyright and for any and all renewals and extensions under any present or future laws throughout the world.

Royalties

4.1 *PUBLISHER*, shall pay WRITER a royalty of fifty percent (50%) of any and all sums actually received by PUBLISHER, or its authorized agent, less the following:

  a. All outstanding recoupable advances between PUBLISHER and WRITER hereunder;
  b. any administration or collection fees, foreign taxes, etc. actually charged to PUBLISHER, its authorized agents, or foreign subpublishers;
  c. royalties payable to cowriters; and
  d. any other sums or monies due PUBLISHER by WRITER of any nature.

4.2 *Royalties* on foreign sales shall be payable by PUBLISHER only after PUBLISHER, or an authorized agent of PUBLISHER, has received payment in the United States in United States currency, and shall be payable at the same rate of exchange as received by PUBLISHER.

4.3 *Notwithstanding anything herein to the contrary, WRITER* shall not be entitled to any portion of any sums received by PUBLISHER with respect to public performance of any Original Works from any performing rights society which pays a portion of performance fees directly to writers or composers and WRITER shall receive the writer's share of public performance royalties directly from the performing rights society.

4.4 *Such royalties shall be paid in instances where WRITER* is or was the sole author, composer, and arranger of the Original Works. With regard to Original Works where WRITER is or was only one of two or more authors, writers, or collaborators, WRITER shall be paid only the portion of such royalties equal to WRITER's percentage of authorship of said Original Work.

4.5 *PUBLISHER* or any of its affiliates, subsidiaries, or parent companies shall have the right to cross collateralize this Agreement and any previous exclusive songwriter agreement, single-song agreement, arranger's agreement, or similar type of agreement between WRITER and PUBLISHER or any of its affiliates, subsidiaries, or parent companies. PUBLISHER or any of its affiliates, subsidiaries, or parent companies shall also have the right to cross collateralize Original Works created hereunder.

4.6 *In the event PUBLISHER* creates or causes to be created a separately copyrighted arrangement or translation of any Original Work, or in the event PUBLISHER causes to be created lyrics for an instrumental composition, WRITER agrees that PUBLISHER may pay to the author or composer of such arrangement or translation, a portion, but not exceeding one-half (1/2), of the royalties due WRITER for said Original Work. In such event WRITER shall receive such royalties due WRITER for said Original Work less such portion.

4.7 *No royalties will be earned or payable on Original Works* contained in any uses pursuant to paragraphs 4.6 herein;

a. given away or furnished on a "no charge" or "service charge" basis for promotional purposes or furnished as a sales inducement or otherwise to distributors, subdistributors, dealers, and others otherwise distributed as part of promotional activities; or

b. sold as cutouts, scrap, salvage, overstock, or closeout units.

Writer's Name and Likeness

PUBLISHER shall have the perpetual, nonexclusive worldwide right to use and to permit others to use: WRITER's name, both legal and professional and whether presently or hereafter used by WRITER; WRITER's approved likeness; WRITER's facsimile signature; and other identification and biographical material concerning WRITER. Such use shall be for purposes of trade and otherwise, without restriction, in connection with Original Works produced hereunder.

Writer's Warranties

6.1 *WRITER* represents and warrants that WRITER has the authority to enter into and perform all the terms of this Agreement and that WRITER is under no disability, restriction or prohibition, whether contractual or otherwise, with respect to WRITER's right to execute this Agreement, to grant the rights granted by WRITER to PUBLISHER hereunder and to perform each and every term and provision hereof.

6.2 *WRITER* warrants that all of the lyrics and music or Original Works delivered by WRITER to PUBLISHER hereunder shall be WRITER's own original compositions or arrangements and that no part thereof shall be an imitation or copy of any other copyrighted work.

6.3 *WRITER* further covenants and agrees to protect and defend the right, title, and interest of PUBLISHER in said Original Works hereby sold and assigned to PUBLISHER, to the fullest extent and to hold PUBLISHER free and harmless of and from all loss, liability, and damage in any action brought against PUBLISHER by reason of the inclusion in said works of material owned or copyrighted by others or on account of WRITER's violation of any warranties contained herein. All

525

costs, fees, and expenses paid or incurred by PUBLISHER in defending, protecting, or perfecting its title to the Original Works, and the copyright thereon, shall be charged against WRITER and deducted from all sums due or becoming due from PUBLISHER to WRITER.

Royalty Payments

7.1 *PUBLISHER*, or an authorized agent, will compute royalties and payments due on or about forty-five (45) days after the end of a calendar quarter, for that quarter, and if royalties or other payments are due WRITER, PUBLISHER will make such payments less any then non-recouped advances paid to WRITER or on WRITER's behalf pursuant to this Agreement. If WRITER has accounts of any nature with PUBLISHER or PUBLISHER's affiliates and if WRITER owes PUBLISHER or PUBLISHER's affiliates money in such accounts(s), PUB- LISHER may apply all royalties or payments which might be due and payable to WRITER to reduce the balance in WRIT- ER's account(s).

7.2 *WRITER*, or a Certified Public Accountant representing WRITER, shall have the right, during PUBLISHER's nor- mal business hours, and at WRITER's expense, after giving PUBLISHER, or an authorized agent of PUBLISHER, at least thirty (30) days advance notice in writing, to examine PUB- LISHER's, or an authorized agent of PUBLISHER, books and records insofar as they pertain to the royalties or payments payable to WRITER.

7.3 *Any payments to be made to WRITER* hereunder shall be made to WRITER at the address set forth at the beginning of this Agreement. PUBLISHER shall make payment by a single check made payable to WRITER, or such agent as WRITER may designate in writing in accordance with the provisions of this Agreement governing notices.

7.4 *All statements and all accounts rendered to WRITER* hereun- der shall be binding upon WRITER and not subject to any objection for any reason whatsoever unless specified objec- tions in writing, setting forth the basis thereof, is given to PUBLISHER, or PUBLISHER's authorized agent, within two

(2) years from the date the statement is rendered. No action, audit, or proceeding of any kind or nature may be instituted or maintained by WRITER with respect to any statements recorded hereunder unless such action or proceeding is commenced within one (1) year after delivery of such written objection by WRITER to PUBLISHER.

Suits and Legal Actions Involving Original Works

PUBLISHER shall have the sole right to prosecute, defend, settle and compromise all suits and actions respecting the Original Works, and generally to do and perform all things necessary to prevent and restrain the infringement of copyrights therein or other rights with respect to the Original Works. If PUBLISHER recovers any moneys as a result of a judgment or settlement, the moneys shall be apportioned between PUBLISHER and WRITER fifty percent (50%) for PUBLISHER and fifty percent (50%) for WRITER prorated by WRITER's portion of ownership, if any, in the Original Works, after first deducting the expense of obtaining the moneys, including attorney's fees. WRITER shall have the right to obtain counsel for WRITER, but at WRITER's own expense, to assist in any such matter. Any judgments against PUBLISHER and any settlements by PUBLISHER of claims against it, respecting the Original Works, together with costs and expenses including attorney's fees, shall be covered by the indemnity provisions.

Indemnification

WRITER hereby agrees to and does hereby indemnify, save, and hold PUBLISHER harmless from any damages, liabilities, costs, losses, and expenses including legal costs and attorney's fees arising out of or connected with any claim, demand, or action by a third party which is inconsistent with any of the warranties, representations, or covenants made by WRITER in this Agreement. WRITER agrees to reimburse PUBLISHER, on demand, for any payment made by PUBLISHER at any time with respect to any such damage, liability, cost, loss, or expense to which WRITER's indemnity applies. PUBLISHER shall notify WRITER of any such claim, demand, or action promptly after PUBLISHER has been formally advised thereof. Pending the determination of any such claim, demand, or action, PUBLISHER shall have the right, at PUBLISHER's election, to withhold payment of any moneys otherwise payable to WRITER under this or any other Agreement between WRITER and any of PUBLISHER's affiliates.

Notices

All notices to be given to WRITER or PUBLISHER hereunder shall be addressed WRITER or PUBLISHER at the addresses set forth on page 1 or at such other address as WRITER or PUBLISHER shall designate in writing from time to time. All notices shall be in writing and shall either be served by personal delivery, mail, or telegraph, all charges prepaid. Except as otherwise provided herein, such notices shall be deemed given when personally delivered, mailed, or delivered to a telegraph office, all charges prepaid, except that notices of change of address shall be effective only after actual receipt thereof.

Entire Agreement

This Agreement sets forth the entire understanding between WRITER and PUBLISHER concerning the subject matter hereof. No modification, amendment, waiver, termination, or discharge of this Agreement or of any of the terms or provisions hereof shall be binding upon either party unless confirmed by a written instrument signed by WRITER and by a duly authorized officer of PUBLISHER. No waiver by WRITER or PUBLISHER of any term or provision of this Agreement or of any default shall affect WRITER's or PUBLISHER's respective rights thereafter to enforce such term or provision or to exercise any right to remedy in the event of any other default, whether or not similar.

If any portion of this Agreement shall be held void, voidable, invalid, or inoperative, no other provision of this Agreement shall be affected as a result thereof, and, accordingly, the remaining provisions of this Agreement shall remain in full force and effect as through such void, voidable, invalid, or inoperative provision had not been contained herein.

Notice of Breach

Before PUBLISHER or WRITER may assert that the other is in default in performing any obligation contained herein, the party alleging the default must advise the other in writing of the specific facts constituting default and the specific obligation breached. The other party shall be allowed a period of thirty (30) days after receipt of such written notice to cure the default. No breach of any obligation shall be deemed to be incurable during such thirty (30) day period.

Relationship of Parties

Nothing herein contained shall constitute a partnership or a joint venture between WRITER and PUBLISHER. Neither party here to shall hold itself out contrary to the terms of this paragraph and neither WRITER or PUBLISHER shall become liable for any representation, act, or omission of the other contrary to these provisions. This Agreement shall not be deemed to give any right or remedy to any third party whatsoever unless said right or remedy is specifically granted by PUBLISHER in writing to such third party.

Binding Effect and Benefit of Agreement

This Agreement shall be binding upon and shall inure to the benefit of PUBLISHER and its designees, successors, and assigns, as well as to WRITER or his heirs, executors, administrators, personal representatives, and/or assigns.

Assignment

PUBLISHER may assign, license, or otherwise transfer to any other entity any or all of PUBLISHER's rights, privileges, and property under this Agreement including, without limitation, the right to WRITER's services. In the event of any assignment, the obligations of PUBLISHER shall be binding upon any such assignee or assignees for the express benefit of PUBLISHER and WRITER.

Controlling Law

This contract shall be deemed to have been made in the State of Tennessee, and its validity, construction, and effect shall be governed by the laws of said state. Any disputes arising between the parties shall be brought in the state and/or federal courts located in Nashville, Davidson County, Tennessee.

IN WITNESS WHEREOF, the parties hereto have executed this Agreement and the accompanying Schedules and Appendices as of the date first written above.

Publisher: _____

Writer: _____

Name: _____

Social Security No: _____

Date of Birth: _____

Copyright Assignment

For and in consideration of the mutual covenants, promises, and undertakings set forth in separate publishing agreements between _____ (WRITER) and _____, with respect to the composition(s) described below, WRITER assigns, transfers, sets over, and conveys to _____, all WRITER's right, title, and interest in and to the following musical composition(s):

Title _____   Writer(S) _____   Percentages _____

The within assignment, transfer, and conveyance include, without limitation, the lyrics, music, and title of said composition(s) any and all works derived therefrom, the United States and worldwide copyright therein, and any renewals or extensions thereof, and any and all other rights that WRITER now has or to which WRITER may become entitled under existing or subsequently enacted federal, state, or foreign laws, including, without limitation, the following rights: to reproduce the composition(s) in copies or phonorecords, to prepare derivative works based upon the composition(s), to distribute copies or phonorecords of the compositions, and to perform and display the compositions publicly. The within grant further includes any and all causes of action for infringement of the compositions, past, present, and future, and all proceeds from the foregoing accrued and unpaid and hereafter accruing.

THIS ASSIGNMENT IS EFFECTIVE THE _____ DAY OF _____, _____.

WRITER: _____

STATE OF TENNESSEE

COUNTY OF DAVIDSON

Personally appeared before me, a Notary Public in and for said county and state, WRITER, the within named publisher, with whom I am personally acquainted, and who acknowledge execution of the foregoing Assignment of Copyright for the purposes therein contained.

Witness my hand and official seal at office on this _____ day of _____, _____.

My commission expires: _____.

Notary Public

Exclusive Recording Artist Agreement

This EXCLUSIVE RECORDING ARTIST AGREEMENT (this "Agreement") is made and entered into as of _____, by and between _____, whose address is _____ _____ ("Company") and _____ ("you"), whose address is _____ _____.

Exclusive Services

Company hereby engages your exclusive personal services as a recording artist in connection with the production of Records and you hereby accept such engagement and agree to exclusively render such services for Company in the Territory during the Term and all extensions and renewals thereof. (You are sometimes called "Artist" below.)

Term

a. The term of this Agreement shall consist of an "Initial Period" and the "Option Periods" set forth below as may be exercised by Company pursuant hereto.

b. The "Initial Period" shall commence on the date hereof and shall end on the date that is twelve (12) months after Delivery and acceptance by Company of the Recording Commitment (as defined below) for the Initial Period.

c. You hereby grant to us four (4) separate, consecutive, and irrevocable options, each to extend the Term for additional contract periods (each herein called an "Option Period") upon the same terms and conditions applicable to the Initial Period, except as otherwise provided herein. Each such Option Period shall run consecutively beginning on the expiration of the Initial Period or the previous Option Period, as the case may be, and shall end on the date twelve (12) months after Delivery and acceptance by Company of the Recording Commitment for such Option Period. The Initial Period and the Option Periods are sometimes referred to herein as "Contract Periods," as the same may be suspended or extended as provided herein. Company shall exercise each such option by notice to you at any time prior to the expiration of the then current Contract Period.

d. Notwithstanding anything to the contrary contained in paragraph 2(a)–(c), if, as of the date when the current Contract Period would otherwise have expired, Company has neither exercised its option to extend the Term for a further Contract Period nor notified you that Company does not wish to exercise such option, then: (i) you shall immediately notify Company that its option has not yet been exercised (an "Option Warning"); (ii) Company shall be entitled to exercise its option at any time before receiving the Option Warning or within ten (10) business days thereafter; and (iii) the current Contract Period shall be deemed to have continued until Company exercises its option or until the end of such ten business day period (whichever shall occur first).

Recording Commitment

a. During each Contract Period you shall record and Deliver to Company sufficient Masters to constitute the Record specified in the following schedule (the "Recording Commitment"):

CONTRACT PERIOD	RECORDING COMMITMENT
Initial Period	One Album (the "First Album")
First Option Period	One Album (the "Second Album")
Second Option Period	One Album (the "Third Album")
Third Option Period	One Album (the "Fourth Album")
Fourth Option Period	One Album (the "Fifth Album")

b. (i) The First Album shall be Delivered to Company within ninety (90) days following commencement of the Initial Period; (ii) the Recording Commitment in respect of each Option Period shall be Delivered to Company within ninety (90) days following commencement of the applicable Option Period; and (iii) unless an authorized officer of Company shall otherwise agree in writing, and without limiting any of the other provisions herein, you shall not Deliver any Album within nine (9) months following the date of the initial commercial release of the immediately preceding Album in the United States. If any Recording Commitment is Delivered between October 15th and December 31st of a particular year, then Delivery of the Recording Commitment concerned will be deemed to have occurred on January 2nd of the succeeding year.

c. No multiple Albums, "theme" Masters (e.g., Christmas Masters), "live" performances, instrumental Masters, joint recordings, or spoken-word Masters shall be recorded or Delivered hereunder in satisfaction of the Recording Commitment without Company's prior written consent, which may be withheld in Company's sole discretion; provided however, if Company consents to the Delivery of a multiple Album hereunder, then such multiple Album shall be deemed a single Album for the purposes of Artist's Delivery obligations under this Agreement. If Artist Delivers and Company accepts Masters consisting of "theme," "live," instrumental, joint, or spoken-word recordings, then such Masters shall not be deemed to be in partial or complete fulfillment of any of Artist's obligations hereunder.

d. During the Term, Company shall have one (1) option ("Greatest Hits Sides Option"), to require Artist to record and Deliver up to two (2) Sides recorded after Company's exercise of such Greatest Hits Sides Option (the "New Greatest Hits Masters"). Each such New Greatest Hits Master shall embody a Composition not previously recorded by Artist and shall be intended for initial release on a "Greatest Hits" or "Best Of" Album (a "Greatest Hits Album"). Artist shall deliver such New Greatest Hits Masters no later than sixty (60) days after Company's exercise of the Greatest Hits Sides Option. New Greatest Hits Masters shall not be deemed to fulfill any of Artist's obligations hereunder with respect to Recording Commitments.

Recording Elements and Procedures

a. Prior to the commencement of recording any Recording(s), you and Company shall mutually agree on each of the following, in order, before you proceed further: (i) selection of, and compensation for, individual producer(s) (including, without limitation any producer advance [or fee] and producer royalty, if any); (ii) selection of material, including the Compositions to be recorded; and (iii) the dates of recording and mixing and studios where recording and mixing are to take place. Artist shall be responsible for engaging and paying all producers of each Master ("Producer") (except that Company may engage Producer(s) at its election and deduct any sums paid to such Producer(s) from any sums payable hereunder). Company shall have the right to have a representative attend each recording session. For the first album,_____is approved as record producer.

b. Upon the reasonable request of Company, Artist shall re-record any selection until a Master commercially and technically satisfactory to Company has been obtained. Company may refuse to accept, and may require Artist to deliver substitute Masters for, compositions Company deems patently offensive or which, in its judgment, violates any law, violates the rights of any person, or subjects Company to material liability for any reason.

c. It is of the essence of this Agreement that Artist timely supply Company with all of the information Company needs in order: (i) to make payments due or required in connection with the Masters; (ii) to comply with any and all other obligations Company may have in connection with recording the Masters, and (iii) to release Records embodying the Masters. Subject to the provisions of this Agreement, Company will pay Recording Costs incurred in connection with any Master(s) Delivered hereunder not to exceed the applicable Recording Fund with respect to such Master(s). Without limiting the foregoing, Artist shall deliver to Company within forty-eight hours after each recording session the following documents pertaining to such session: all union contract forms or report forms and all necessary payroll forms (including without limitation all I-9 forms and related documentation and all W-4 forms). Artist shall deliver all other invoices, receipts, vouchers, and documents within one week after the related expense is incurred. Artist shall be solely responsible for and shall pay any penalties and/

535

or interest charges incurred by Company for late payments by reason of Artist's failure to comply with the terms of this paragraph (or Company may deduct any resulting penalty or interest charges from any and all monies payable under this Agreement or any other agreement between you or Artist and Company or its affiliates). In addition, Artist shall deliver promptly complete label copy, any liner credits, and any information required to be submitted to unions, guilds, or other third parties.

d. Your submission of Masters to Company shall constitute your representation and warranty that you have obtained all necessary licenses, approvals, consents, and permissions including, without limitation, written clearance(s) from the copyright owner(s)/publisher(s) for any and all "first-use" compositions, licenses for sampled material, etc.; provided that, unless Company has specifically requested that Artist do so, Company will secure the actual Mechanical Licenses for compositions recorded hereunder; provided that you are solely responsible for obtaining and you hereby represent, warrant, and covenant that you will obtain written clearance(s) from the copyright owner(s)/publisher(s) for any and all "first-use" compositions prior to recording any such composition and promptly provide Company with copies of all such clearances. You shall be solely responsible for and pay any and all costs, fees, and expenses in connection with any and all sample licensing and authorization, and all such sums (including, without limitation, royalties and any "rollover" payments) to the extent not paid by you, shall be deducted from any and all monies payable under this Agreement or any other agreement between you or Artist and Company or its affiliates. Company shall have the right to approve or disapprove (in its unrestricted discretion) all terms and conditions of any such sample license prior to embodiment on any Master. Notwithstanding the foregoing, Company may elect to arrange directly for such authorization and licensing in which event, you shall nevertheless be solely responsible for and pay any and all costs, fees, and expenses in connection with such sample licensing and all such sums (including, without limitation, royalties and any "rollover" payments) to the extent not paid by you, shall be deducted from any and all monies payable under this

Agreement or any other agreement between you or Artist and Company or its affiliates.

e. Nothing in this Agreement shall obligate Company to continue or permit the continuation of any recording session, even if previously approved hereunder, if Company reasonably anticipates that the Recording Costs attributable to the recording session shall exceed the Advances/Recording Fund for that Album or other applicable Recording Commitment or that the Recordings being produced will not be technically and commercially satisfactory in accordance with the provisions of this Agreement.

f. Unless Company requests or approves otherwise, your performances hereunder shall be reasonably consistent in concept and style, and the Masters will be similar in general artistic concept and style, to Masters recorded and accepted by Company as satisfying your Recording Commitment for the Initial Period hereof. You will obtain Company's written consent prior to performing in any new concept or style.

Grant of Rights

a. All Masters, Video Masters, and other Recordings embodying Artist's performances made during the Term from the inception of the recording thereof and all reproductions derived there from, together with the performances embodied thereon (but excluding musical compositions embodied therein), shall be the property of Company, free from any claims whatsoever by Artist or any person deriving rights or interests from or through Artist. Without limiting the generality of the foregoing, Company shall have the exclusive and unlimited right to all the results and proceeds of Artist's recording services rendered during the Term, including, without limitation, the exclusive, unlimited, and perpetual right throughout the Territory: (i) to manufacture, advertise, sell, lease, license, distribute, or otherwise use or dispose of, in any or all fields of use by any method now or hereafter known (including but not limited to any form of digital or electronic transmissions as well as sales through Company's website), Records embodying Masters (including accompaniment tracks without the vocal

537

tracks), or to refrain therefrom; (ii) to use and publish and to permit others to use and publish Artist's name (including any professional name currently utilized or hereafter adopted by Artist), approved photographs and likenesses, and approved biographical material concerning Artist for advertising and trade purposes in connection with the sale of Artist's Recordings and the exploitation, in accordance with the terms hereof, of all Masters and Video Masters produced during the Term. The materials approved by Artist that contain the Artist's name or likeness or photograph or biography may also be used by Company's foreign affiliates and licensees; (iii) to obtain copyrights and renewals thereof in sound Recordings and Video Masters (as distinguished from the musical compositions embodied thereon) recorded by Artist during the Term, in Company's name as owner and "employer-for-hire" of such sound recordings and Video Masters. Artist acknowledges that Artist's services hereunder are rendered as Company's employee-for hire for the purposes of copyright ownership of the sound Recordings and Video Masters and any related artwork made hereunder. If any such Recording or artwork is determined not to be a "work-made-for-hire," it will be deemed transferred and assigned to Company by this Agreement, together with all rights in it, including without limitation the worldwide copyright therein and all renewals and extensions thereof. You and Artist hereby irrevocably and unconditionally waive any and all moral and like rights (including, droit morale) that you and Artist have in such Recordings and in the performances embodied therein and related artwork and hereby agree not to make any claim against Company or any party authorized by Company to exploit such Recordings or artwork based on such moral or like rights; (iv) to release Records derived from Masters recorded during the Term by Artist under any name, trademark or label which Company may from time to time elect; and (v) to perform such Records and to permit performances thereof by means of radio broadcast, television, Internet, or any other method or medium now or hereafter known or devised.

b. Company shall have the perpetual right, without any liability to any party, to use and to authorize others to use your name and the names (including any professional names heretofore or hereafter adopted), and any likenesses (including photographs, portraits, caricatures, and stills from any Videos made hereunder) and biographical material relating to Artist and any producer of Masters hereunder, on and in the packaging of Records hereunder, for purposes of advertising, promotion, and trade and in connection with the making, exploitation, promotion, marketing, and publicity of Records hereunder, the writing and publishing of articles by Company or third parties, in general goodwill advertising (advertising designed to create goodwill and prestige for Company and not for the purpose of selling any specific product or service), and including, without limitation, purposes collateral to such permitted purposes (e.g., MTV's advertising and promotion), without payment of additional compensation to Artist or any other person. You warrant and represent that you own the exclusive right to so use such names, likenesses, and biographical materials and that the use of same will not infringe upon the rights of any third party. If any third party challenges Artist's right to use a professional name, Company may, at its election and without limiting Company's rights, require Artist to adopt another professional name approved by Company without awaiting the determination of the validity of such challenge. During the Term, Artist will not change the name by which Artist is professionally known without Company's prior written approval.

c. Company shall have the right to maintain an artist website (including the right to incorporate Artist's name, likeness, photographs, and biographical material therein) in accordance with Company's standard policies (including, without limitation, incorporating links to other sites). Artist may also maintain his/her own website; provided that (i) during the Term of this Agreement, Company shall have approval rights regarding all aspects of such website and (ii) Artist may not sell recorded product on such website. No Person other than Company and Artist shall have the right to maintain an artist website regarding Artist during the Term hereof.

Marketing Restrictions

Company agrees that it will not release any Album delivered in fulfillment of the Recording Commitment as a Budget Record in less than twelve (12) months; or as a Mid-Price Record in less than six (6) months after initial release of the Album in the United States. If Company releases any such Album that is a Mid-Price Record or Budget Record, as applicable, prior to the expiration of the applicable time period without your consent, your sole remedy shall be that Company shall not reduce your royalty rate pursuant to paragraph 8(i) below for sales of such Album made during such period.

Advances and Recording Fund

Company shall pay to Artist the following sums as Advances:

a. In connection with Artist's Delivery to Company of the applicable Album(s) hereunder, Company will pay Artist an Advance in the amount by which the applicable sum below (each herein called a "Recording Fund"), as reduced pursuant to paragraph (b) below, exceeds all Recording Costs paid or incurred by Company for such Album(s). Such Advance shall be payable to Artist promptly following the Delivery of the applicable Album(s). (i) With respect to the First Album, the Recording Fund shall be $_____, payable as follows:

   A. The sum of $_____ shall be payable to the Artist upon the execution of this Agreement.
   B. The balance shall be payable upon Delivery of the Master Recordings. (ii) With respect to each of the Second Album through the Fifth Album, if any, the Recording Fund shall be an amount equal to sixty-six (66%) percent of the royalties earned by Artist under paragraph 8 hereof on paid Net Sales through United States normal retail distribution channels of the previously released Recording Commitment Album for twelve (12) months from USA release of such previously released Recording Commitment Album. In no event shall any such

Recording Fund be less than the "Minimum" or more than the "Maximum" set forth below for the Album in question:

C.

Applicable Album/Recording Commitment		
	MINIMUM	MAXIMUM
Second Album	$150,000	$250,000
Third Album	$200,000	$300,000
Fourth Album	$250,000	$400,000
Fifth Album	$300,000	$500,000

b. Any monies paid by Company to third parties for independent record promotion and marketing shall constitute an Advance to Artist.

c. In calculating each Advance hereunder, the applicable Recording Fund shall be reduced by all Anticipated Costs, partial payments of such Recording Fund, and all charges and other Advances deductible therefrom (each such Recording Fund, as so reduced, is sometimes referred to herein as the "Available Fund"). As used herein, "Anticipated Costs" means any costs Company reasonably anticipates will be paid or incurred by Company for recording, mastering, mixing, or remixing the Masters concerned and all costs which Company reasonably anticipates are necessary to clear "samples" on such Masters. Any Anticipated Costs which are deducted from a Recording Fund but are not paid or incurred by Company as set forth in the previous sentence shall be remitted to Artist. Artist agrees that the Recording Fund includes the prepayment of session union scale to Artist as provided in the applicable union codes and Artist shall complete any documentation required by the applicable union to implement this sentence. It is understood and agreed that Company shall not be responsible for paying any charges or fees for arrangements or orchestrations supplied by Artist.

d. Each Recording Fund set forth in paragraph 7(a) above is inclusive of all Recording Costs for the production of the Masters comprising the applicable Album. All Recording Costs incurred by Company with your approval and/or all Recording Costs incurred by you or your representatives which are in excess of the aforementioned Recording Funds shall be your sole responsibility, and you hereby agree to forthwith pay and discharge all such excess Recording Costs. In the event Company elects to pay any such excess Recording Costs on your behalf (which Company shall have the right but not the obligation to do), you shall, upon demand, reimburse Company for such excess Recording Costs or, in lieu of requesting reimbursement: (i) with respect to excess Recording Costs incurred by Company (with your approval), Company shall have the right to deduct such excess Recording Costs from any monies otherwise due you under this Agreement except for Mechanical Royalties and (ii) with respect to excess Recording Costs incurred by you or your representatives, Company shall have the right to deduct such excess Recording Costs from any monies otherwise due you under this Agreement, including without limitation Mechanical Royalties.

e. (i) All sums paid to you or Artist or on your or Artist's behalf, or at your written request to anyone on your or Artist's behalf, or to or on behalf of any person, firm, or corporation representing you or Artist, other than royalties payable hereunder and (ii) any and all Recording Costs paid or incurred by Company hereunder, shall constitute Advances. Company may recoup Advances from any and all royalties (excluding Mechanical Royalties, except as otherwise specified herein) accruing hereunder.

f. If any Album is not delivered within ninety (90) days following its due date, Artist shall, upon Company's written demand, repay Company any amounts previously paid by Company for or in connection with such Album.

Royalties

Company shall credit to your royalty account royalties as described below. Royalties shall be computed by applying the applicable royalty percentage rate specified below to the applicable Royalty Base Price in respect of top-line Net Sales of the Record concerned:

a. (i) The royalty rate (the "Basic U.S. Rate") in respect of Net Sales of Records (other than Audiovisual Records) consisting entirely of Masters made hereunder during the respective Contract Periods specified below and sold by Company through Normal Retail Channels in the United States ("USNRC Net Sales") shall be as follows:

TYPE OF RECORD	BASIC U.S. RATES
First Album	12%
Second and Third Albums	13%
Fourth Album and Fifth Albums	14%
Singles, EPs, and Twelve-Inch Singles	10%

(ii) Notwithstanding anything to the contrary, a sale of a Digital Download shall result in Company crediting to your royalty account a royalty rate of 10% prorated based upon the number of downloads. By way of example, should Company sell an Album via Digital Download and the retail price to the consumer is $9.99, you will receive 10% or 99.9 cents per album download.

b. The Basic U.S. Rate will escalate prospectively solely in respect of Net Sales of any particular Album constituting your Recording Commitment in excess of the following number of units: 1% at 500,000 USA SoundScan units, and an additional 1% at 1,000,000 USA SoundScan units.

c. The royalty rate on Records sold for distribution through normal retail distribution channels outside of the United States shall be the following percentages of the Basic U.S. Rate, applied to the applicable Royalty Base Price, and, if sold by a licensee not owned or controlled by Company, shall be paid to Artist upon the same number of Records for which Company is paid:

Territories Percentage of Basic U.S. Rate	
Canada	75%
EU, Australia, New Zealand, and Japan	66.67%
Rest of the World	50%

d. With respect to Records sold by Company through mail order or a record club or in conjunction with a television advertising campaign, and for Records sold other than through normal retail distribution channels, Artist's royalties shall be computed at fifty percent (50%) of the applicable rate set forth above. With respect to Records licensed by Company for sale through mail order or a record club or licensed by Company for sale in conjunction with a television advertising campaign, and with respect to Company's licenses of the Masters or Video Masters, Company shall credit Artist's royalty account with fifty percent (50%) of Company's Net Royalty Receipts, or fifty percent (50%) of the Net Amount Received by Company, as applicable, from such sales and licenses. No royalties shall be payable with respect to Records received by members of any Club Operation in an introductory offer in connection with joining it or as a result of the purchase of a required number of Records including, without limitation, Records distributed as "bonus" or "free" Records, or Records for which the Club Operation is not paid.

e. On Records sold as premium merchandise, Artist's royalties shall be computed at fifty percent (50%) of the applicable rate and the Royalty Base Price shall be the price the distributing record company receives for the premium Record.

f. Notwithstanding any provision to the contrary herein contained and without limiting any of Company's rights hereunder, Company shall have the right to license Masters for all types of use (visual and non-visual) on a flat fee or royalty basis, in Company's discretion, for any uses referred to in this subparagraph, and as to any such license Company may credit to Artist's royalty account, in lieu of any other royalty, fifty percent (50%) of the Net Royalty Receipts from that license.

g. Notwithstanding anything to the contrary contained herein, the royalty rate on any Record in a New Technology Configuration shall be seventy-five percent (75%) of the otherwise applicable royalty rate. Audio-only compact discs are not a New Technology Configuration. It is specifically acknowledged that Company's actual out-of-pocket costs incurred directly in connection with the development and production (but not the manufacturing) of Records hereunder in any

New Technology Configuration shall constitute Recording Costs.

h. With respect to Records sold directly to consumers by Company in the United States or by a Principal Licensee outside the United States, other than by the methods described in subparagraph 8(d) (e.g., without limitation, telephone, satellite, cable, direct transmission over wire or through the air, and online computer sales) (collectively, "Direct Transmissions"), the royalty rate shall be seventy-five percent (75%) of the royalty rate that would otherwise apply if the Record concerned was sold through Normal Retail Channels. With respect to Records licensed by Company for sale directly to consumers by means of a Direct Transmission, Company shall credit Artist's royalty account with fifty percent (50%) of Company's Net Royalty Receipts, or fifty percent (50%) of the Net Amount Received by Company, as applicable, from such licenses.

i. The royalty rate on a Budget Record or on any Record sold to the United States government, its subdivisions, departments, or agencies, through military exchange channels, or to educational institutions or libraries shall be fifty percent (50%) of the applicable royalty rate set forth above; on a Mid-Price Record, seventy-five percent (75%) of the applicable royalty rate set forth above.

j. No royalties shall be payable to Artist in respect of Records sold or distributed for promotional purposes; as surplus, overstock, or scrap; as cutouts after the listing of such Records has been deleted from the catalog; as "free," "no charge," or "bonus" Records other than Albums; for Records distributed to radio stations; or for Records distributed for use on transportation carriers or for use in juke boxes. As to Records sold at a discount to "one-stops," rack jobbers, distributors, or dealers, whether or not affiliated with Company, in lieu of the Records given away or furnished on a "no charge" basis as provided above, the applicable royalty rate otherwise payable hereunder with respect to such Records shall be reduced in the proportion that said discount wholesale price bears to the usual stated wholesale price.

k. If any Master is recorded by an Artist jointly with another artist or musician to whom Company is obligated to pay a royalty for the Master, Artist's applicable royalty rate therefor shall be divided by the number of persons (including Artist) to whom Company is obligated to pay a royalty in respect of the Master. For purposes of the immediately preceding sentence, a group of artists to whom Company is obligated to pay one "all-in" royalty shall constitute one person.

l. The royalties provided for in this Agreement are inclusive of all royalties payable to producers and all third parties (other than Mechanical Royalties) for sale of Records and use of the Masters.

m. As to a Record not consisting entirely of Recordings delivered hereunder, the royalty to be paid hereunder shall be prorated in the proportion the Masters (or Video Masters, as applicable) that are embodied on the Record bear to the total number of royalty-bearing master recordings (or Video Masters, as applicable) embodied on that Record.

Accounting

a. Statements as to royalties payable hereunder shall be sent by Company to Artist on or before the September 30th and March 31st for the respective semiannual period ending the preceding June 30th and December 31st, together with payment of accrued royalties due to Artist, if any, on sales and licenses for which Company has received payment by the end of the semiannual accounting period involved. No statements need be rendered by Company for any accounting period after the expiration of the Term for which there are no sales, or other exploitations of Records derived from Masters hereunder. Company may withhold from Artist's royalty account a reasonable reserve against anticipated returns, rebates and credits. With respect to sales of singles and other Record configurations, such reserves shall be held in accordance with Company's reasonable business judgment. Company will liquidate such reserves within four (4) accounting periods following the period in which such reserves were initially established.

b. Artist shall be deemed to have consented to all royalty statements rendered by Company to Artist, and they shall not be subject to any objection by Artist for any reason, unless specific objection in writing, stating the basis thereof, is given by Artist to Company within two (2) years from the date the statement is rendered to Artist (the "Objection Period"). Any action, suit, or proceeding action relating to any royalty statement (rendered hereunder) must be commenced by the Artist within one (1) year after expiration of the Objection Period for that statement.

c. Company shall maintain books of account concerning the sale, distribution, and exploitation of Records and Masters made hereunder. A certified public accountant or attorney on Artist's behalf may, at Artist's expense, once per calendar year, examine Company's books as same pertain to the sale, distribution, and other exploitation of Records hereunder, at Company's office, during usual business hours and upon no less than thirty (30) days' prior written notice. Company's books relating to activities during any accounting period may only be examined as aforesaid during the Objection Period.

d. Royalties hereunder for sales and licenses derived from sources outside the United States shall be computed in the national currency in which Company is paid therefor, shall be credited to Artist's royalty account hereunder at the same rate of exchange as Company is paid or credited, and shall be proportionately subject to any foreign income, withholding, added value, transfer, or comparable taxes which may be imposed upon Company's royalties for those sales and licenses. If Company cannot collect payment in the United States in U.S. Dollars, Company shall not be required to account to you for the sale, except as provided in the next sentence. Company shall, at your request and at your expense, deduct from the monies so blocked, and deposit in a foreign depository, the equivalent in local currency of the royalties which would be payable to you on the foreign sales concerned, to the extent such monies are available for that purpose, and only to the extent to which your royalty account is then in a fully recouped position. All such deposits shall constitute royalty payments to you for accounting purposes. To the extent possible, Company will allow you to select the foreign depository referred to in this paragraph (d).

e. Company shall have the right to deduct from any amounts payable to Artist hereunder (1) such portion thereof as may be required to be deducted by any governmental authority, and (2) any amount payable by Company with respect to Artist's royalties under any union or guild agreement applicable to the Records and Masters hereunder. Artist agrees to execute and deliver promptly to Company such forms and other documents that may be required in connection with this paragraph.

f. In the event Company makes any claim, or brings any action, suit, or proceeding, that recovers any sums with respect to any Master, and in the event royalties are payable to Artist under this Agreement from such sums received by Company, all costs and expenses of such recovery (including, without limitation, reasonable attorneys', accounting, and auditing fees and expenses) shall be deducted from the gross sums so recovered.

Notices

Except as otherwise specifically provided herein, all notices under this Agreement shall be in writing and shall be given by courier or other personal delivery, by overnight delivery by an established overnight delivery service (e.g., Federal Express, Airborne Express, DHL, or United Parcel Service), or by registered or certified mail, return receipt requested, at the applicable address below, or at a substitute address designated by written notice by the party concerned.

TO YOU AT: _____

WITH A COURTESY COPY TO: _____

TO COMPANY AT: _____

WITH A COURTESY COPY TO: _____

Notices shall be deemed given when mailed or deposited into the custody of an overnight delivery service for overnight delivery, or, if personally delivered, when so delivered, except that a notice of change of address shall be effective only from the date of its receipt.

Licenses for Musical Compositions

a. Artist hereby grants to Company an irrevocable license under copyright to reproduce each Controlled Composition for uses as contemplated hereunder. Company shall pay Mechanical Royalties on Controlled Compositions at 100% of the respective minimum statutory rates (without regard to playing time) for the United States and Canada, determined at the date effective on the earlier of (i) the date such Masters are Delivered to Company hereunder or (ii) the date such Masters are required to be Delivered to Company hereunder, as applicable.

b. There shall be an aggregate Mechanical Royalty cap of twelve (12) times such rate per Album; two (2) times such rate per Single; three (3) times such rate per Twelve-inch Single; and five (5) times such rate per EP. To the extent that the aggregate Mechanical Royalty rate would exceed the cap in any Record, the excess will be deducted from any and all monies payable hereunder including without limitation, the Mechanical Royalties payable for the Controlled Compositions recorded under this Agreement. If a Composition recorded hereunder is embodied more than once on a particular Record, Company shall pay Mechanical Royalties in connection therewith at the applicable rate for such Composition as though the Composition were embodied thereon only once. All Mechanical Royalties payable hereunder shall be paid on the basis of Net Sales of Records hereunder for which royalties are payable pursuant to this Agreement. No Mechanical Royalties shall be paid on Records described in subparagraph 8(j). Company may maintain reasonable reserves with respect to payment of Mechanical Royalties. Notwithstanding anything to the contrary contained herein, Mechanical Royalties payable in respect of Controlled Compositions for sales of Records for any use described in subparagraphs 8(d), 8(e), and 8(i) shall be seventy-five percent (75%) of the otherwise applicable rate in the United States or Canada, as the case may be. Any assignment of the ownership or administration of copyright in any Controlled Composition shall be made subject to the provisions hereof and any inconsistencies between the terms of this Agreement and mechanical licenses issued to and accepted by Company shall be determined by the terms of this Agreement. If any Single, Maxi-Single, EP, or Album contains Compositions which are not Controlled Compositions, you will obtain for Company's benefit mechanical licenses covering such Compositions on the same terms and conditions applicable to Controlled Compositions pursuant to this paragraph 11.

c. In respect of all Controlled Compositions, Company is hereby granted the irrevocable perpetual worldwide right to reprint the lyrics on the jackets, sleeves, and other packaging of Records derived from Masters hereunder. Company is hereby granted the irrevocable right throughout the Territory to recreate the title and/or lyrics to any Composition embodied in a Recording delivered hereunder in the so-called "text mode" of digital compact cassettes, interactive compact discs, or any other New Technology Configuration embodying such Recording, without payment to any person or entity. If Company is required to pay any monies for the exercise of any of the rights granted to it under this subparagraph 11(c), then Company shall have the right to demand reimbursement, therefore from you and Artist (and you and/or Artist shall immediately make such reimbursement) and/or the right to deduct such costs from any monies payable under this Agreement or any other agreement for Artist's services.

d. In respect of all Controlled Compositions performed in Videos, Company is hereby granted an irrevocable perpetual worldwide license to record, synchronize, and reproduce such Compositions in such Videos and to distribute and perform such Videos including, without limitation, all Audiovisual Records thereof, and to authorize others to do so. In addition, in respect of all Controlled Compositions performed in Webcasts, Company is hereby granted an irrevocable perpetual worldwide license to record, synchronize, and reproduce such Compositions in such Webcasts, and to distribute and perform such Webcasts and to authorize others to do so. Company will not be required to make any payment in connection with the uses referred to in the immediately preceding two sentences, and those licenses shall apply whether or not Company receives any payment in connection with those uses. Notwithstanding the immediately preceding sentence, following Company's full recoupment of all costs in connection with a Video, Company shall negotiate in good faith with you regarding a royalty to be paid prospectively with respect to Controlled Compositions embodied in such Video in connection with the commercial exploitation of such Video. Simultaneously with your and Company's selection of creative elements of each Video produced hereunder, you shall furnish Company with a written acknowledgment from the person(s) or entity(ies) controlling the copyright in each non-Controlled Composition to be embodied in any Video confirming the terms upon which said person(s) or entity(ies) shall issue licenses in respect thereof and in respect of Webcasts. Upon Company's request therefore, you shall cause said person(s) or entity(ies) to forthwith issue to Company (and its designees) licenses containing said terms and such other terms and conditions as Company (or its designees) may require. Royalties in connection with licenses for the use of non-Controlled Compositions pertaining to Videos and Audiovisual Records are included in the royalties set forth in paragraph 17 hereof. If the copyright in any Controlled Composition is owned or controlled by anyone else, you will cause that person, firm, or corporation to grant Company the same rights described in this paragraph 11, on the same terms.

Events of Default

a. In the event Artist fails to fulfill any of Artist's recording commitments hereunder in accordance with all of the material terms and conditions of this Agreement, then, in addition to any other rights or remedies available to Company, Company shall have the right, upon notice to Artist at any time prior to the expiration of the then current Contract Period (i) to terminate this Agreement without further obligation to Artist as to unrecorded or unfinished Masters or Video Masters, or (ii) to extend the then current Contract Period for the duration of such default plus one hundred and fifty (150) days, with the times for the exercise by Company of its options to extend the Term and the dates of commencement of subsequent Option Periods deemed extended accordingly. Company's obligations hereunder, other than the obligation to pay earned royalties to Artist, shall be suspended for the duration of any such default. The provisions of this subparagraph shall not result in an extension of the Term for a period in excess of the period permitted by applicable law, if any, for the enforcement of personal service contracts.

b. Company reserves the right, at its election, to suspend the operation of this Agreement for the duration of any of the following contingencies, if, by reason of any such contingency, Company is materially hampered in the performance of its obligations under this Agreement or its normal business operations are delayed or become impossible or commercially impracticable: Act of God, fire, catastrophe, labor disagreement, acts of government, its agencies or officers, any order, regulation, ruling, or action of any labor union or association of artists, musicians, composers, or employees affecting Company or the industry in which it is engaged, delays in the delivery of materials and supplies, or any other cause beyond Company's control. No suspension under this subparagraph shall exceed six (6) months during any Contract Period unless such contingency is industry-wide, in which event Company shall have the right to suspend the applicable Contract Period for the duration of such contingency.

c. If Company refuses without cause to allow you to fulfill your Recording Commitment for any Contract Period and if, not later than sixty (60) days after that refusal takes place, you notify Company of your desire to fulfill such Recording Commitment, then Company shall permit you to fulfill said Recording Commitment by notice to you to such effect, such notice to be given, if at all, within forty-five (45) days of Company's receipt of your notice. Should Company fail to give such notice as aforesaid, you shall have the option to terminate the Term by notice given to Company within thirty (30) days after the expiration of the forty-five (45) day period; on receipt by Company of such notice, the Term shall terminate and all parties will be deemed to have fulfilled all of their obligations hereunder except those obligations that survive the end of the Term (e.g., warranties, Re-Recording Restrictions, rights of approval, and obligation to pay royalties), and you shall have no other remedy for such refusal without cause by Company to allow you to fulfill your Recording Commitment.

Injunctive Relief

Artist expressly acknowledges that Artist's services hereunder are of a special, unique, and intellectual character which gives them a peculiar value, and that in the event of a breach or threatened breach by Artist of any term, condition, or covenant hereof, Company shall incur immediate irreparable injury. Artist expressly agrees that Company shall be entitled to injunctive and other equitable relief, as permitted by law, to prevent a breach or threatened breach of this Agreement by Artist, which relief shall be in addition to any other rights or remedies, for damages or otherwise, available to Company.

Collective Bargaining Agreements

During the Term, Artist warrants and represents that, if Company so requests, Artist shall become and remain a member in good standing of any labor unions with which Company may have agreements lawfully requiring such union membership, including, but not limited to, the American Federation of Musicians and the American Federation of Television and Radio Artists.

Warranties and Representations: Indemnities

a. You and Artist warrant and represent that: (i) Artist is over the age of eighteen (18) and neither you nor Artist is under any disability, restriction, or prohibition, whether contractual or otherwise, with respect to your right to execute this Agreement or your and Artist's rights to perform its terms and conditions. Without limiting the foregoing, you specifically warrant and represent that no prior obligations, contracts, or agreements of any kind undertaken or entered into by Artist, will interfere in any manner with the complete performance of this Agreement by Artist or with Artist's right to record any and all compositions hereunder.

You further warrant and represent that Company shall not be required to make any payments of any nature for, or in connection with, the rendition of your or Artist's services or the acquisition, exercise, or exploitation of rights by Company pursuant to this Agreement, except as specifically provided herein. As of the date hereof, Artist is not a resident of the State of California. Artist shall notify Company immediately in the event that Artist becomes a resident of the State of California. (ii) (1) Artist shall enter into a valid and binding contract with each producer (each, a "Producer Contract") prior to the rendering of each such producer's services hereunder. Each Producer Contract shall grant to Artist all rights necessary for Artist to fulfill all of Artist's obligations hereunder. Artist will fully and promptly perform its obligations under the Producer Contracts. Without limitation of the foregoing, Artist shall cause each producer to execute and deliver a document in the form of Exhibit A prior to commencement of recording of the applicable Master. If Artist does not enforce any of Artist's rights under a Producer Contract, Company may without limitation of Company's rights enforce such rights in Artist's name and/or the name of Company. If Artist breaches any Producer Contract, then Company may cure such breach on Artist's behalf and at Artist's expense. No modification of or amendment to a Producer Contract will be made which would directly or indirectly diminish any of Company's rights hereunder. Artist will upon Company's request furnish Company with a complete copy of any or all Producer Contracts and/or any modification of the Producer Contracts.

b. You shall be solely responsible for and shall pay promptly all monies becoming payable to Artist, each individual producer of Recordings hereunder and all other parties rendering services on or in connection with such Recordings, both in connection with such individuals' services and also the exploitation by Company and its designees of the results of such services; provided that Company shall, in accordance with all of the terms hereof, pay Mechanical Royalties becoming payable to the copyright proprietors of Compositions

embodied on Masters and monies required to be paid to the AFM. Music Performance Trust Fund and Phonograph Record Manufacturers Special Payments Fund in connection with the manufacture and sale of Records derived from Masters.

(i)   No materials, or any use thereof, will violate any law or infringe upon or violate the rights of any third party. "Material," as used herein, includes:

(ii)   all musical compositions and other intellectual property, embodied in the Masters,

(iii)   the name_____as used in connection with the Masters, and

(iv)   all other ideas, other intellectual property or elements contained in or used in connection with the Masters, or the packaging, sale, distribution, advertising, publicizing, or other exploitation of Records embodying the Masters.

(v)   No changes in the individuals comprising Artist will be made without Company's prior written consent. Neither you nor Artist shall have the right, so long as this Agreement is in effect, to assign Artist's professional name as mentioned on page 1 hereof or any other name(s) utilized by Artist in connection with Masters or to permit the use of said name(s) by any other individual or group of individuals without Company's prior written consent, and any attempt to do so shall be null and void and shall convey no right or title. You hereby warrant and represent that you are and shall be the sole owner of all such professional name(s), and that no other person, firm, or corporation has the right to use said name(s) or to permit said name(s) to be used in connection with Records, and that you have the authority to grant Company the exclusive right to use said name(s) in the Territory in accordance with all of the terms and conditions of this Agreement, and Company shall have the exclusive right to use said professional name as aforesaid.

(vi)   Except as otherwise specifically set forth in this Agreement, during the Term, Artist shall not perform for the purpose of making Records for anyone

555

other than Company for use in the Territory and neither you nor Artist shall authorize the use of Artist's name, likeness, or other identification for the purpose of distributing, selling, advertising, or exploiting Records for anyone other than Company in the Territory.

(vii) Artist shall not perform any Composition recorded hereunder for anyone other than Company for use in the Territory on Records for a period of (i) five (5) years after the initial date of release of the respective Record containing such selection or (ii) two (2) years after the expiration or other termination of the Term, whichever is later ("Re-recording Restriction").

(viii) The Masters made and/or Delivered hereunder shall be produced in accordance with the rules and regulations of the American Federation of Musicians, the American Federation of Television and Radio Artists, and all other unions having jurisdiction. All persons rendering services in connection with such Masters shall fully comply with the provisions of the Immigration Reform Control Act of 1986 and any other applicable laws.

(ix) There exist no previously recorded Recordings embodying Artist's performances except those that have been sold, transferred, or otherwise assigned to Company.

c. (i) Artist agrees to and does hereby indemnify, save, and hold Company harmless of and from any and all liability, loss, damage, cost, or expense (including without limitation reasonable attorneys' fees) arising out of or connected with any breach, threatened breach, or alleged breach of this Agreement or any claim which is inconsistent with any of the warranties or representations made by you or Artist in this Agreement, provided that such claim has been settled with Artist's consent, or has been reduced to final judgment. Artist agrees to reimburse Company on demand for any payment made or incurred by Company with respect to any liability or claim to which the foregoing indemnity applies. Pending final determination of any claim involving such alleged breach or failure, Company may withhold sums due Artist hereunder in an amount reasonably related to the amount of such claim.

If no action is filed within one (1) year following the date on which such claim was first received by Company, Company shall release all sums withheld in connection with such claim, unless Company, in its reasonable business judgment, believes an action will be filed thereafter. Notwithstanding the foregoing, if, after such release by Company of sums withheld in connection with a particular claim, such claim is reasserted, then Company's rights under this paragraph will apply in full force and effect. Artist shall have the right to participate in the defense of any action instituted on a claim for which Artist is responsible to indemnify Company, with counsel of Artist's choice and at Artist's expense; however, Company shall have the right at all times to maintain control of the conduct of the defense.

(ii) Notwithstanding anything to the contrary contained herein, Company shall have the right to settle without your consent any claim involving sums of Seven Thousand Five Hundred Dollars ($7,500) or less, and this indemnity shall apply in full to any claim so settled; if you do not consent to any settlement proposed by Company for an amount in excess of Seven Thousand Five Hundred Dollars ($7,500), Company shall have the right to settle such claim without your consent, and this indemnity shall apply in full to any claim so settled, unless you post a surety bond with a surety acceptable to Company in its sole discretion. The surety bond must name Company as the beneficiary and assure prompt, unconditional payment to Company of all expenses, losses, and damages (including without limitation costs and reasonable attorneys' fees) that Company may incur in connection with said claim.

Approvals

Whenever in this Agreement Artist's approval is required, it shall not be unreasonably withheld or delayed. Artist must send written notice of approval or disapproval within five (5) days after Artist receives Company's request for the approval. Failure to give due notice of approval or disapproval shall be deemed to be approval.

Videos

a. Company shall have the right to require Artist to perform at such reasonable times and places as Company designates for the production of Video Masters featuring Artist's performances of Compositions embodied in the Masters and Artist agrees to perform to the best of Artist's ability thereon. Company shall be the exclusive owner throughout the Territory in perpetuity of such Video Masters and all rights therein, including, without limitation, all copyrights and renewal of copyrights with respect thereto, and the right to use and exploit such Video Masters in any and all forms and media. Company will consult with Artist with respect to the budget, producer, director, and storyboard for each Video Master produced hereunder, but Company shall make the final decisions thereon.

b. Company shall pay all Video production costs incurred in connection with the Videos consistent with the approved production budget. All sums paid by Company in connection with the production of the Video Masters shall constitute Advances to be charged against and recouped from Artist's royalties (excluding Mechanical Royalties) under this Agreement, subject to the following sentence. Fifty percent (50%) of the aggregate amount of Video production costs shall not be recoupable from royalties payable pursuant to paragraph 8 above, but shall nevertheless be one hundred percent (100%) recoupable from Video royalties hereunder. All Video production costs in excess of the approved budget that have been incurred due to your or Artist's acts or omissions shall be your sole responsibility, and you hereby agree to forthwith pay and discharge all such excess costs. In the event that Company agrees to pay any such excess costs on your behalf, you shall, upon demand, reimburse Company for such excess costs or in lieu of requesting reimbursement, Company may deduct such excess costs from any monies otherwise due you under this Agreement or any other agreement between you or Artist and Company or its affiliates (including without limitation Mechanical Royalties). In the event that Artist fails to appear at locations and/or on dates which have been mutually approved by you and Company, without reasonable excuse, the costs of cancellation of the shoot shall be fully deductible from all monies payable to you under this or any other agreement between you or Artist and Company or its affiliates (including without limitation Mechanical Royalties).

c. In the event that Company decides to commercially release any Video Masters produced hereunder for sale, Company and Artist shall negotiate in good faith an artist royalty payable to Artist hereunder in connection with such sales of such Video Masters. Any such artist royalty negotiated with respect to sales of Video Masters hereunder is inclusive of any third-party payments (including without limitation any payments to music publishers with respect to non-Controlled Compositions) required to be made by Company in connection with the manufacture and commercial exploitation of Video Masters hereunder.

Marketing and Publicity

a. Artist shall be available from time to time to appear for photography, poster, and cover art, and the like, under the direction of Company or its nominees and to appear for interviews with representatives of the communications media and Company's publicity personnel.

b. Artist shall be available from time to time at Company's request to perform for the purpose of recording for promotional purposes by means of film, videotape, or other audio-visual media performances of Compositions embodied on Masters.

Group Provisions

Artist's obligations under this Agreement are joint and several. All references in this Agreement to "Artist" include all members of the Group, inclusively, and each member individually, unless otherwise specified. If any member of the Group ceases to perform as a member of the Group, that member will be deemed to be a "Leaving Member" and the following shall apply in that circumstance:

a. The remaining members of Artist will notify Company within thirty (30) days after a member has left, that a member has left, and the Leaving Member will be replaced by a new member, if Artist and Company so agree, to be substituted as a party to this Agreement in the place and stead of the Leaving Member. Artist agrees to cause the new member (and any new member joining the Group after the date hereof, even if not replacing a Leaving Member) to execute and deliver to Company such instruments as Company may require to accomplish that substitution (hereinafter collectively referred to as the "Substitution Instruments"). Thereafter, the Leaving Member will have no further obligation to perform under this Agreement but will continue to be bound by the other provisions of this Agreement, and Company will continue to have the right to use the name of the Group (and any other name hereafter used for the Group) pursuant to this Agreement. Artist agrees that no person will be permitted to perform in place of the Leaving Member under this Agreement unless that performer has executed and delivered to Company his/her Substitution Instruments. Company will continue to have the right to use the name of the Group (and any other name hereafter used for the Group). No Leaving Member will make any use of the Group's name in any circumstances, nor authorize or permit anyone other than Company to use or trade upon the Group's name for any purpose. For the avoidance of doubt, it is specifically agreed that the Leaving Member shall not make or authorize any reference to the Group's name for or in connection with the Leaving Member's performances, live or recorded or taped or filmed, nor make or authorize any reference in any packaging of phonorecords or videograms, any artwork or stickers used in connection with phonorecords or videograms, or any advertising, marketing, promotion, or publicity in connection with phonorecords or videograms.

b. Company will have the right to terminate the Term with respect to the remaining members of the Group by notice to be given to Artist at any time before expiration of ninety (90) days after Company's receipt of the notice referred to in subparagraph (a) above. In the event of such termination, all members of the Group will be deemed to be Leaving Members as of the date of such termination notice, and subparagraph (c) will apply to all of them, collectively or individually as Company may elect.

c. Artist grants to Company an option to engage the exclusive recording services of each Leaving Member (hereinafter referred to as a "Leaving Member Option"). Each Leaving Member Option may be exercised by Company at any time within ninety (90) days after Company's receipt of notice under subparagraph (a), or within ninety (90) days after the date of Company's termination notice pursuant to subparagraph (b) above, whichever shall be later. If Company exercises a Leaving Member Option, the Leaving Member concerned will be deemed to have entered into a new agreement with Company containing the same provisions as this Agreement, except as follows:

   (i) The new agreement will apply only to that Leaving Member, and all references to "Artist" in that new agreement will be deemed to refer to the Leaving Member;

   (ii) The term of the new agreement will commence on the date Company exercises that Leaving Member Option and may be extended by Company (exercisable in the same manner as provided in paragraph 2 above) for the same number of Option Periods that remained under paragraph 2 hereof at the time Company exercises that Leaving Member Option (but Company shall have at least 2 such Option Periods in any event);

   (iii) The Recording Commitment for each Contract Period of such Term shall be one (1) Album;

   (iv) If Artist's royalty account under this Agreement in an unrecouped position at the date Company exercises a Leaving Member Option, a pro-rata portion of the amount of that unrecouped balance will be recoupable from the royalties payable by Company under the new agreement. That portion shall be determined by a fraction, the numerator being one and the denominator being the number of members constituting the Artist prior to Company's receipt of the notice referred to in subparagraph (a) above (e.g., 25% of that unrecouped balance if there are 4 members of Artist and one of them becomes a Leaving Member for whom Company has exercised a Leaving Member Option). To the extent that unrecouped portion is recouped under the new agreement, it shall be credited toward recoupment under this Agreement, and to the extent that unrecouped portion is recouped under this Agreement it shall be credited toward recoupment of charges against royalties under the new agreement.

(v) As to any Leaving Member who has a songwriter's agreement or copublishing agreement with Company or any of Company's affiliates, the term of that agreement shall be coterminous with the new recording agreement, as same shall be substituted for, extended, and renewed.

Definitions

For the purposes of this Agreement, the following definitions and terms shall be:

a. *Advance*—a prepayment of royalties. Company may recoup Advances from all royalties to be paid or accrued to or on behalf of Artist pursuant to this Agreement. Mechanical Royalties shall not be chargeable in recoupment of any Advances unless otherwise expressly provided.

b. *Album*—one (1) or more audio-only Records, at least forty-five (45) minutes in playing time and embodying at least ten (10) Masters of different compositions sold in a single package.

Mid-Price Album or Record—a Record which is sold by Company or its Principal Licensee(s) at a price that is below Company's or the applicable Principal Licensee's then-prevailing top-line suggested retail list price, which price is consistently applied by Company to such Records and which Records are sold by Company or its Principal Licensee(s) as mid-priced Records.

Multiple Record Set—two or more Records packaged and/or marketed as a single unit.

Budget Album or Record—A Record which is sold by Company or its Principal Licensee(s) at a price which is below Company's or the applicable Principal Licensee's then-prevailing top-line suggested retail list price, which price is consistently applied by Company to such Records and which Records are sold by Company or its Principal Licensee(s) as budget Records.

Single—a vinyl, audio-only Record not more than seven (7) inches in diameter, or the equivalent in non-vinyl configurations.

Twelve-Inch Single—an audio-only record, which contains not more than four (4) Recordings of different compositions.

Extended Play Record or EP—an audio-only Record embodying thereon either five (5) Masters or six (6) Masters, but does not constitute an Album.

Audiovisual Record—a record that embodies, reproduces, transmits, or otherwise communicates visual images, whether or not the interaction of a consumer is possible or necessary for the visual images to be utilized or viewed.

c. *Base Price*—for Records and Audiovisual Records, the Suggested Retail List Price. "Royalty Base Price" is the Base Price less all excise, sales and similar taxes included in the Base Price and less the applicable Container Charges and less distributors' discounts and rebates, if any.

d. *Container Charge*—the applicable percentages of the Base Price specified below for sales by Company:

(i) Analog cassette Records—twenty percent (20%);

(ii) Compact disc Records and Records in all other configurations—twenty-five percent (25%);

(iii) Analog vinyl Records—twenty-five percent (25%); and

(iv) Electronic phonorecord delivery—no deductions.

e. *"Controlled Composition"*—a composition wholly or partly written, or directly or indirectly owned or controlled, by Artist or a producer of Masters or by any representative of Artist or a producer hereunder.

f. *"Digital Download"*—a type of record by which a Master is transmitted or otherwise communicated to a consumer on demand via digital distribution over the Internet which results in that particular Record being placed in storage on that consumer's computer or other device such that the consumer owns the Record in perpetuity (as opposed to a limited rental or license). Company will not deduct a Container Charge from your portion of Digital Download royalties.

g. *"Delivery," "deliver," or "Delivered"*—the actual receipt by Company of completed, fully mixed, leadered, and edited Masters comprising the applicable Recording Commitment, technically and commercially satisfactory in Company's opinion and ready for the manufacture of Records, together with all materials, consents, approvals, licenses, and permissions Artist is required to supply to Company hereunder.

h. *"Master"*—an individual sound Recording recorded hereunder that embodies Artist's performance(s) as the featured recording artist(s).

"Video" or "Video Masters"—videocassettes, Videodiscs, or any other devices, now or hereafter known or developed (including, without limitation, any visual and/or audiovisual work using digital technology), that enable motion pictures and other audiovisual works that have a soundtrack substantially featuring the performances of Artist to be perceived visually, with sound, when used in combination with or as part of a piece of electronic, mechanical, or other apparatus.

i. *"Mechanical Royalties"*—royalties payable to any person for the right to reproduce and distribute copyrighted compositions on Records, including, without limitation, Audiovisual Records.

j. *"Merchandise Rights"*—any use, reproduction, or other exploitation in any manner, media, or formats of the name, photographs, likenesses, biographical material, trademarks, and/or any other identification utilized by you (other than in connection with Records hereunder), including, without limitation "fan clubs," any products, endorsements, sponsorships, and/or the sale of merchandise (e.g., T-shirts and other apparel, caps, posters, biographical books).

k. *"Net Amount Received"*—the gross, earned, non-returnable amount received by Company (excluding royalties that are based upon a percentage of a retail or wholesale or other price and excluding advances against such royalties), less any out-of-pocket costs or expenses which Company is contractually required to make to third parties, including, but not limited to, payments to a trustee or fund to the extent required by any agreement between Company and any labor organization or trustee, and less Mechanical Royalty payments. An example of a Net Amount Received is an earned payment (not an advance) under a flat fee license or flat rate license.

l.   *"Net Royalty Receipts"*—company's gross, earned royalty receipts less any out-of-pocket costs or expenses which Company is contractually required to make to third parties, and less any amounts included in the receipts that are for payments to a trustee or fund required by any agreement between Company or Artist and any labor organization or trustee, and less Mechanical Royalty payments.

m. *"Net Sales"*—eighty-five (85%) percent of Gross Sales of Albums, and seventy (70%) percent of Gross Sales of Singles and EPs, sold, paid for, and not returned. "Free goods" that are given by Company or Company's distributor in lieu of discounts shall be included in "Gross Sales" under this paragraph, except for additional free goods given by Company or its distributor pursuant to special "impact" marketing programs of limited duration.

n.   *"New Technology Configurations"*—records in the following configurations: mini-discs, digital compact cassettes, digital audio tapes, DVD, laser discs, CD-ROM, and other Records embodying, employing, or otherwise utilizing any non-analog technology (whether or not presently existing or hereafter created or developed), but specifically excluding audio-only compact discs.

o.   *"Record"*—any form of reproduction, transmission, or communication of Recordings, now or hereafter known, manufactured, distributed, transmitted, or communicated primarily for home use, school use, juke box use, or use in means of transportation, including, without limitation, Records embodying or reproducing sound alone and Audio-visual Records.

p.   *"Recording"*—every recording of sound, whether or not coupled with a visual image, by any method and on any substance or material, or in any other form or format, whether now or hereafter known, which is used or useful in the recording, production, and/or manufacture of Records or for any other commercial exploitation.

q. *"Recording Costs"*—all amounts paid or incurred in connection with the production of Masters hereunder. Recording Costs include, without limitation, all union scale payments required to be made to the Artist in connection with Masters recorded hereunder, all costs of instrumental, vocal, and other personnel in connection with the recording of the Masters, travel, rehearsal, and equipment rental expenses, per diems, advances to producers, studio and engineering charges and personnel, all other amounts required to be paid pursuant to any applicable law or any collective bargaining agreement between Company and any union representing persons who render services in connection with the Masters, and all costs of mixing, remixing, mastering, and re-mastering. Recording Costs do not include the costs of producing metal parts, but include all studio and engineering charges or other costs incurred in preparing Masters for the production of metal parts and/or final production Masters. (Metal parts include lacquer, copper, and other equivalent masters.)

r. *"Suggested Retail List Price"*—(i) with respect to Records sold for distribution in the United States, Company's suggested retail list price in the United States during the applicable accounting period for the computation of royalties to be made hereunder, it being understood that a separate calculation of the suggested retail list price shall be made for each price configuration of Phonograph Records manufactured and sold by Company; and (ii) with respect to Records sold hereunder for distribution outside the United States, Company's or its licensees' suggested or applicable retail price in the country of manufacture or sale, as Company is paid, or, in the absence in a particular country of such suggested retail list price, the price as may be established by Company or its licensee(s) in conformity with the general practice of the recording industry in such country, provided that Company shall not be obligated to utilize the price adopted by the local mechanical copyright collection agency for the collection of Mechanical Royalties. Notwithstanding anything to the contrary contained herein, (A) the Suggested Retail List Price for premium Records shall be Company's actual sales price of such Records and (B) the Suggested Retail List Price with respect to Videos manufactured and distributed by Company shall be Company's published wholesale price as of the commencement of the accounting period in question.

s. *"Territory"*—the Universe.

t. *Sales "through normal retail distribution channels"*—sales made to retail stores, or for resale to retail stores, through Company's record distributors, or, outside the United States, through Company's foreign licensees, or by Company itself, to retail stores or for resale to retail stores.

Assignment

Company shall have the right to assign this Agreement in whole or in part to any subsidiary, parent company, or affiliate of Company or to any third party acquiring all or a substantial portion of Company's assets or equity interests or pursuant to an initial public offering. Neither you nor Artist shall have the right to assign this Agreement.

Assignment of Publishing Interest in Recorded Songs

With respect to any and all musical compositions written in whole or in part by any or all members of Artist which is recorded and commercially released on any Master or Album recorded hereunder (a "Recorded Composition"), each such member of Artist hereby assigns and transfers to Company or its publishing designee, its successors and assigns, fifty percent (50%) of all of such Artist member's right, title, and interest in and to each and every Recorded Composition, or portion thereof, as described above, for the entire Universe, including but not limited to such Artist member's interest in the copyrights therein and any renewals or extensions thereof. Each

Artist member agrees, upon Company's request, to execute a publishing agreement with Company's affiliated publishing company with respect to each such Recorded Composition, promptly following the recording thereof by Artist and commercial release of same on any Master or Album hereunder. The parties agree that Company's affiliated publishing entity will administer and license 100% of such Recorded Composition.

Merchandising

In consideration for Company's development of your name, brand, identity, and services as a recording and touring artist, you grant Company exclusive Merchandise Rights during the term of this Agreement. After deducting from gross income received from the sale of merchandise, the costs of manufacturing, shipping, and fulfillment, costs associated with a merchandise salesperson (if required) such as per diems, salary, transportation and lodging, insurance, venue or hall fees, product design fees, and other costs customarily associated with selling artist merchandise, the parties shall equally divide all net receipts. Company shall account to you monthly for your portion of net receipts, which shall not be subject to recoupment of Advances hereunder, other than advances that specifically constitute prepayment to you of a portion of your merchandise net receipts. Company, or its designee, shall maintain books and records for the purpose of verifying the accuracy of statements rendered by Company pursuant to this paragraph, which books and records are available for your examination upon reasonable notice and at Company's principal office.

Confidentiality

Each of Artist and Company agree to keep strictly confidential the monetary terms of this Agreement; provided however that this restriction shall not apply to information which:

a.   is required to be disclosed by law or by any order, rule, or regulation of any court or governmental agency; paragraph 25.

b.   is disclosed to Artist's or Company's attorney(s) or manager(s) in the ordinary course of business so long as such persons have agreed to be subject to restrictions identical to those imposed upon Company and Artist under this provision.

Miscellaneous

a.   This Agreement sets forth the entire agreement between the parties with respect to the subject matter hereof. No modification, amendment, waiver, termination, or discharge of this Agreement shall be binding upon Artist or Company unless confirmed by a written instrument signed by an authorized person of Company and Artist. A waiver by either party of any term or condition of this Agreement in any instance shall not be deemed or construed as a waiver of such term or condition for the future, or of any subsequent breach thereof. All of Company and Artist's rights, options, and remedies in this Agreement shall be cumulative and none of them shall be in limitation of any other remedy, option, or right available to Company. Wherever possible, each provision hereof shall be interpreted in such manner as to be effective and valid under applicable law, but in case any one or more of the provisions of this Agreement shall, for any reason, be held to be invalid, illegal, or unenforceable in any respect, that provision shall be ineffective to the extent, but only to the extent, of such invalidity, illegality, or unenforceability without invalidating the remainder of that provision or any other provisions of this Agreement. It is agreed that all accountings and royalty payments required herein, and all grants made herein, shall survive and continue beyond the expiration or earlier termination of this Agreement. No breach of this Agreement by Company or Artist (except for Artist's breach of the exclusivity provisions hereof) shall be deemed material unless notice is given specifying the nature of the breach and the recipient of the notice fails to cure such breach, if any, within thirty

(30) days after receipt of the notice; notwithstanding the foregoing, Artist expressly acknowledges that Company may obtain injunctive relief hereunder immediately, and Company is not required to delay for thirty (30) days or any other period.

b. This Agreement is made in the State of Tennessee and its validity, construction, and performance shall be governed by the laws of the State of Tennessee applicable to agreements made and to be entirely performed in Tennessee, without regard to any conflicts of laws principles. The Federal and State courts in Davidson County shall have exclusive jurisdiction of any dispute arising under or concerning this Agreement. Service of process pursuant to this paragraph may be made, among other methods, by delivering the same via overnight mail or mailing by certified air mail, return receipt requested, in the same manner as giving other notices under this Agreement, and shall be effective upon sending the process. Such service is deemed to have the same force and effect as personal service within the State of Tennessee.

c. This Agreement has binding legal effect, and grants certain exclusive rights to Company for Artist's services. Artist acknowledges that Company has requested Artist to consult with and be represented by an attorney who is knowledgeable about the subject of this Agreement and the record and music and entertainment industries, to be advised about the content and effect of the provisions of this Agreement, and to follow Artist's attorney's advice about entering into this Agreement.

d. This Agreement shall not become effective until it is executed by all parties.

IN WITNESS WHEREOF, the parties hereto have executed this Agreement on the day and year first above written.

COMPANY: _____

By: _____

ARTIST: _____

By: _____

Date: _____

The Personal Management Agreement

This AGREEMENT (the "Agreement") made and entered into the _____ day of _____, _____ by and between _____ (referred to as "Artist" or "your"), whose address is _____ and _____ ("Manager"), whose address is _____.

WHEREAS, Artist wishes to obtain Manager's exclusive advice, guidance, counsel, and direction to promote and develop Artist's Career as defined herein; and

WHEREAS, Manager wishes to provide such services on the terms and conditions set forth herein.

NOW, THEREFORE IN CONSIDERATION of the mutual covenants and agreements set forth below, and other good and valuable consideration, the receipt and sufficiency of which are hereby acknowledged, the parties agree as follows:

Term

The initial term of this Agreement (the "Initial Term") shall be for a period of two (2) years commencing on the date first written above, and terminating as of the end of the last day of the second year thereafter. Artist hereby grants to Manager three (3) additional one-year option periods (each year referred to as the "Option Period") automatically exercisable by Manager unless Manager on or before the end of the preceding Initial Term and/or Option Period, whichever is applicable, notifies the Artist in writing of its desire to terminate this Agreement.

Personal Manager's Services

Throughout the Term of this Agreement, Manager agrees

- To advise, guide, direct, and counsel Artist in any and all matters pertaining to employment, publicity, public relations, advertising, the selection of musical material, and all other matters pertaining to Artist's Career which are not specifically excepted herein.

- To advise, guide, direct, and counsel Artist with relation to the adoption of the proper format for presentation of Artist's talents in the determination of proper style, mood, setting, and characterization in keeping with Artist's talent and best interest.
- To advise, guide, direct, and counsel Artist in the selection of artistic talents to assist, accompany, or embellish Artist's artistic presentation.
- To advise, guide, direct, and counsel Artist concerning compensation and privileges for his talent and similar artistic talent.
- To advise, guide, direct, and counsel Artist concerning the selection of booking agencies, artists' agents, artists' managers, and persons, firms, and corporations who will counsel, advise, seek, and procure employment and engagements for Artist.
- As pertinent to Artist's Career, to advise, guide, direct, and counsel with regard to general practices in the entertainment, music, and recording fields and with respect to compensation and terms of contracts related thereto.
- To advise, guide, direct, and counsel Artist regarding hiring publicists, marketing consultants, advertising agencies, and similar consultants and service providers hired to further Artist's Career.
- To meet with Artist when reasonably requested by Artist.

For purposes of this Agreement, "Artist's Career" shall mean and refer to Artist's career worldwide in the entertainment and related businesses, including without limitation, work done by Artist in the recording, acting, literary, theatrical, music publishing, music composing, personal appearance, advertising, entertainment, amusement, music, music performance, music video, television, radio, motion picture, motion picture sound track, commercials, endorsements, video, Internet and merchandising fields, and otherwise related to Artist's career in the entertainment field, as now known or hereinafter devised in which Artist's artistic talents and/or name, voice, likeness, and/or public image are developed and exploited.

Non-Exclusivity and Territory

Manager's services under this Agreement are nonexclusive. Manager shall at all times have the right to render the same or similar services to others whose talents may be similar to or may be in competition with Artist, as well as engage in any and all other business activities; however, Manager agrees to be reasonably available to render the services to Artist hereunder. Artist's engagement of Manager under this Agreement is exclusive throughout the world.

Agencies and Publicity

Artist may from time to time enter into agreements with talent agencies, theatrical agencies, and employment agencies whose function and obligation shall be to procure employment and engagements for Artist. Any compensation (which Artist may be required to pay to these agencies) shall be at the sole cost and expense of Artist. If requested by Artist, Manager agrees to supervise and screen the selection and activity of such agency. All expenses of persons of companies specifically retained by Artist to do publicity, public relations, or other work on behalf of Artist shall be at the sole cost and expense of Artist.

Manager's Authority

Consistent with the services Manager is engaged to provide as set forth herein, and subject to any specific limitations set forth in this Agreement:

Artist hereby grants to Manager the right to approve and permit any and all publicity and advertising and to approve and permit the use of Artist's name, pre-approved photograph(s), pre-approved likeness(es), pre-approved voice, pre-approved sound effects, pre-approved caricature(s), and pre-approved artistic and musical materials, for purposes of advertising and publicity in the promotion and advertising of Artist's products and services. Manager agrees to consult with Artist about tour plans and other activities on a regular basis. To the extent the foregoing requires Artist's signature or written approval, Manager shall reasonably attempt to obtain Artist's signature directly or via facsimile (considering Artist's actual availability and the turnaround time with respect to the particular matter concerned), or Manager shall attempt to obtain Artist's written approval via email which, solely if granted by Artist, shall confer in Manager the limited power of attorney to render the corresponding approval on behalf of Artist in respect of the particular matter concerned. In the event Manager makes any such decision with which Artist disagrees, then Artist's decision controls and the parties agree to use their best efforts to implement Artist's preference, as may be practical and reasonable under the circumstances then existing. Notwithstanding any other provision contained in this Agreement, Manager does not have the authority to execute or approve, and shall not execute or approve (pursuant to the authority granted to Manager in this paragraph 5 or otherwise) on behalf of Artist any recording, music, or book publishing, booking agency, video, endorsement, merchandising, or any other contract or agreement of any kind whatsoever, whether written or oral. Any such contract or agreement so executed or approved by Manager shall be null and void, and Manager indemnifies Artist with respect thereto. The authority granted to Manager pursuant to this paragraph 5 above is coupled with an interest and shall be irrevocable during the Term of this Agreement.

Receipt of Artist's Compensation

Artist shall engage at its expense a business manager or appropriate bookkeeping personnel to receive all Gross Income for Artist and shall account and remit to Manager in each month when there is either Compensation or expenses due Manager out of Gross Income including the period provided for in Section 7(c) below. Such business manager or bookkeeping personnel shall be available to Manager during regular business hours to confirm the receipt of monies and the payment of expenses. At the request of Artist, Manager shall assist Artist in the selection of a business manager or bookkeeping personnel. Manager and Artist (or Artist's business manager or appropriate bookkeeping personnel, as the case may be) shall keep accurate and complete books of account and records with respect to all amounts received by Artist and Manager in connection with Artist's Career, which books may be inspected during regular business hours, by a certified public accountant designated either by Artist or Manager, upon reasonable notice to the other. Both Artist and Manager may audit up to three years at a time but may audit any given year only once. Each quarter during the Active Term and the period provided for in Section 7(c) below, Artist shall cause to be delivered to Manager a full statement of account showing the monthly Gross Income accruing to and received by Artist and any compensation and expense reimbursement due to Manager along with a payment for any additional amount due, if any, Manager may request, during the Active Term, monthly internal statements from Artist's business manager or appropriate bookkeeping personnel and other information upon reasonable request.

Manager's Compensation

Since the nature and extent of the success or failure of Artist's Career cannot be predetermined the parties desire that Manager's compensation be determined in such manner as will permit Manager to accept the risk of failure and to benefit from Artist's success. Therefore, as compensation for Manager's services, Artist agrees to pay Manager during and through the Term of this Agreement (and any modifications, extensions, replacements, renewals, or substitutions of this Agreement), the following in consideration of all of Manager's services hereunder ("Fee"): a sum equal to fifteen percent (15%) of all Gross Income (as defined herein in the attached and incorporated Glossary) received or accrued (as of the last day of the Term) directly or indirectly by Artist or by any other person or entity on Artist's behalf, including without limitation, Artist's heirs, executors, administrators, and assigns, regarding Artist's Career.

a. For purposes of this Agreement, "Gross Income" shall include, without limitation but subject to the exclusions stated in paragraph 7 of the Agreement, all of the following relating to Artist's Career: salaries, earnings, fees, royalties, advances against royalties, merchandise, gifts (excluding bona fide gifts as defined by the Internal Revenue Code), bonuses, shares of profits, shares of stock, residuals, repeats or return fees, partnership interests, share of profits, percentages, and the total amount paid for a package television or radio program (live or recorded) or other entertainment package, received directly or indirectly by Artist or by any other person or entity on Artist's behalf, including the aggregate amount paid to bands led by Artist (but only for Artist's services in connection therewith). It is understood that, for purposes hereof, no expense, cost, or disbursement paid or incurred by Artist in connection with the receipt or earning of "Gross Income" (including salaries, shares of profits, or other sums paid to individuals participating in Artist's presentations, recordings, videos, or other forms of performances) shall be deducted therefrom prior to the calculation of Manager's compensation hereunder except only for the items excepted in paragraph 7(b) above. In the event that Artist receives, as all or part of his compensation for activities within Artist's Career, stock or other equity interests, or the right to buy stock or other equity interests, in any corporation or entity or that Artist becomes the packager or owner of all or part of an entertainment property, whether as an individual proprietor, stockholder, partner, joint venture, or otherwise, Manager's Compensation shall apply to said stock or equity, right to buy stock or equity, individual proprietorship, partnership, joint venture, or other forms of interest, and Manager shall be entitled to Manager's Compensation thereof. Should Artist be required to make any payment for such interest, Manager will pay his percentage share of such payment, except if Manager does not want his Fee thereof, the reasonable value of such interest (taking into account Artist's payment therefore) shall be deemed to be Gross Income for the purposes of this Agreement.

b. The following are specifically excluded from Gross Income: (i) tour support funds to the extent such funds cover reasonably incurred actual tour losses (or any tour support funds to the extent actually paid for or incurred on behalf of Artist by any record company or tour/live performances sponsor to cover tour/live performance losses); (ii) funds or other considerations, which are received by Artist or on Artist's behalf and which are subsequently refunded to the payee (example: deposits received by booking agencies on live engagements which are not subsequently performed by Artist, and therefore the deposit must be refunded); (iii) funds which are paid by Artist or by third parties on Artist's behalf to third parties which are expended on actual video cost, recording costs (including, but not limited to, those costs, expenditures, and advances remitted or defined as recording costs or the equivalent thereof in any recording agreement entered into by Artist), and production costs, writers and publishers, producer, mixers, engineers, studio facilities, and salaried musicians; (iv) sound and light, opening acts, support acts, each with respect to live performances or other personal appearances by Artist; (v) actual publisher, subpublisher, and administrator fees under agreements related to music publishing income; (vi) the expenses of Artist for engaging musical groups other than Artist in the entertainment industry to the extent Artist is a "packager" of an entertainment property but such expense shall not reduce Gross Income by more than the Gross Income created by such entertainment package unless Manager has approved such package arrangement in advance in writing (such approval not to be unreasonably withheld); (vii) union pension and welfare plan benefits paid to Artist and on Artist's behalf with respect to recording sessions; (viii) verifiable reimbursed expenses; (ix) monies earned by Artist from passive investments; (x) gifts received by Artist to the extent not in lieu of compensation; and (xi) loans to Artist.

c. After the Term, Manager shall be paid, in lieu of the Fee, a sum (the "Post-Term Commission") equal to fifteen (15%) percent of all Gross Income (except for items excluded under the paragraph 7 above) received or accrued by Artist for the first five (5) years after the Term has ended, which Post-Term Commission shall be reduced to ten (10%) percent for the five (5) years thereafter, and then further reduced to five (5%) percent for the next five years. For purposes of clarity,

Manager shall not be entitled to any Post-Term Commission (or other compensation of any kind) beyond fifteen (15) years after the termination of this Agreement.

Personal Manager's Expenses

In addition to the compensation provided in paragraph 7 above, Artist shall reimburse Manager for all reasonable expenses necessarily and actually incurred on Artist's behalf directly in connection with Manager's rendition of services under this Agreement during the Term including, but not limited to, overnight, certified or registered mail postage long distance telephone calls, facsimile transmissions, and travel expenses, but excluding office expenses and overhead. No Individual expense in excess of $500 shall be incurred by Manager without first having obtained Artist's prior written consent. Reimbursable expenses in any month shall not exceed an aggregate of $1,000 without Manager having first obtained Artist's prior written consent. Expenses shall only be reimbursed out of Gross Income actually earned by Artist from Artist's Career. Artist agrees that, in the event that Manager renders any services under this Agreement to Artist outside of the metropolitan area in which Manager's office is located, Manager shall be entitled to reimbursement from Artist for reasonable travel expenses (subject to applicable pre-approval hereunder). In the event that Manager must travel by air, Manager agrees to purchase airline tickets at the lowest fare available at the time of booking and, when applicable, for the same "class" that Artist travels (further subject to applicable pre-approval hereunder). All expenses charged or incurred by Manager shall be prorated with respect to Artist if Manager is simultaneously rendering services for or on behalf of other clients or parties. Subject to the foregoing, reimbursement of expenses shall be due Manager within thirty (30) days after receipt of itemized statements setting forth the nature and amount of each expense. During the Term of this Agreement and for a period of (3) years following, Manager agrees to keep and maintain reasonable documentation of each expense from which he requested reimbursement by Artist and all documentation shall be provided to Artist or his representative promptly after Artist's reasonable request.

Expenses, Loans, and Advances

Except as otherwise stated herein, Artist shall be solely responsible for the payment of all expenses which may arise in connection with Artist's Career, including, without limitation the cost of material, equipment, facilities, transportation, lodging and living expenses, costumes, makeup, promotion, publicity, accounting, and legal fees, and Manager shall not have any liability whatsoever in such connection and therefore, such expense shall not reduce Gross Income subject to Manager's commission hereunder except only as set forth in paragraph 7 above. Manager shall not be obligated to lend to or advance money to Artist, but if it does so, Artist shall repay such amount(s) promptly from Gross Income (and only from Gross Income); Manager agrees to provide Artist a written tabulation of any such loan or advance actually made upon request by Artist.

Offers of Employment

Artist shall advise Manager of all offers of employment related to Artist's Career submitted to Artist and will refer all inquiries concerning Artist's Career to Manager, so that Manager may advise Artist whether they are compatible with Artist's Career. Reciprocally, Manager shall inform Artist of any offers or prospective employment opportunities of or concerning Artist's Career which Manager receives.

Career

Artist agrees at all times to attend to Artist's Career, and to do all things reasonably necessary and desirable and consistent with this Agreement to promote Artist's Career and earnings. With Manager's advice, guidance, direction, and counsel, Artist shall at all times endeavor to engage and utilize proper theatrical agents and booking agents to obtain engagements and employment for Artist.

Notice of Breach

In order to eliminate misunderstandings between the Artist and Manager, Artist and Manager agree to advise each other in writing of the specific nature of any claimed breach of this Agreement. The party receiving such notice shall have thirty (30) days after the receipt of the notice in which to cure the claimed breach. The written notice shall be deemed a condition precedent to the commencement of any arbitration and shall be sent by certified mail, return receipt requested. Notice to Artist shall be given to the address set forth in the first paragraph of this Agreement, with a copy to Artist's legal counsel, _____. Notice to Manager shall be given to the address set forth in the first paragraph of this Agreement, with a copy (which shall not constitute notice) to Manager's legal counsel, _____.

Business Entities

Artist's Career may be developed and exploited primarily through certain corporate, limited liability company or similar entities. Therefore, references herein to Artist, such as, without limitation, reference to income earned by Artist or received by Artist, shall be deemed to include the income earned or received by these entities. However, there shall be no double compensation hereunder, and therefore any distributions to Artist from such entities, including without limitation, dividends and salaries which are derived from gross revenues that have already been compensated hereunder, shall not be commissioned again upon payment to Artist.

Manager's Other Businesses

Artist acknowledges that from time to time during the Term, either alone or with others, Manager or individuals or entities affiliated with Manager may act as promoter of an event at which Artist shall perform. Manager agrees that in any such arrangements Artist will be compensated in the same manner as if Manager or such individuals or entities affiliated with Manager were not involved in the promotion and after full and adequate disclosure of the circumstances; however, Manager agrees to waive the Fee and Post-Term Commissions in respect of any such events to avoid a conflict of interest hereunder.

Parties Free to Enter Into Agreement

Manager and Artist warrant that: they are under no disability restriction or prohibition with respect to their right to execute this Agreement and form its terms and conditions; no act or omission by either hereunder will violate any right or interest of any person or firm, or will subject either party to any liability, or claim of liability to any person or firm; and Artist warrants that he has the right to and shall collect any and all Gross Income derived from Artist's Career. Manager and Artist agree to indemnify the other and to hold the other harmless against any damages, cost, expenses, fees (including reasonable attorneys' fees) incurred by the other party in any third-party claim, suit or proceeding instituted against the other party in which any assertion is made which is inconsistent with any warranty, representation, or covenant of them provided such claim is finally adjudicated or settled with their consent (which consent shall not be unreasonably withheld).

Assignment

Manager shall not have the right to assign Manager's rights or responsibilities under this Agreement to any entity without Artist's prior written approval, which approval shall not be unreasonably withheld.

Reviews of Counsel

The Artist and Manager acknowledge that they have had an opportunity to retain independent counsel of their own choosing to review this Agreement and advise them regarding its terms and conditions.

Age

Artist represents and warrants that Artist has attained the age of eighteen (18) years.

Binding Effect

This Agreement shall be binding upon Artist and Manager, and, to the extent applicable, shall also be binding upon their heirs, executors, permitted successors, and permitted assigns.

Controlling Law

This Agreement shall be deemed to be executed in the State of Tennessee and shall be construed in accordance with the internal laws of the State of Tennessee.

Status of Parties

This Agreement shall not be construed to create a partnership or joint venture between Manager and Artist. Manager is an independent contractor hereunder.

Modification

No modification or alteration of this Agreement shall be valid unless it is made in writing and signed by both parties.

IN WITNESS WHEREOF, the parties hereto have executed this Agreement as of the day and date first written above.

ARTIST: MANAGER:

_____   _____

Date Signed   Date Signed

Note

1. Hicks, Rush. *Music Industry Contracts*. 2015. A Sample Single Song Contract, Exclusive Recording & Artist Management Deal, Belmont University, Nashville, TN 37212.

APPENDIX B

An Insider's Guide to Entertainment Industry Lingo

If you want to be in the entertainment and music business, listen to what's being played, streamed, and viewed to see what you're going to be competing against. Surviving in the industry also requires you to be able to speak the language.

A *2-D barcode* is read by a digital scanner horizontally and vertically. A 2-D barcode is generated when an applicant completes Form CO on the U.S. Copyright Office's website. The 2-D barcode is unique for each work registered. It contains information that the applicant has entered on the blank form, but in encrypted format.

A & B Schedules are terms found on the forms sent to SAG-AFTRA by record labels, detailing the total scale payments made to label recording artists (designated as "A" vocalists/artists) and to nonroyalty artists (designated as "B" vocalists/artists).

Accounting statements are provided by music publishers quarterly or semiannually to songwriters, who are their clients. The accounting statement lists the money the publisher paid the songwriter in advance, plus chargeable expenses against the money credited or due to the songwriter from the royalties.

Acoustics can make or break the quality of a recording. All the talents of the recording artists, studio musicians, audio engineer, and producer are only as good as the studio's acoustics. Poor acoustics can destroy the best creative efforts of world-class musicians. Floating walls, ceilings, and floors are used to isolate sound from room to room in a performance studio and control room. Grooves are cut in the floors, walls are mounted on rubber tubes, and ceilings are often spring-loaded. Windows between the performance studios and control room are made of thick, double-pane glass with 5.5 inches of air space between each panel. Soundproof doors complete the package.

Administrative rights determine who controls the potential revenue-generating assets of a song. Administrative rights include the right to assign or transfer the copyright at any time; the right to control licensing; and the right to collect royalties for the use of the song.

Advances are all monies advanced for the recordings, marketing, promotion, and publicity of a product and an artist. However, actual money is often provided at the time an artist signs a contract to help the artist with living expenses for a short period of time (usually during recording sessions). The financial advances are 50%–100% recoupable paid from the artist's royalties (based on each unit sold) to satisfy the total debt to the label before the artist is paid for units sold.

Ambiance is the "feel" or "sound" of a room that is created by its mixture of hard and soft walls, floors, ceiling, and so forth.

Amplitude is "how loud" a sound wave is and is measured in decibels (dBs).

Amps or *amplifiers* increase the loudness of a sound wave or the amplitude (volume) of an electronic signal. Examples include guitar amps, monitor speakers, and operational amps inside the console.

Analog tape machines record both the amplitude (loudness) and frequency (number of vibrations) of the electronic signal sent by the mics to the tape recorder through the console.

Application forms refer to those provided by the Copyright Office for use in applying for registration of a claim to copyright. Registration makes a public record of the basic facts of a particular copyright.

Artist accounts or in-house accounts are set up in the artist's name by the record label to fund all approved projects, including recordings, advances, and tour support, that are recoupable at 100%. The marketing, promotion, and publicity are recoupable at 50%; however, most labels run them through tour support so that all expenses will become 100% recoupable.

Assignment means the appointment of the manager to represent an artist. The term also covers the assignment of the exclusive rights and copyright to a music publisher in a single songwriter deal.

Audiovisual works consist of a series of related images (pictures, etc.), which are intrinsically viewed, or shown by machines or devices, such as projectors, viewers, or electronics equipment, together with accompanying sounds. Mediums include films or tapes in which the works themselves are embodied.

Audits unlike those associated with the IRS, in the entertainment industry are typically included in artists' contracts. They allow the artist's representation to notify the labels approximately sixty days before the audit that permits an accountant to analyze the label's books to confirm correct royalty payments. Audits are conducted annually only.

Authority occurs when artists transfer their decision making for performing, interviews, and so forth over to the manager, who will determine which offer is in the best long-term interest of the artist. This also means that the artist may not accept movies, concert opportunities, or record deals without the advice of the manager.

Authorship refers to who wrote or owns the song. If the songwriter is a work-for-hire employee, then the publisher is usually considered the "author" of a song. If the songwriter is an independent contractor, staff writer, or one who owns a publishing company (as in co- or subpublishing), then he or she is usually considered the author of a song. A work-for-hire songwriter may also retain authorship and ownership if he or she has a signed agreement stating that the authorship and ownership will remain with the original songwriter(s).

Automated consoles are included in the standard recording technology today. Some of the older boards were computer-assisted, using VCAs (voltage control amplifiers) to scan the settings on the console and record their positions onto one of the tape machine tracks or a computer hard disk. Most of these functions are now part of Pro Tools and Logic Pro software.

Bern Convention is an international treaty, called the Convention for the Protection of Literary and Artistic Works, initially signed at Bern, Switzerland, on September 9, 1886, and all acts, protocols, and revisions thereto. The United States acceded to the Bern Convention effective March 1, 1989. Also shown in print as the Berne Convention.

The *best edition* of a work is the edition published (distributed) in the United States at any time before the date of deposit, which the Library of Congress determines to be most suitable for its purposes.

Best efforts is used here in the context of music publishers agreeing to use their best efforts to exploit the copyright material across all forms of distribution in the industry, including artists, labels, film and television production companies, advertisement agencies, digital media, mass media, and other companies that use music in their businesses. Best efforts can also be found in contracts between artists, personal managers, and tour promoters, among others, indicating that they will do "their best" in the business arrangement and agreements. This is the equivalent of good-faith efforts in contracts and is challenging to prove should an artist or other entertainment personnel need to go to court.

Bidirectional pickup patterns allow sound waves from the front and back of the mic and cancel out sound waves on the sides of the mic.

Billing is a term that refers to the location or placement of the act's or artist's name (logo) on tickets, advertisements, and press releases. Top billing means that the name of the artist will be on top of the venue's billboard or marquee and appear in larger print than other acts' or artists' names on tickets, and first mention on radio promotion spots.

Bonuses are negotiated, nonrecoupable monies paid to songwriters (and often recording artists and producers) for albums that reach various sales levels, usually gold (500,000 units), platinum (1,000,000 units), or multiplatinum.

Break-even point is the point in financial statements where the income (monies gained) from the event (concert, recording contract, etc.) equals the cost or total expenses for the event (or cost of recording, promotion, marketing, etc., of the recording).

Cardioid pickup patterns cancel the sound from the sides and rear of the mics. They sense sound waves that are only directly in front of the mic. Most mics in the recording studio use cardioid pickup patterns to avoid instruments' music from leaking into other instruments' mics.

Cartage is an extra payment (paid by the session producer) for musicians (or a private company) to haul instruments and amplifiers to and from the recording sessions. Union rules require additional insurance on harps, timpani, and other types of instruments that may be damaged more easily than others.

Catalog is a collection of the songs (copyrights) and recordings (usually demos) a music publisher owns and controls.

Circulars are informational brochures published by the Copyright Office for distribution to the public. Each circular deals with an aspect of the copyright law—for example, duration of copyright, copyright registration of musical compositions, and copyright notice.

Collaboration happens when two or more people work together to create a song, recording, and so forth. It is suggested that all the specific items of the agreement be in writing, including exceptions to the traditional royalty splits, permission to change each other's lyrics, and a listing of who will pay demo recording, legal, and promotion expenses. However, once the working arrangement is decided, a written agreement concerning the actual copyright is not required. The Copyright Law calls the songs written by more than one individual a "joint work."

A *collective work* is a work, such as a periodical issue, anthology, or encyclopedia, in which a number of contributions (which are separate and independent by themselves) have been assembled into a new collective work.

Commission base is the negotiated artist's revenue sources that will earn the manager a commission.

Commission rate is the percentage of the commission base that is paid to the manager for his or her service.

Commitment album is the only album that the label is committed to financing. Labels may offer a seven-year, seven-album deal, but they have to pay only for the recording, marketing, and promotion of the first album during the initial period (usually one year), and then each additional album *if* they pick up the corresponding option (usually based on album sales figures from the previously released recording).

Compensation is the manager's percentage of income as stated in the contract. The percentage often ranges from 15%–20% of monies resulting from all direct and indirect artist activities to very specific percentages based on defined revenue sources.

A *compilation* is a work formed by the collection and assembling of preexisting materials or of data that are selected, coordinated, or arranged in such a way that the resulting work as a whole constitutes an original work of authorship. Compilations include collective works to which different artists contribute to create an album.

Compression is used to reduce the amplitude (amount of volume) of a signal to record a proper signal on the tape track. Compression is also used to change the quality of the sound of an instrument or vocalist. Most consoles have four compression ratios, various attack times (how fast the compression is applied), and release from compression times. It is often used to "fatten up" the sound of a kick drum or bass guitar.

Compressors are outboard equipment used to reduce the amplitude (loudness) of a signal. They allow the audio engineer to record the proper amount of signal and to change the quality of the sound. They are often used on bass guitars and kick drums to make them sound "punchier" and on singers to help the engineer "catch" increases in amplitude.

Comp ticket or complementary ticket to a concert is often provided to radio stations and local important music insiders to promote an event. The number of tickets is usually supplied by the promoter and approved by the artist's manager.

Condenser mics have two plates that hold a static electronic charge: one is a permanent plate, and the other is a moveable plate. Sound waves vibrate the moveable plate, changing the distance between the two plates, which generates the electrical output of the mic. Condenser mics need batteries or phantom power supplied from the console.

Consoles or *"boards"* act as electronic traffic lights, dividing and directing the microphone and electrical instrument signals to various destinations.

Contingency scale payments are made to SAG-AFTRA nonroyalty artist members (who recorded the album).

Contractors are in charge of union members and are required by those unions to be hired. SAG-AFTRA requires one of the background vocalists (BGVs) to be designated as the contractor when more than three singers are hired. Further, this contractor is paid an additional 50%. The AFM requires a contractor pay the leader or lead musician double when more than 12 musicians are called for in a session.

Control room speakers (monitors) are used in the control room to convert microphone signals back into sound waves. To insure the speakers and control room acoustics are providing a correct sound (flat frequency response), room equalizers are used to compensate for the "hype of the sound" created by the speakers and control room acoustics. "Hype of the sound" refers to the change in the sound of the monitors caused by the acoustics of the control room. Reflective sound from the acoustics can phase add, causing the speakers to sound as if they are reproducing more bass or treble.

Creative representation is the term used for the contract between the artist, who grants power of attorney to managers, and the manager. Creative representation is a mutual agreement to exhibit common interest in the development of the artist's creative talents, recording, performing, and professional career.

Cross collateralization is the financial process through which the labels/music companies recoup past debts (on previous releases) by transferring income from currently profitable projects to cover previous debts. Labels cross collateralize record sales, while music companies cross collateralize record sales, concert tickets grosses, merchandise, and sometimes corporate sponsorships in their 360 deals.

Crossover networks are in speakers to separate the frequencies (bass, midrange, and treble) and send them to the appropriate speakers.

Cue systems allow the musicians and recording artists to hear themselves in their headphones. The mic signals sent to the console are simply mixed and returned to the artists and musicians through the cue systems and their headphones. Headset boxes allow the musicians and artists to control the amplitude (loudness) of their own headphones. Most consoles have a minimum of two stereo or four mono cue systems.

DART is an acronym for digital audio recording technology, which refers to digital audio recording devices and media covered by the Audio Home Recording Act of 1992 (P.L.102–563), the first statutory license to grant royalties to copyright owners for home copying.

Delivery date is the actual date the artist is to present an acceptable digital master of his or her recording to the label (usually the vice president of A&R). Also, this is known as the street date, when record labels release products to the public for sale and streaming. It is now usually on a Saturday.

Deposit (noun) refers to the copy, copies, or phonorecords of an original work of authorship that are placed in the Copyright Office to support the claim to copyright in the work or to meet the mandatory deposit requirement of the 1976 Copyright Act. Deposits become part of the public record and may be selected by the Library of Congress for its collections.

Deposit (verb) is to send to the Copyright Office a copy, copies, or phonorecords of an original work of authorship to support a claim to copyright or to meet the mandatory deposit requirement of the 1976 Copyright Act.

Deposit account refers to money kept in a special account set up in the Copyright Office by individuals or firms and from which copyright fees are deducted.

Direct boxes may be used instead of mics to convert the high impedance signal of the musical instruments into a low impedance signal acceptable to the console when musicians connect their guitars and keyboards directly to the console. Direct boxes eliminate the sound of one electrical musical instrument leaking into another open mic. However, they also limit the musicians' ability to control their own instrument's sound quality as the guitar amp's tone controls are also removed from the signal flow.

To *display a work* means to show a copy of it either directly or by means of a film, slide, television image, or any other device or process, or in the case of motion picture or other audiovisual work, to show individual images nonsequentially.

Document, in copyright terms, is a paper relating to the ownership of a copyright or to any other matter involving a copyright. Documents may be recorded in the Copyright Office for the public record.

Dry/wet refers to recording a signal on the tape with or without reverb or "effects" while listening to the signal in the control room monitor speakers and musicians' and recording artists' headsets. Dry is no effect; wet is when you are recording with the effect.

Dynamic mics have a coil of wire in the element (or top) of the mic. Sound waves vibrate a plastic diaphragm, which moves the coil in and out of a permanent magnetic field. This movement converts the sound waves into an electronic signal.

Dynamic range is the increase or decrease in amplitude of a sound wave. As an example, the dynamic range (loudness) of music is 120 dB.

Editing in the old days was the process of removing unwanted sounds or musical notes by cutting the analog tape with a razor blade and then taping the parts back together. Now it is usually done on computer software programs.

Effects, usually called outboard equipment (not located on the console), are "patched" into the console to change the quality of the signal. Examples include digital reverb, echo and delay units, noise gates (which cut off the signal at various amplitude levels used to prevent noise), compressors and limiters (which reduce the amplitude or loudness of the signal), and harmonizers (which are used to change the pitch).

Effect sends and returns are controls on each module of the console that are used to send a signal to the effects equipment to change the quality of the sound.

EQ or *equalization* refers to the bass, midrange, and treble controls on the console used to change the quality of the sound. Similar examples include the tone controls on your home stereo or car radio.

Exclusivity applies to artist contracts and means to "the exclusion of all others." You cannot record for another record company, assign songs to another music publisher, or use another manager.

Faders are an amplitude control on the console that moves up and down (instead of left to right, like the volume control on your car radio) to control the amount of signal coming in to a speaker. One hand can manipulate several faders at the same time. Faders control the monitor speakers, cue, the signal being sent to the recording artists' and musicians' headsets, and the amount of signal being sent to the master tape machine or the computer.

Favored nations clause is mostly used in film soundtracks and compilation albums. Mainly used as a negotiating tool, it states that the newest artist or lesser-paid artist negotiates to get paid the same amount of money as the highest-paid artist being featured.

Firm offer is a signed contracted agreement, which includes the concert dates, time of appearance, negotiated price for the act's performance, and a deposit consisting of a cashier's check or money order that will be placed into a secured account.

Float is a term used when a writer assigns performance royalties to be paid through a music publisher (performance royalties are usually paid directly to the songwriter by the performance rights organization). The publisher will often "float" the payment in order to earn interest off of the monies. This is a practice that is not supported by songwriter organizations as the music publisher may recoup expenses from the writers' royalties and may also make interest from the writers' royalties before they are paid.

A *food service* or *drinking establishment* is a restaurant, inn, bar, tavern, or any other similar place of business, in which the public or patrons assemble for the primary purpose of being served food or drink. Any such establishment in which the majority of the gross square feet of space is nonresidential may be used for that purpose.

Formula is usually a predetermined percentage of the profits or losses that will determine the artist's cash advances for the next album if the label picks up the "option." If the artist is to receive an advantage of 20% of profitable sales of, let's say, $1 million, then the advance would be $200,000. Formulas are often based on a series of issues; however, the key is for the artists to know the actual amount will have a floor (guaranteed low amount and a possible high ceiling), which creates a range for the advance to fall within. An example is a range of $50,000–$200,000. If the record loses money the artist is guaranteed to receive at least the bottom floor of $50,000 on his or her next album advance. However, let's say that the artist is to receive 30% of a million-dollar profit. The ceiling is $200,000, so the artist will receive only the $200,000 ceiling figure instead of the higher $300,000.

Four walls are the most basic levels of service a venue has available to the concert promoter and act listed under the term "four walls" or "four walling." The amount of electricity available (amps), heating, air conditioning, lights, and personnel, including house security, is listed on the contractual agreement between the venue and concert promoter.

Frequency is the number of complete cycles (sound waves or vibrations) per second, measured in "Hz" (or Hertz, after the inventor of the concept). We hear 20 to 320 Hz as bass, 320 to 5,120 Hz as midrange, and 5,120 Hz to 20 kHz as treble. High end is treble.

Gates are electronic switches (thresholds) used to eliminate a signal based on the amplitude. Gates are used to "block" tape hiss heard between musical notes as well as ambiance, rumble, and leakage from other microphones. They are regularly used on drums and overdub vocals to "tighten" the sound.

Gross monthly earnings are usually the total monies made by the artist from recording royalty points, bonuses or salaries, profits from concerts and merchandise sales, music publishing and songwriting royalties, commercial endorsements, movie and television performances, and profits from shares in capital ventures.

Harmonizing offers a variety of special effects, including the doubling of an input signal, delay of signal, and changing of the pitch. Doubling allows the engineer to make one instrument or vocalist sound like two. Delay provides echo or reverberation. Changing the pitch allows the engineer to detune or tune instruments and vocalists.

Headliner is the main attraction or most famous act to perform at a concert. Headliners are paid more money to perform than supporting acts.

"In the mud" is a term that refers to recording an analog signal too low, causing the playback to have too much noise.

"In the red" is a term that refers to recording an analog signal too "hot" on the tape machine, causing distortion.

Key person clause refers to a contract clause (verbal or written) that states a singer or group can negotiate to leave a management company or record company if the one key person who the artist has designated at the management company or record company ceases to be with the contracting company. It can be a manager or record company executive you believe is essential to your career.

Limiter is a compressor at maximum compression. The amplitude of the output is "limited" to the manually adjusted level of the threshold.

Loudspeakers are one of the several types of performance studio monitors that are used in the control room to determine the quality of the sound through large to small speakers.

Microphone pickup patterns are areas where mics are most sensitive to sound.

Modules allow for quick repairs and smaller and lighter consoles. They are another benefit of the space race: specifically, the design of plug-in modules.

Moral rights concern the protection of the reputation of the author. According to Stephanie C. Ardito of *InformationToday*, "Moral rights" is the English translation of the French phrase droit moral. Moral rights differ from copyright. Copyright protects property rights, which entitles authors to publish and economically benefit from their published works. Moral rights safeguard personal and reputational rights, which permit authors to defend both the integrity of their works and the use of their names. In countries that legally recognize moral rights, authors have redress to protect any distortion, misrepresentation, or interference of their works that could negatively affect their honor. Moral rights are often described as "inalienable." French law recognizes perpetual moral rights. In Germany, moral rights end when the author's copyright expires (70 years after he or she dies), while in other countries, moral rights terminate with the author's death.[1]

Negotiation period for an artist with a management agreement may include a performance clause (e.g., a requirement for the manager to secure a major recording contract) for the act. If the manager fails to achieve the stated performance clause within a certain time (often 12–24 months), the artist may cancel the agreement by sending a written statement to the manager. The agreement is usually concluded within 30 days after the manager receives the cancellation notice. These can also be considered "goals" for the manager and thus are not usually found in an artist/manager agreement.

Omnidirectional pickup patterns allow sound waves to enter from all directions into the mic at approximately the same loudness level.

Op amps are the many small operational (op) amplifiers that power a console, instead of one large, powerful amplifier. This technology was developed by the space race when it was determined that many small one-watt amplifiers were lighter to lift into outer space than one huge amp. The additional benefits to the audio recording business include a more compact size as well as less noise in the final recording.

Originality in a song contract means the writer has to state that a creative work is original and not plagiarized from another person's creative efforts. This statement provides legal proof the publisher can use to protect itself (sometimes against the writer) if sued by other writers who claim authorship.

Oscillator is a frequency or tone generator that was used to record signals on the tape, align tape machines, and provide special effects when mixed with musical instruments in the old days.

Pan pots are controls on the console that place a signal into the left-side or right-side monitor speaker or in the center of the mix.

Patch bays allow an audio engineer to reroute the signal to a different module, special effect, or other equipment.

Payment of services refers to all monies from the artist's performances, record sales, and commercial endorsements paid to the business manager or the artist's accountant as approved by the manager. The manager usually collects his or her negotiated percentage of the commission base and then has the business manager invest or place the remaining portion of the monies collected into the artist's accounts or businesses.

To *perform a work* means to recite, render, play, dance, or act it, either directly or by means of any device, or in the case of a motion picture or other audiovisual work, to show its images in any sequence or to make the words accompanying it audible.

To *publish* a work is to distribute copies or phonorecords of the work to the public by sale or other transfer of ownership, or by rental, lease, or lending. Publication also includes offering to distribute copies or phonorecords to a group of persons for purposes of further distribution, public performance, or public display. A public performance or display of a work does not of itself constitute publication.

Representation refers to the provision that if both the publisher and writer agree to a contract, it shall be subject to any existing agreements between the parties and their following performance rights organizations (ASCAP, BMI, SESAC), mechanical rights organization (the Harry Fox Agency), and digital rights management agency (SoundExchange). Representation is also a term used by recording artists to refer to management and booking agencies. Actors also are represented by union agents representing unions (e.g., SAG-AFTRA; musicians are represented by the AFM).

Reverberation is sound reflections from many sources (several hard surfaces) heard more than ten milliseconds after a direct sound wave. Reverb is an effect that makes the vocalists or musicians sound as if they were in a large auditorium (a hard-surfaced, empty acoustic room) instead of the smaller, acoustically correct recording studio.

Reversion clause is a provision in the songwriter–music publisher's contract that states if the publisher does not obtain a commercial sound recording (or some other form of written performance) then it will return the song's copyright to the songwriter within a specified time.

Ribbon mics operate on the same principle as dynamic mics, except they have flexible, metallic ribbons. Sound waves vibrate in and out of the permanent, magnetic force field. Because the ribbon is flexible, the quality of sound generated is often considered "warmer" and "smoother."

Rooms in a recording studio have many different names and functions. The recording artists and musicians perform in rooms that are collectively called "the studio." Producers and audio engineers supervise and record the artists' and musicians' performances from the control room. An equipment room isolates the noise of the tape machines from the control and performance studios, and a storage area is used for keeping microphones, music stands, headsets, cords, direct boxes, and other recording equipment that are not in use. The studio's acoustics are used to emphasize and match the feel of the song to the image and vocal characteristics of the recording artist. Placement of the performers and their microphones in the studio adds the final acoustics for the recording.

Seeking employment is not always allowed by artist managers, as in some states they are barred from seeking employment for the acts they represent. They usually limit their role to negotiating with agents and labels for various types of deals, and consulting the artists on hiring, firing, and employment opportunities.

Sound recording is a work that results from the fixation of a series of musical, spoken, or other sounds, regardless of the nature of the material objects in which they are embodied. A sound recording does not include the sounds accompanying a motion picture or other audiovisual work. Copyright in a sound recording protects the particular series of sounds embodied in the sound recording. Copyright registration for a sound recording alone is not the same as registration for the musical, dramatic, or literary work recorded. The underlying work may be registered in its own right apart from any recording of the performance.

Statements, either monthly or quarterly, are provided by the business manager to the personal manager and artist(s) concerning financial matters.

Suspension of artists in the music industry can be from label deals and artist management agreements if they become sick, refuse to perform, breach any part of their legal contracts, or break written moral clauses or social norms by committing a criminal act.

Term is the actual length of the initial period of an agreement (usually one year) followed by six one-year options. In a songwriting agreement it includes a period of time and number of required acceptable songs. In a reading deal it's usually a period of time and number of recordings, and in a management agreement it often refers to the initial period being 2–3 years and then 4–5 one-year options.

Ticket manifest is a computerized list of every seat (and ticket price) in a venue for a specific act or artist for each concert date.

A *transfer* of copyright ownership is an assignment, mortgage, exclusive license, or any other legal conveyance of a copyright or any of the exclusive rights comprised in a copyright, regardless of whether it is limited in time or place of effect, not including a nonexclusive license.

Transfer of copyright and all exclusive rights happens when the writer assigns, transfers, and delivers to the publisher an original musical composition written and/or composed by the named songwriter(s). The writer also assigns to the publisher the title, words, and music thereof, and the right to secure the copyright ownership registration throughout the world.

And just when you thought there were no more terms to learn . . .

- Trade-offs
- Opportunity costs
- Markets
- Law of demand
- Demand schedule
- Demand curve
- Change in quantity demanded
- Determinants of demand
- Change in demand
- Normal good
- Inferior good
- Substitute good
- Complementary goods
- Consumer sovereignty
- Price elasticity of demand
- Revenue
- Supply schedule
- Supply curve
- Law of supply
- Change in quantity supplied
- Change in supply

- Technology
- Backward bending supply curve
- Income effect
- Substitution effect
- Equilibrium
- Equilibrium price
- Equilibrium quantity

Note

1. Stephanie C. Ardito, "Moral Rights for Authors and Artists: In Light of the Tasini Ruling, Is the Next Step to Advocate for Legislation?", *InformationToday*. Accessed May 11, 2013. http://www.infotoday. com/it/jan02/ardito.htm.

ILLUSTRATIONS

Figures

Tables

INDEX

Note: Italicized page numbers indicate a figure on the corresponding page. Page numbers in bold indicate a table on the corresponding page.